GENETICALLY MODIFIED PLANTS

Genetically Modified Plants

Timm Wolf
and
Jonas Koch
Editors

Nova Science Publishers, Inc.
New York

For permission to use material from this book please contact us:
Telephone 631-231-7269; Fax 631-231-8175
Web Site: http://www.novapublishers.com

Library of Congress Cataloging-in-Publication Data

Genetically modified plants / Timm V. Wolf and Jonas P. Koch (editors).
p. cm.
ISBN 978-1-60456-696-3 (hardcover)
1. Transgenic plants. 2. Plant genetic engineering. I. Wolf, Timm V. II. Koch, Jonas P.
SB123.57.G483 2008
631.5'233--dc22 2008014985

Published by Nova Science Publishers, Inc. ✢ New York

CONTENTS

PREFACE

This book provides research from around the globe on Genetically modified plants which is a plant genetically engineered to contain one or more genes of another species. The aim is to introduce a new trait to the plant species which does not occur naturally in this species, for example resistance to certain pests, diseases or environmental conditions, or the production of a certain nutrient or pharmaceutical agent. Genetically modified plants are often called "transgenic plants", as they contain one or more transgenes from other organisms, however, this term also includes plants in which the transgene was integrated by naturally occurring processes.

Chapter 1 - Monitoring changes in Bacillus thuringiensis (Bt) resistance allele frequencies in field insect populations is essential for sustainability of the transgenic Bt crops. Resistance monitoring for Bt crops should be able to provide information on early shifts in resistance allele frequency so that proactive measures can be implemented to delay and even restrict the occurrence of widespread control failures. However, early detection of a change in the Bt resistance frequency in an insect population is often difficult and costly, especially if information is lacking or unavailable on resistance mechanism(s). Since 2004, a coordinated monitoring program for detecting a low frequency of Bt resistance in field populations of the sugarcane borer, Diatraea saccharalis (F.), a major target lepidopteran pest of Bt corn in the mid-southern region of the United States, has been implemented in Louisiana. This monitoring program includes 1) planting field corn after the optimum planting dates at multi-locations in the primary corn production areas of the state to attract and maintain late-season pest populations; 2) collecting larvae and pupae from these areas on multiple dates to ensure sufficient sample sizes for developing family-lines; 3) rearing field-collected larvae to pupal stage and sexing all pupae within a location; 4) establishing two-parent family-lines for an F_2 screen or hybridizing field-derived moths with those from a Bt resistant strain that were used in an F_1 screen; 5) detecting Bt resistance alleles using F_2/F_1 screens on Bt corn leaf tissues; 6) confirming Bt resistance for potential positive family-lines identified in the F_2/F_1 screens; and 7) estimating Bt resistance allele frequencies. Each year since the implementation of this monitoring program, 251-560 feral sugarcane borers were examined for Bt resistance alleles. Major resistance alleles that allowed resistant insects to complete larval development on Bt corn plants were detected in three populations sampled during 2004, 2006, and 2007, but not in 2005. The estimated Bt resistance allele frequency ranged from 0-0.0039 with an overall rate of 0.0013 and a 95% confidence interval of 0.0003-0.0028 during the monitoring period. The estimated time required for examining two feral insects (a two-parent family-line for F_2

screen or two lines for F_1 screen) was 0.91 hours using the F_2 screen and 0.66 hours using the F_1 screen. These results indicate that the Bt resistance allele frequency in Louisiana populations of sugarcane borer appears somewhat higher compared to that estimated for other corn stalk boring species targeted by Bt corn. No significant shift in the Bt resistance allele frequency has been observed during this four-year period. Furthermore, the resistance frequency observed in these populations does not require remediation or support a proactive change in the current resistance management strategies for Bt field corn.

Chapter 2 - The generation of therapeutic antibodies and fusion proteins for medical application is one of the fastest growing areas of the pharmaceutical industry with more than 150 therapeutic antibodies and fusion proteins currently either in clinical trial or use. At the same time, the use of virus-like particles has become an interesting tool in the fight against viral infections. Thus, some devastating high-incidence diseases such as HIV or cancer are currently chosen as clear targets for this type of therapeutical strategy. However, the high production cost of the current manufacturing systems of these molecules is a latent hurdle to overcome. With the advent of biotechnology, transgenic plants have emerged as a more economical new strategy for recombinant protein production. Antibodies and virus-like particles have been demonstrated to be well expressed in plants. In addition, the achieved protein expression level of most of them in the plant system has been reported to be compatible with that established for commercial viability. These facts make the use of plants for the generation of these types of recombinant molecules a very promising strategy to the development of lower cost biopharmaceuticals. In consequence, it could lead to exert important economical and medical implications as being affordable for developing countries where the incidence of infectious diseases is the highest. The development and production of these therapeutic molecules in plants is reviewed in this chapter, and the medical implications, advantages and limitations of both the plant-system and plant-derived molecules for practical use are discussed.

Chapter 3 - Lolium and Festuca species are the most important forage and turf grasses, and are cultivated in temperate zones around the world. Since most of these species are outcrossing plants, genetic diversity is observed even within a single cultivar developed by means of conventional breeding. Thus, genetic engineering is expected to be an attractive and powerful tool for improving the sophistication of breeding programs and ensuring that even within a genetically diverse cultivar, it is possible to incorporate useful genes in all individuals; as a result, this approach has been experimentally adopted in both groups of grass species. In this review, the authors present the current state of genetic modification that has been achieved in Lolium and Festuca species, and discuss how we can produce practical transgenic plants of these species.

Chapter 4 - The main focus of first-generation GE (Genetically Engineered) plants were agronomic traits, such as herbicide tolerance or pest resistance, which benefited mainly farmers. The benefits for human or animal health, when present, were indirect, for instance a reduced mycotoxin content, due to a lower plant colonization by mycotoxigenic fungi, or a reduction in commodity prices, due to higher yield or improved farming practices, translating in increased food availability.

At present GE plants with nutritionally interesting traits are becoming available, both for feed and food. Forages with improved digestibility, due to a reduction in lignin, have been obtained and also soybean and cereals with increased content of lysine, methionine and tryptophan have been developed. Other interesting GE plants in the pipeline are corn and soybean with increased resistance to mycotoxins-producing fungi or able to degrade mycotoxins. Another important research trend is the reduction of anti nutritional factors such as phytate, which reduces phosphorus availability and increases environmental impact of livestock, particularly of pigs. Oleaginous crops with altered oil composition are available since many years but, while in previous years the increase of oleic acid content was the main target of genetic engineering, recent research focused to improve the content of the cardiovascular protective omega 3 and 6 fatty acids in seed oil.

Other important topics of importance both for human and animal nutrition are the increased content of vitamins (Vit. E, B_9 and C), pro-vitamin (β-carotene) and flavonoids, an important component of a health promoting diet. The development of oilseed crops with increased sterol content could be an interesting way for reducing cholesterol uptake in humans.

The alteration of starch composition or the accumulation of sugars in different polysaccharides could be used in order to obtain processed foods with a more suitable glycemic index, an important characteristic, especially considering the increasing incidence of diabetes.

Chapter 5 - "Hairy root" – an infectious disease caused by the soil bacteria Agrobacterium rhizogenes – is a natural phenomenon, which has existed for centuries. The development of molecular biology and genetic engineering tools at the end of the 1970s and in the early 1980s revealed the principles and mechanisms of infection and allowed scientists to "copy" this phenomenon. Today, genetic transformations using the "natural genetic engineer" Agrobacterium rhizogenes are widely used for induction of the so-called "hairy root" plant in vitro systems. Such systems are valuable for the production of economically important biologically-active substances. Furthermore the opportunity to express and integrate foreign genes into plant cells using Agrobacterium rhizogenes plasmids has allowed the mass production of desirable phytochemicals, medicinally important emzymes and foreign proteins either in in vitro conditions (bioreactors) or through plant regeneration (in greenhouses or in the field).

This chapter summarizes recent progress in the techniques and methods relating to the genetic transformation of plant cells and the further maintenance and cultivation of hairy root cultures, as well as recent results from the authors' laboratories on using flow cytometry for investigating the polysomaty of transformed root in vitro systems, and the bioreactor systems (submerged and temporary immersion) used for the cultivation of hairy roots.

Chapter 6 - Mycotoxins are acutely toxic or carcinogenic compounds produced primarily by Aspergillus and Fusarium molds that infect seeds of high oil content in the field, such as maize, cottonseed, peanuts, wheat and tree nuts. Damage by insects facilitates entry of the molds, and maize hybrids that express the Bacillus thuringiensis (Bt) crystal protein at sufficiently high levels to provide near immunity to the European corn borer can have greatly reduced levels of mycotoxins such as fumonisin in the presence of these pests. However, the number of different species of ear-damaging insect pests requires many different Bts, and the recent reports of field collected European corn borers with high levels of resistance to Bt CryIAb suggest nonBt strategies for broad spectrum insect control may be valuable in the future.

The authors approach toward solving the mycotoxin problem involves the investigation of plant-derived genes, alone and in combination with each other, that have the potential to provide high levels of resistance to insects, including those genes that can be used as selectable markers. The authors have investigated directly active proteins, such as ribosome-inactivating proteins and N-acetyl hexosaminidases, and demonstrated significant effects against both beetles and caterpillars when these genes were expressed transgenically. Broad spectrum resistance to species in several different insect orders has been noted for plants producing enhanced levels of peroxidases, including in the field. A model anti-cell-death protein enhanced insect resistance when expressed in transgenic plants. Robotics has been used to develop antiinsectan peptides with enhanced oral activity over original forms. The authors have recently begun to introduce combinations of genes for directly active proteins in different functional classes, and have identified plants with greatly enhanced resistance over that seen with introductions of individual genes.

Chapter 7 - Mycotoxins are toxic or carcinogenic compounds produced by fungi, including Aspergilllus and Fusarium molds that colonize seeds of maize, cotton, peanuts or tree nuts. Insect herbivory of plant tissue often enhances mold infection. Transgenic maize expressing the Bacillus thuringiensis (Bt) toxin produce ears with lower mycotoxin levels when the target insect pest is effectively controlled. However, Bt toxins are generally species specific and therefore new transgenic strategies for broad insect control need to be developed. The vast diversity of plant secondary biochemicals reflects the strategy of non-motile plants to defend themselves from pathogens and animal herbivores. Research in recent years revealed that many of these plant biochemicals are synthesized by a number of enzymes that are coordinately regulated at the gene level by transcription factors. Altering the expression of these transcription factors by genetic engineering opens the possibility that these secondary biochemicals may be synthesized in specific tissues to combat insect herbivory and reduce mycotoxin contamination. Alternatively, well studied secondary biochemical pathways may be modified or transferred from one plant species to another and tested for effective insect resistance. Transgenic manipulation of secondary metabolism must be carefully considered because redirection of plant energy reserves may hinder plant yield. The advent of genetic engineering and advances in plant biochemistry has opened up the exciting possibilities of utilizing the diversity of secondary biochemicals for protecting valuable crop commodities.

Chapter 8 - Gene expression levels of transgene 35S-gshI (γ-glutamylcysteine synthetase) cloned from E. coli, and the endogenous gene gsh1 of poplar (Populus x canescens) were up-regulated by the DNA demethylating agent DHAC (5,6-dihydro-5'-azacytidine hydrochloride) (10^{-4} M for 7 days) in aseptic leaf discs cultures. Two 35S-gshI-transgenic (6lgl and 11ggs) and wild type (WT) poplar clones were used. The efficiency of gene upregulation was also analyzed under herbicide paraquat stress (4 x 10^{-7} M). Levels of gshI-mRNA and gsh1-mRNA were determined by RT-qPCR (reverse transcriptase quantitative PCR) after cDNA synthesis. For internal control, the constitutively expressed housekeeping poplar genes α-tubulin and actin were used, and the $2^{-\Delta\Delta Ct}$ method was applied for data analysis. In long term DHAC treatment (21 days), a morphogenetic response of de novo root development was observed on leaf discs in a wide concentration range of DHAC (10^{-8} to 10^{-6} M). Adventitious shoots (11ggs clone) also emerged from leaf discs after a combined treatment with DHAC (10^{-4} M) and paraquat (10^{-7} M). Shoots were dissected, rooted and transplanted in glass houses for further analyses for phytoremediation capacity. Since DNA methylation patterns are inherited (epigenetic memory), these poplar plants with increased gene expression levels of both transgene 35S-gshI and endogenous gene gsh1 provide novel plant sources for in situ application.

Chapter 9 - Soy is a food rich in proteins, fibers, vitamins, minerals and isoflavones, compounds with a very similar chemical structure to estrogen, being able to perform similar activities as this hormone. Studies report that its use decreases the risk of cardiovascular diseases by the reduction of cholesterol and triglycerides serum levels. 24 female Wistar rats were divided into three groups (N = 8): organic soy group (OSG) receiving an organic soy-based diet, genetically modified soy group (GMSG) receiving a transgenic soy-based diet, and a control group (CG) receiving a casein-based. All animals with water, isocaloric and isoproteic diets Ad libtum during all their lives until old age. It was observed that the consumption of organic and transgenic soy seems to promote a reduction in the levels of cholesterol and triglycerides compared with control group. It was also observed that the genetically modified soy behaved in a similar way to the organic soy in relation to lipids levels, no inserted DNA or any other component of the transgenic soy had any adverse effect that could imply the transgenic food is not adequate for consumption.

Chapter 10 - Canola is one of the biotechnologically developed crops in the world. Biotech canola is a GMO (genetically modified organism) in food and an LMO (living modified organism) in environment. Several kinds of biotech canola have been commercialized in United States and Canada. Among them, two kinds of biotech canola, GT73 and Ms8/Rf3, were mainly cultivated in the world.

The detection method is an essential element for the GMO labeling system or LMO management about the biotech crops. In this study, the specific primer pairs and probes were developed to the qualitative and quantitative PCR methods for GT73 and Ms8/Rf3. The designed primers were confirmed in specificity and sensitivity. The single PCR product was obtained from GT73 or Ms8/Rf3 by the specifically designed primer pair, respectively. Limits of detection (LODs) of the qualitative method were all 0.03% for these biotech canola events.

On the other hand, the quantitative method was developed by using a TaqMan real-time PCR. As a reference molecule, a synthetic plasmid was constructed from a taxon-specific DNA sequence of canola and construct-specific DNA sequences of GT73 and Ms8/Rf3. The validation of this quantitative method was performed using six levels of mixing samples, 0.1 to 10.0%, with two biotech canola. As a result, the biases from the true value and the relative standard deviations were almost within the range of ±30%. Limits of quantitation (LOQs) of the quantitative method were all 0.1%. Consequently, the authors report that the proposed detection methods were applicable for qualitative and quantitative analysis for the biotech canola GT73 and Ms8/Rf3.

In: Genetically Modified Plants
Editors: Timm Wolf and Jonas Koch

ISBN 978-1-60456-696-3

Chapter 1

DETECTION AND MONITORING OF *BACILLUS THURINGIENSIS* RESISTANCE ALLELES IN THE SUGARCANE BORER, *DIATRAEA SACCHARALIS* (F.)

Fangneng Huang and B. Rogers Leonard
Department of Entomology, Louisiana State University Agricultural Center, Baton Rouge, Louisiana 70803, U.S.A

ABSTRACT

Monitoring changes in *Bacillus thuringiensis* (Bt) resistance allele frequencies in field insect populations is essential for sustainability of the transgenic Bt crops. Resistance monitoring for Bt crops should be able to provide information on early shifts in resistance allele frequency so that proactive measures can be implemented to delay and even restrict the occurrence of widespread control failures. However, early detection of a change in the Bt resistance frequency in an insect population is often difficult and costly, especially if information is lacking or unavailable on resistance mechanism(s). Since 2004, a coordinated monitoring program for detecting a low frequency of Bt resistance in field populations of the sugarcane borer, Diatraea saccharalis (F.), a major target lepidopteran pest of Bt corn in the mid-southern region of the United States, has been implemented in Louisiana. This monitoring program includes 1) planting field corn after the optimum planting dates at multi-locations in the primary corn production areas of the state to attract and maintain late-season pest populations; 2) collecting larvae and pupae from these areas on multiple dates to ensure sufficient sample sizes for developing family-lines; 3) rearing field-collected larvae to pupal stage and sexing all pupae within a location; ; 4) establishing two-parent family-lines for an F_2 screen or hybridizing field-derived moths with those from a Bt resistant strain that were used in an F_1 screen; 5) detecting Bt resistance alleles using F_2/F_1 screens on Bt corn leaf tissues; 6) confirming Bt resistance for potential positive family-lines identified in the F_2/F_1 screens; and 7) estimating Bt resistance allele frequencies. Each year since the implementation of this monitoring program, 251-560 feral sugarcane borers were examined for Bt resistance alleles. Major resistance alleles that allowed resistant insects to complete larval development on Bt corn plants were detected in three populations sampled during 2004,

2006, and 2007, but not in 2005. The estimated Bt resistance allele frequency ranged from 0-0.0039 with an overall rate of 0.0013 and a 95% confidence interval of 0.0003-0.0028 during the monitoring period. The estimated time required for examining two feral insects (a two-parent family-line for F_2 screen or two lines for F_1 screen) was 0.91 hours using the F_2 screen and 0.66 hours using the F_1 screen. These results indicate that the Bt resistance allele frequency in Louisiana populations of sugarcane borer appears somewhat higher compared to that estimated for other corn stalk boring species targeted by Bt corn. No significant shift in the Bt resistance allele frequency has been observed during this four-year period. Furthermore, the resistance frequency observed in these populations does not require remediation or support a proactive change in the current resistance management strategies for Bt field corn.

Keywords: *corn, transgenic plants, resistance monitoring, Cry proteins, Bt, F_1/F_2 screen, IRM, corn borer.*

INTRODUCTION

Transgenic corn varieties expressing *Bacillus thuringiensis* (Bt) endotoxins have become the most important tools for managing corn stalk boring pests in the United States and several other countries. During 2007, Bt corn cultivars were planted on more than 18.4 million hectares and represented approximately 49% of the total corn hectarage planted in that year (NASS 2007). The widespread acceptance of transgenic Bt corn will place significant selection pressure on corn borer populations and could accelerate the development of resistant populations; thereby increasing concerns about the long-term sustainability of Bt corn as an effective insect pest management tool (Ostile et al. 1997, Gould 1998). To prolong the effectiveness of Bt corn, a mandatory insect resistance management (IRM) plan has been implemented for planting Bt corn in the United States and Canada (US EPA 2001, Baute 2004).

Field corn is a major crop and contributes significant value to agriculture in Louisiana, as well as, in other mid-southern U.S. states. During 2007, producers across Texas, Louisiana, and Mississippi planted >1.5 million hectares of field corn. Recently and during the foreseeable future, the urgent need for alternative energy sources has greatly increased the value of corn. Producers across this region will likely continue to plant significant corn hectarage because of the high price of this commodity.

A corn stalk borer complex consisting of the sugarcane borer, *Diatraea saccharalis* (F.) and southwestern corn borer, *D. grandiosella* Dyar are major yield limiting pests across this region. Yield losses up to 30% on conventional non-Bt corn can be attributed to this pest complex (Sankula and Blumenthal 2004). Bt corn cultivars adapted to the mid-southern region were commercialized during 1999 and adoption of this technology has increased rapidly. Bt corn has significantly reduced yield losses from all corn borer species and is currently the primary strategy for managing these pests. This technology has nearly saturated the marketplace and accounts for about 50% of the total corn acreage in this region. Given the recognized value of this technology by producers, additional hectarage would be planted if IRM guidelines would allow them to do so.

Managing or delaying sugarcane borer resistance to Bt corn in the mid-southern region appears to be even more critical than that proposed for the two most important corn borer species across the North Central and Mid-Western Regions of the US Corn Belt, European corn borer, *Ostrinia nubilalis* Hübner, and southwestern corn borer. The sugarcane borer, a key pest of sugarcane (Reagan 2001), has expanded its geographic range during the late 1990's and has now replaced the southwestern corn borer as the dominant corn borer species in Louisiana and some areas of Texas (Castro et al. 2004a, Porter et al. 2005, Huang et al. 2006a). In Louisiana, significant field corn hectarage was seriously damaged by late-season populations of sugarcane borer during 2002 and 2003 (Castro et al. 2004a). In many non-Bt corn fields, yield losses from sugarcane borer infestations exceeded 30%. A four-year survey (2004 – 2007) of Louisiana corn stalk borer species found that sugarcane borer accounted for more than 80% of the total corn borer populations across the major production areas (Huang et al. 2006a; FH, BRL unpublished data). During 2005 across some areas of Texas, field corn was heavily damaged by sugarcane borer (Porter et al. 2005). Economic infestations were reported from Corpus Christi to Victoria, extending northwards to the Dallas area. Field samples collected during 2006 and 2007 showed that sugarcane borer accounted for >99% of the total corn borer populations in these areas (RP, CC, FH, unpublished data). Sugarcane borer infestations have also been reported in Mississippi and southern Arkansas corn fields. Sugarcane borer has been documented to successfully overwinter in corn stubble across northeast Louisiana and in much of the mid-South. This species is now occurring in regions of corn where no sugarcane is planted (Huang et al. 2006a). The explanations for this increase in the pest status of sugarcane borer across the region have largely remained unknown.

Because of the increasing sugarcane borer problem on corn in the mid-southern region, this boring species was recently listed as a target pest of Bt corn in the United States (US EPA 2005a, b). In addition, laboratory bioassays showed that sugarcane borer is inherently less susceptible to the Cry1Ab protein in Bt corn compared to European corn borer and southwestern corn borer (Huang et al. 2006b). Two previous studies from Louisiana also showed that Bt corn performance against sugarcane borer varied among Bt corn varieties (Castro et al. 2004b, McAllister et al. 2004). The widespread adoption of Bt corn, present and future value of this technology to sustainable corn IPM, and increased incidence of sugarcane borer across the mid-southern region demand an effective monitoring program that can detect early shifts in Bt resistance allele frequency in field borer populations. These data are necessary to support implementation of the high dose/refuge strategy for managing sugarcane borer resistance to Bt corn across this region.

A systematic monitoring program for detecting early changes of Bt resistance allele frequencies in field populations of sugarcane borer was coordinated among scientists across Louisiana in 2004. Based on the principles of an F_2 screen method proposed for detecting Bt resistance in European corn borer (Andow and Alstad 1998), a cost-effective modification of an F_2 screening technique was developed for detecting rare Bt resistance alleles in sugarcane borer populations sampled from corn fields across the major Louisiana corn production areas during 2004-2005 (Huang et al. 2007a, b). During 2006-2007, an F_1 screening procedures for identifying resistance, simplified from the F_2 screening method, was developed and used to screen F_1 progeny of crosses between feral individuals and a known resistant strain of sugarcane borer. This paper outlines the F_2/F_1 screening procedures, reports the results of the four-year monitoring, and discusses the costs of the monitoring program.

Procedures of F2/F1 Screens for Detecting Rare Bt Resistance Allels in Sugarcane Borer

Insect samples. The first step of the monitoring program to detect Bt resistance in sugarcane borer populations required collections of insects from field of corn across the sampling regions (Figures 1 and 2). During 2004, larvae and pupae of first and second generations of sugarcane borer were collected from corn fields near Winnsboro, Louisiana (Franklin Parish) (Huang et al. 2007a). To facilitate the insect collection during 2005 to 2007, corn was planted beyond optimum planting dates at multiple locations in the primary production areas to attract and maintain late-season corn borer populations. Larvae and pupae of sugarcane borer were collected from these field areas on several dates to ensure sufficient sample sizes for developing family-lines to detect Bt resistance alleles. Field-collected larvae were reared individually in 30-ml plastic cups (Fill-Rite, Newark, NJ) containing 5-10 ml of a meridic diet (Bio-Serv, Frenchtown, NJ), and maintained in environmentally-controlled chambers within the conditions of 20 - 28 °C, with 50-65% RH, and a 16:8 (L:D) h photoperiod. Pupae, collected from fields or derived from field-collected larvae, were separated according to sex before adult emergence and were segregated until adult emergence for pair matings in the development of family-lines for screening Bt resistance.

F_2 screen. A cost-effective F_2 screening method was developed to detect rare Bt resistance alleles in field populations of sugarcane borer sampled during 2004 and 2005 (Figure1.) (Huang et al. 2007b). Two-parent family-lines were established by pairing newly emerged male and female adults derived from field-collected samples in 450-ml paper containers (Huhtamaki Foodservice, De Soto, Kansas) (Huang et al. 2007b). The paper containers were then placed in growth chambers maintained at 27-28 °C, 50-65% RH, and a 16:8 (L:D) h photoperiod for adult mating and oviposition. F_1 eggs produced from each single-pair mating were collected after 3-4 days. F_1 larvae (60-150) of each single-pair mating were reared individually in 30-ml plastic cups each containing a meridic diet in environmental growth chambers maintained at 27°C, 50% RH, and a 16:8 (L:D) h photoperiod. F_1 adults within each single-pair mating were sib-mated in 3.8-liter cardboard cartons (Neptune Paper Products, Newark, NJ) to produce an F_2 generation. F_2 progeny produced from a single-pair mating were considered as a two-parent family-line and used to screen for Bt resistance.

F_2 screens for detecting Bt resistance alleles were performed in 8-well trays (Bio-Smart-8, C-D International, Pitman, NJ) containing corn leaf tissue, dissected from fully expanded leaves of V5-V8 stages of a commercial Bt (Cry1Ab) corn hybrid, DKC 69-70 (Monsanto, St. Louis, MO) (Huang et al. 2007b). For each two-parent family-line, 150-300 F_2 neonates were screened in 3-5 wells in environmental chambers maintained at 27-28°C, 50% RH, and a 16:8 (L:D) h photoperiod. Larval survival was recorded after 5, 10, and 15 days. . If one of the two parents for a family-line carried a major recessive Bt resistance allele, 6.25% of F_2 larvae were expected to be homozygous for Bt resistance and should survive in the F_2 screen (Andow and Alstad 1998). A line was considered a potential positive line expressing Bt resistance alleles if F_2 neonates survived for more than 10 days in the F_2 screen.

Resistance confirmation for the F_2 screen. To validate if the initial positive lines identified in the F_2 screen actually possessed resistance alleles, non-selected F_3 neonates (>240 neonates/ family-line) produced from the potential positive lines were placed into the whorls of vegetative stage plants of DKC69-70 Bt and DK697 non-Bt plants in the greenhouse for larval growth and development (Figure 1). DK697 is a conventional non-Bt corn hybrid that is genetically similar to the DKC69-70 Bt corn hybrid, both commonly planted in Louisiana. Larval survival on Bt and non-Bt corn plants were recorded after 27 days in 2004 and 21 days in 2005. Similarly, if one of the two parents for a line carried major recessive Bt resistance alleles, 6.25% of F_3 larvae were expected to be homozygous for Bt resistance and should survive on Bt corn plants (Huang et al. 2007b).

A family-line (line 55) derived from samples collected during 2004 was identified as a potential positive line in the F_2 screen. Larvae of this family line survived on intact Bt corn plants in greenhouse tests for >27 days. To evaluate if line 55 could complete larval development (from neonate to pupal stage) on Bt corn plants, survivors were recovered from Bt corn plants after 27 days, and were backcrossed with a Bt-susceptible laboratory strain (Bt-SS strain) (Huang et al. 2007b). Second generation neonates of this backcross were re-exposed to Bt corn plants in the greenhouse and evaluated for larval survival and development. Approximately 80 neonates from the second generation of the backcross were exposed to each of 75 Bt plants. Insect survival was recorded when the first pupa from a Bt plant was observed.

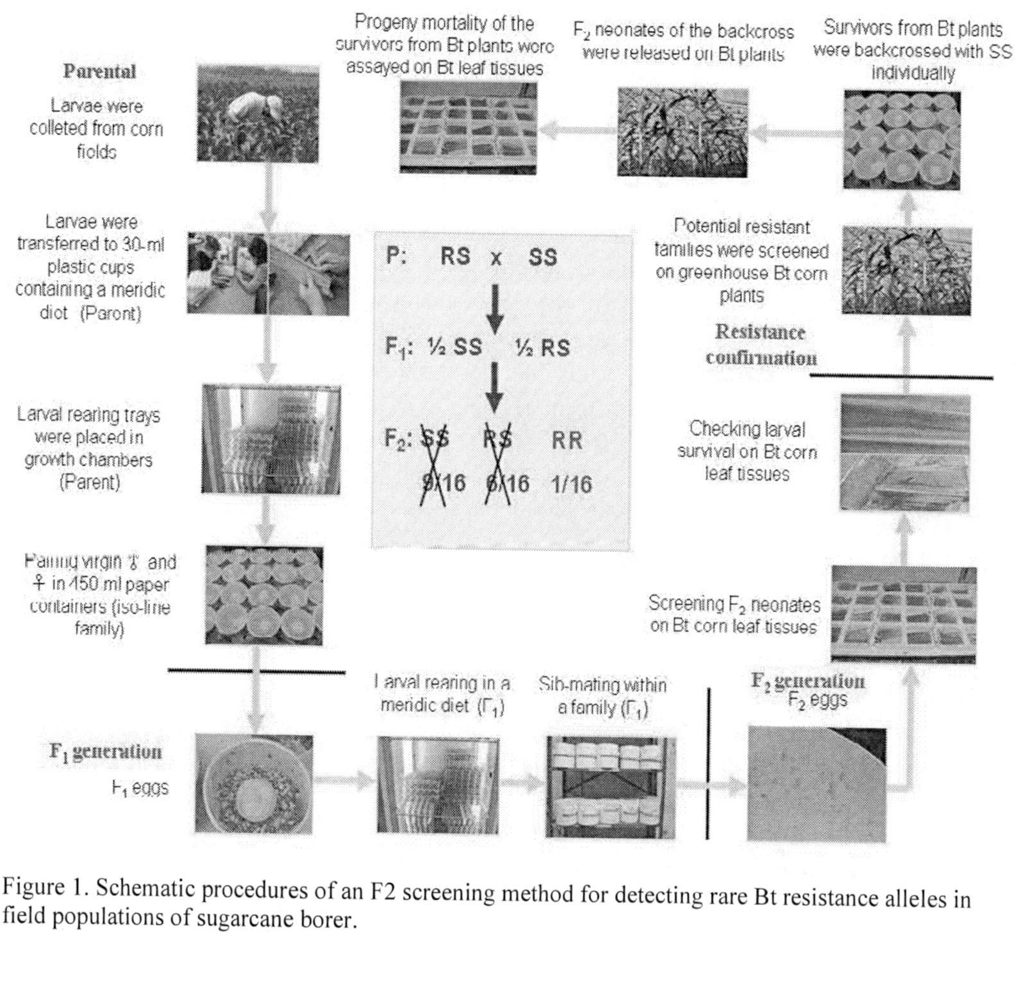

Figure 1. Schematic procedures of an F2 screening method for detecting rare Bt resistance alleles in field populations of sugarcane borer.

If survivors of the non-selected F_3 insects were homozygous for Bt resistance, 25% of the second generation larvae of the backcross were expected to be homozygous for Bt resistance and should survive on Bt corn plants in the resistance re-verification study.

Estimation of Bt resistance allele frequencies for the F_2 screen. For the F_2 screen, a family-line was defined to possess a major resistance allele(s) if F_2 larvae of the line could survive for 10 days on Bt corn leaf tissues and its offspring could complete larval development on Bt corn plants. A line was considered to carry a minor resistance allele(s) if F_2 larvae of the line survived for 10 days but its offspring could not survive on Bt corn plants for ≥21 days. The expected Bt resistance allele frequencies (E[q]) in the combined populations collected each year were estimated using a Bayesian analysis (Andow and Alstad 1998),

$$E[q] = (S + 1)/[4(n + 2)] \qquad [1]$$

where n = the number of two-parent family lines, and S = the number of positive two-parent family-lines possessing Bt resistance alleles. The 95% confidence interval of the estimated resistance frequency was calculated using F distributions as described by Andow and Alstad (1999). The probability (or detection power) that a resistance allele could be identified in a family-line (if present) was estimated according to the methods described in Stodola and Andow (2004).

F_1 screen. Field samples of sugarcane borer collected during 2006 and 2007 were examined for Bt resistance alleles using an F_1 screening method (Figure 2). For Bt resistance traits expressed in a recessive manner, F_2 screening procedures can be simplified by screening F_1 progeny of the crosses between field-collected individuals and Bt resistant insects (hereafter referred as F_1 screen) (Gould et al. 1997). Using the previously described F_2 screen method, line 55 collected in 2004 was identified to possess a major resistance allele to Cry1Ab in Bt corn. A Bt resistant strain (Bt-RR strain) was developed from pupae and large larvae (5th intars) of this line that survived on Bt corn plants during the resistance re-verification tests. Resistance in this sugarcane borer strain most likely was controlled by a recessive or near completely recessive gene (Huang et al. 2007b). The availability of this Bt resistant strain allowed detection of Bt resistance alleles in field populations of this insect species using F_1 screening procedures. If the feral parent in a single-pair cross possesses the same resistance allele as the resistant parent, 50% of the F_1 progeny are expected to be homozygous for Bt resistance, and the homozygous Bt-resistant insects should survive in the F_1 screen (Gould et al. 1997).

Two types of mating (single pair and group) were used to develop F_1 progeny for screening Bt resistance in field collections of sugarcane borer. In the single pair-mating procedure, newly emerged virgin males and females of field-derived moths were paired individually with Bt-resistant insects in 450-ml paper containers for adult mating and oviposition as described in the single-pair matings for the F_2 screen (Huang et al. 2007b). In the group mating procedure, five newly emerged virgin females of field-derived moths and four virgin males of the Bt-RR strain were placed into a 1-liter cardboard carton (Neptune Paper Products, Newark, NJ) for adult mating and oviposition to produce F_1 generations for each group.

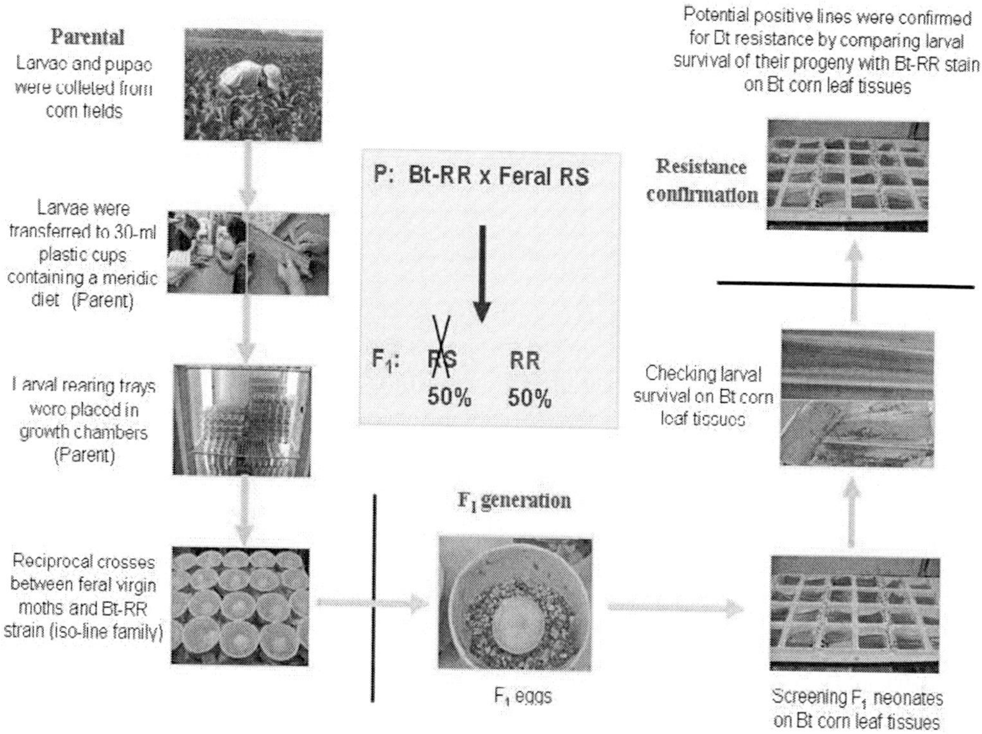

Figure 2. Schematic procedures of an F1 screening method for detecting rare Bt resistance alleles in field populations of sugarcane borer.

In the resistance screen, 50-F_1 neonates were released on Bt corn leaf tissues in each of two wells for each single-pair mating (n = 100) or four wells for each group mating (n = 200 neonates) as described in the F_2 screen (Figure2, Huang et al. 2007b). Larval survival was evaluated after 5 and 10 days during 2006 or after 6 and 12 days during 2007. Chi-square (χ^2) tests were used to determine if the observed survival rates were different from the expected values at the $\alpha = 0.05$ level. The expected survival rates were adjusted to the baseline survival of the Bt-RR strain and the heterozygous genotype (Bt-RS) on Bt corn leaf tissues before the data were subjected to statistical analyses. Baseline survival levels of the three genotypes (Bt-SS, Bt-RS, and Bt-RR) on Bt and non-Bt corn leaf tissues were determined using the same procedures as described in the F_1 screen. A line was considered as a potential positive line for a Bt-RS feral parent if its observed larval survival rate after 10 or 12 days in the F_1 screen was not considerably lower (e.g. >10%) than the expected rate.

Resistance confirmation for the F_1 screen. To validate the presence of resistance alleles previously identified in the F_1 screen, F_1 survivors were transferred into 30 ml plastic cups containing a meridic diet and reared to the pupal stage. Pupae derived from the survivors were placed in 3.8-liter cardboard cartons for adult emergence, mating, and oviposition. Upon eclosion, larvae (20 neonates) representing the 1st generation progeny of the survivors in the F_1 screen and the Bt-RR strain were placed on Bt and non-Bt corn leaf tissues in each of 6-8 wells (n =120-160) (Figure 2). Larval survival was examined after 10 and 15 days or after 5 and 10 days for 2006 and 2007, respectively. Differences in larval survival between the two strains was compared using *t*-tests at the $\alpha = 0.05$ level. A suspected positive line was validated if the survival rates from the F_1 screen were similar to that of the Bt-RR strain.

Estimation of Bt resistance allele frequencies for the F_1 screen. For the F_1 screen, the expected frequency of Bt resistance alleles ($\hat{q}$) for insect samples collected in each year was estimated using the equation (Wenes et al. 2006),

$$\hat{q} = 1 - [1 - (S + 1)/(N_{tot} + 2)]^{1/2} \quad [2]$$

where S = the number of positive lines with Bt-RS feral parents, and N_{tot} = the number of single parent lines screened plus the number of females in the group screens that contributed to the F_1 generation. A female was considered successfully mated and a contributor to the F_1 generation in a group mating if the female contained empty spermatophores in the bursa copulatrix. F_1 neonates hatched from eggs of each single-pair or group mating were used in the F_1 screen for Bt resistance. The 95% confidence interval was calculated based on the equation [15] from Andow and Alstad (1999). The detection power in an F_1 screen was estimated using the equation as described in Wenes et al. (2006).

Estimated BT Resistance Allele Frequencies in Sugarcane Borer

F_2 screen for insect samples collected during 2004. A total of 213 two-parent family-lines of sugarcane borer derived from field collections near Winnsboro, Louisiana (Franklin Parish) were screened for Cry1Ab resistance on Bt corn leaf tissue using the F_2 screening method (Figure1., Table 1). In the F_2 screening, 37 neonates from four family-lines survived for >10 days on Bt corn leaf tissue. Most of the survivors (27) in the F_2 screening were from a single line (line 52). In the greenhouse confirmation tests, 14 live larvae of the non-selected F_3 generation of line 52 were recovered from 27 Bt plants after 27 days. In the resistance re-verification tests, the first pupa of the second generation of the backcross was observed on a Bt corn plant 55 days after larval inoculation. Therefore, the re-verification tests in the greenhouse were terminated, and insect survival was recorded thereafter. A total of 351 live sugarcane borers including 20 pupae, 48 fifth instars, and 283 second-forth instar larvae were recovered from 55 Bt plants. The survivors were collected in plant stalks, ears, or ear shanks. The other 20 Bt corn plants infested with neonates of the second generation of the backcross were used for monitoring insect growth and development to determine the termination date of the tests. Line 52 survival on Bt corn leaf tissue in the F_2 screen and on Bt corn plants in the resistance confirmation and re-verification tests suggested that one of the parents possessed at least one major Bt resistance allele and was capable of survival on intact Bt corn plants expressing Cry1Ab. Segregation analysis of line 52 suggested that Bt resistance in this family was likely controlled by a near completely recessive allele at a single locus (Huang et al. 2007a). Under this assumption, the estimated major Bt resistance allele frequency in this population of sugarcane borer was 0.0023 with a 95% confidence interval (CI) of 0.0003 to 0.0064.

Table 1. Estimated frequency of *Bacillus thuringiensis* resistance alleles in Louisiana sugarcane borer populations sampled during 2004-2007

Year	F_2/F_1 screen	No. feral individuals screened (lines)	No. individuals possessing minor resistance alleles	No. individuals possessing major resistance alleles	Minor resistance allele frequency (95% CI)*	Major resistance allele frequency and/or 95% CI	Detection probability
2004	F_2	416 (213)	3	1	0.0047 (0.0013-0.0101)	0.0023(0.0003-0.0064)	0.98
2005	F_2	560 (280)	6	0	0.0063 (0.0025-0.0117)	(0 – 0.0027)	0.94
2006	F_1	331 (331)	---	1	---	0.0030(0.0004-0.0084)	0.98
2007	F_1	251 (251)	---	1	---	0.0039(0.0005-0.0110)	0.98
Overall	---	1558	9	3	0.0051(0.0024-0.0087)	0.0013(0.0003-0.0028)	0.96

Frequencies for the combined populations (overall) is re-calculated as a function of the total data set. Overall minor resistance allele frequency was estimated based only the data from F_2 screen during 2004 and 2005.

The other three family-lines demonstrated a low survival rate (< 1.7%) on Bt corn leaf tissue after 10 days in the F_2 screen and were considered to possess a minor resistance allele. Thus, the estimated resistance frequency with minor alleles was 0.0047 with a 95% CI of 0.0013 to 0.0101 in this population. If a resistance allele existed in the population, the F_2 screen had a detection power of 97.5%.

F_2 screen for insect samples collected during 2005. A total of 280 two-parent family-lines of sugarcane borer were examined for Bt resistance using the F_2 screening procedures during 2005 (Table 1). The four populations were collected across primary corn production areas of the state, including three populations (East Carroll, Tensas, and Franklin) from Northeast Louisiana and one population (Rapides) from Central Louisiana. For each population, 54-90 F_2 family-lines were screened for Bt resistance. A total of 84 F_2 larvae from eight family-lines survived on Bt corn leaf tissue after 10 days in the F_2 screen. Six of these lines demonstrated larval survival rates of ≥ 5%, and the other two lines had <1.5 % larval survivorship. None of these families in the F_2 screen completed larval development on Bt corn plants in resistance confirmation tests in the greenhouse. Thus, the F_2 screen suggested that none of the 280 family-lines possessed major resistance alleles to Cry1Ab Bt corn. The estimated frequency for major Bt resistance alleles in the four populations collected during 2005 was less than 0.0027 with a 95% probability. The six family-lines that demonstrated significant survival levels were considered to express minor resistance alleles. Based on Bayesian analysis, the estimated minor resistance allele frequency was 0.0063 with a 95% CI of 0.0025 to 0.0117. The F_2 screen had a detection power of 94% if a resistance allele existed in the populations.

F_1 screen for insect samples collected during 2006. A total of 331 feral individuals of sugarcane borer were collected from the same four locations during 2006 as those during 2005. These insects were crossed with the Bt-RR strain and screened for Bt resistance using the F_1 screening method (Figure 2, Table 1). Insect sample size from each location ranged from 5 (East Carroll Parish) to 223 (Rapides Parish) lines. Two lines (19 and 28) of Rapides populations demonstrated significantly greater larval survival rates (12 and 40%, respectively) than the Bt-RS genotype (6%) on Bt corn leaf tissue after 10 days in the F_1 screen. The observed survival rate of line 28 in the F_1 screen was not significantly different from the adjusted expected rate (37.8%) for a Bt-RS feral parent. The progeny survival rates of F_1 survivors from line 28 on Bt corn leaf tissue were 52% after 10 days and 48% after 15 days, and neither were significantly different from the Bt-RR strain. The results of the F_1 screen and resistance confirmation tests suggested that the feral parent of line 28 possessed the same resistance allele found in the Bt-RR strain. The 12% larval survival rate of line 19 after 10 days in the F_1 screen was considerably lower than that of the expected rate for a Bt-RS feral parent (37.8%). An effort to establish a strain with the survivors of this line for resistance confirmation was unsuccessful because of low survivorship. Therefore, it is likely that the feral parent of this line did not carry resistance alleles. Results of Bayesian analysis suggested the expected Bt-resistance allele frequency for the four Louisiana populations of sugarcane borer collected during 2006 was 0.0030 with a 95% CI of 0.0004 to 0.0084. The F_1 screen had a detection power of 98% if a resistance allele existed in the feral parents.

F_1 screen for insect samples collected during 2007. A total of 251 feral individuals of sugarcane borer were collected during 2007 from three locations (Tensas, Franklin, and Rapides Parishes) in Louisiana. These insects were crossed with the Bt-RR insects and screened for Bt resistance using the F_1 screening procedures (Figure 2, Table 1). The feral parent sample sizes for the F_1 screen were 33, 74, and 144 lines for Tensas, Franklin, and Rapides population, respectively. One line (24) developed from a cross between a field-collected male of the Franklin population and a Bt-RR female demonstrated a 23% survival rate on Bt corn leaf tissue after 12 days in the F_1 screen. The observed survival rate of line 24 was not significantly lower than the expected rate (29%) for a Bt-RS feral parent, and was considerably greater than the survival rate (1%) of the Bt-RS genotype. In the resistance confirmation tests, the corrected progeny survival rates of F_1 survivors from line 24 on Bt corn leaf tissue was 78% after 5 days and 65% after 10 days. These survival levels were not significantly different from that of the Bt-RR strain. The results of the F_1 screen and resistance confirmation tests suggested that the feral parent of line 24 of the Franklin population possessed the same resistance allele as presented in the Bt-RR strain. A different line (22) from a cross between a male collected from the Franklin population and a Bt-RR insect survived the 12-day screen; however the survival rate was only 1%. F_1 neonates of all other lines did not survive on Bt corn leaf tissue after 12 days. Thus, the F_1 screen and the resistance confirmation tests suggested that one (line 24 of the Franklin population) out of the 251 feral parents collected during 2007 possessed the same resistance allele as the Bt-RR strain. Thus, the expected Bt-resistance allele frequency for the three Louisiana populations of sugarcane borer was estimated to be 0.0039 with a 95% CI of 0.0005 to 0.0110. The F_1 screen had a detection power of 98% to identify a resistance allele if one existed in the samples.

Estimation of Labor Costs of the F_2/F_1 Screens for Detecting Rare BT Resistance in Sugarcane Borer

Most of the detecting procedures that have been used by the scientific community to monitor rare Bt resistance alleles in field insect populations are costly and labor intensive (Andow and Alstad 1998, Huang 2006). In particular, the labor cost of an F_2 screen is often considered a significant limitation of this technique. To estimate labor costs, the entire procedure of both screens was divided into individual tasks, 10 for the F_2 screen and 7 for the F_1 screen (Table 2). The time required to complete each task was determined based on a work group consisting of a professor, two undergraduate interns, one M.S. student, and three Ph.D. students. A 70% success rate for single-pair matings was considered in the cost estimation for the F_2 and F_1 screens. A successful single-pair mating means that the pair produced enough F_2 or F_1 progeny for screening Bt resistance.

The time required for each task in examining 100 two-parent family-lines using the F_2 screen ranged from 2 hours in the replacement of leaf tissues to 17 hours for meridic diet preparation (Bio-Serv, Frenchtown, NJ) (Table 2). The total time required to complete the ten tasks of an F_2 screen for screening 100 lines was 90.5 hours (approximately 1 hour/line). If an equivalent number (e.g. 200 lines) of feral individuals were examined for Bt resistance using the F_1 screen, the time required was reduced to 66 hours (Table 2); a reduction of 27% compared to the F_2 screen.

Table 2. Estimated labor costs (time/hours) for examining Bt resistance in sugarcane borer for 100 two-parent family-lines with the F_2 screen or 200 single-pair lines using the F_1 screen

Task	F_2 screen (h)	F_1 screen (h)
Diet preparation	17	1.5
Adult pairing	5.5	13.5
Handling of F_1 and F_2 eggs	12	12
Transfer F_1 larvae	15	0
Recover F_1 pupae	15	0
F_1 sib-mating	5	0
Place F_1 or F_2 eggs into screening trays	7	11
Place leaf tissue for screening	2	4
Replace leaf tissue	5	10
Check results	7	14
Total	90.5	66

* Estimation was based on a personnel team consisting of one professor, two undergraduate interns, one M.S. student, and three Ph.D. students.

Estimations were adjusted with a 70% success rate for single-pair mating. A successful single-pair mating was defined as one that produced sufficient F_2 or F_1 progeny for screening.

Significances of the Present F_2/F_1 Screens for Detecting Rare BT Resistance Alleles in Field Insect Populations

Several methods have been used for monitoring Bt resistance in field insect populations (Andow and Alstad 1998, Huang 2006). Among these, the dose-response and diagnostic concentration techniques (Marçon et al. 2000, Tabashnik et al. 2000) are the two most common procedures because of their relatively simple procedures. However, both methods may be not suitable for monitoring insect resistance to cry proteins in Bt crops at early stages because they are not sufficiently sensitive in detecting rare recessive resistance alleles (Roush and Miller 1986, Andow and Alstad, 1998).

Andow and Alstad (1998) introduced a novel F_2 screening technique for detecting recessive alleles in field populations of European corn borer. This technique involves collecting mated females from fields, inbreeding within each iso-female line, and screening their F_2 progeny for Bt resistance on Bt corn plants or on meridic diet containing Bt toxins. The F_2 screen method is expected to be more effective in detecting rare recessive Bt resistance alleles in field insect populations than the traditional dose-response or diagnostic dose methods. Since 1998, more than 2,000 family-lines of European corn borer and southwestern corn borer have been screened for Bt resistance using the F_2 screening method, but no major resistance alleles have been detected among these populations (Andow et al. 1998, 2000; Bourguet et al. 2003; Stodola et al. 2006; Huang et al. 2007c). A major allele for Bt resistance would allow larval development on the Cry1Ab dose expressed in commercial Bt corn plants. The major Bt resistance allele identified in line 52 of the Louisiana sugarcane borer population collected during 2004 represents the first report of a Bt-resistant corn stalk borer to survive and complete larval development on commercial Cry1Ab corn plants. The successful identification of a major resistance allele to Bt corn in line 52 also suggests that the F_2 screening procedures are effective in detecting rare Bt resistance alleles in field populations of sugarcane borer.

The present F_2 screen can represent several advantages and should significantly reduce the overall cost compared to the F_2 screen proposed for other corn borer species (Huang et al. 2007b). Relative to the F_2 method used for detecting Bt resistance alleles in European corn borer (Andow et al. 1998, 2000; Bourguet et al. 2003), the present F_2 screening procedures have two notable modifications: 1) the use of field-collected larvae or pupae for establishment of family-lines rather than field-collected adult males or mated females and 2) the use of Bt corn leaf tissue for screening F_2 neonates rather than directly screening F_2 larvae on whole Bt corn plants or on diet containing Bt toxins. F_2 screen for detecting Bt resistance alleles needs to rear and maintain each of many family-lines separately for one or more generations after insect samples are collected from fields. In addition, resistance screening should be performed on each family-line. Therefore, the cost (both in labor and materials) can be a significant concern in implementation of an F_2 screen for monitoring Bt resistance (US EPA 2001). The estimated time required for examining one family-line using the present F_2 screen was about one hour. Based on these observations and estimations, a sample of 200-300 two-parent family-lines for a representative field site could be considered a suitable sample size for estimating Bt resistance allele frequencies. This translates to a total of 200-300 hours required to conduct the F_2 screen for these samples, an amount of labor cost that could be affordable in a period of 2-3 months for many laboratories. However, the actual time needed to complete these tasks could vary greatly depending on the experience and skills of personnel conducting the screen. Some tasks such as diet preparation, removing pupae from diet, larval transferring, egg handling, require less experience and skills. However, other activities such as conducting single-pair mating for establishing family-lines and checking larval survival in screening resistance require extensive experience. In addition, temperature conditions for insect rearing could be varied to control insect development so that insect samples could be utilized at different times, which is particularly important when large insect samples were collected.

One major task of an F_2 screen that is not listed in Table 2 involves field collection of insects. The labor costs associated with field collections could vary significantly, depending on insect population densities. The majority of field corn is planted during March and April in Louisiana to maximize yield and reduce insect damage. Corn planted after the optimum planting dates become more attractive to the sugarcane borer late in the growing season. To ensure that sufficient numbers of insects could be collected for the F_2/F_1 screens during 2005-2007, corn was planted late in a season at several selected locations across the primary corn production areas. In most cases, several hundred sugarcane borer larvae and pupae were collected from these late-planted corn fields in a few hours. The use of late-season insects for screening Bt resistance also allows detection of field resistance status at the end of the growing season.

In addition, resistance confirmation tests to validate potential positive lines in an F_2/F_1 screen are necessary. Labor costs associated with the resistance confirmation can be variable depending on the number of lines that require re-testing and the confirmation methods. Based on these experiments, Bt resistance validation tests on intact Bt corn plants in the greenhouse (e.g. method used for line 52 in 2004) cost significantly more compared to the use of Bt corn leaf tissue in the laboratory (e.g. method used for line 28 of the Rapides population collected in 2006 and line 24 of the Franklin population collected in 2007). Our results suggest that Bt resistance confirmation can be done in laboratory bioassays with Bt corn leaf tissue after the larval survival rates of Bt-RR, Bt-RS, and Bt-SS insects (including both on Bt and non-Bt corn) are established. Other than labor costs, material expenditures for an F_2 screen are another concern in monitoring Bt resistance. The F_2 screening procedures using Bt corn leaf tissue could significantly reduce the material costs compared to screening on whole Bt plants. A rough estimate of non-reusable materials required for screening one family-line of sugarcane borer with the current F_2 screening method was about $2 for purchasing insect diet, rearing materials (cups and lids), and plastic covers. The relatively high efficiency of this screening method has allowed this protocol to be adopted with slight modifications for detecting Bt resistance alleles in field populations of southwestern corn borer (Huang et al. 2007c) and the cotton bollworm, *Helicoverpa armigera* (Hübner) (ZX, JLX. unpublished data).

To further facilitate Bt resistance screening in sugarcane borer, an F_1 screening method was evaluated in this study. Two out of 582 feral insect samples collected during 2006-2007 were identified to possess a major resistance allele. The successful identification of resistance alleles in two feral individuals of sugarcane borer suggested that the current F_1 screening procedure is effective and can be used to detect rare Bt resistance alleles in field populations. The estimated labor costs of the F_1 screen were about 30% less than the costs associated with the F_2 screen. The non-reusable materials required for an F_1 screen were estimated to be 50% less ($1 for 2 lines) than the F_2 screen. In addition, the F_1 screening method appears to have a greater detection power than the F_2 screen. In an F_1 screen, 50% of the F_1 progeny in a line are expected to be homozygous for Bt resistance for a feral Bt-RS parent (Gould et al.1997), while only 6.25% individuals in the F_2 generation are expected to be homozygous for Bt resistance if one parent of a family-line possesses a Bt-resistance allele (Andow and Alstad 1998). However, the F_1 screen can only detect the same allele as the one present in the laboratory resistant strain (Bt-RR). The F_1 screen also assumes that the resistance allele in the known Bt resistant strain is the only important genetic factor in resistance development. Therefore, a sufficient number of feral insects need to be examined with the F_2 screening method to identify 'all possible' resistance alleles in target insect populations, before an F_1 screen is used as the primary procedure for detecting rare Bt resistance alleles.

The four-year monitoring program showed that Bt resistance allele frequency in Louisiana populations of sugarcane borer across the primary corn production areas ranged from 0-0.0039 with an overall rate of 0.0013 and a 95% CI of 0.0003 to 0.0028. Bt resistance alleles in Louisiana sugarcane borer may not be as rare as that reported for other corn stalk boring species (Andow et al. 1998, 2000; Bourguet et al. 2003; Stodola et al. 2006; Huang et al. 2007c). However, it appears that the resistance frequency in these populations has not reached a level that would lead to control failures with Bt corn that expresses Cry 1Ab. These results suggest that continuous monitoring of Bt resistance in sugarcane borer is necessary to ensure the sustainable use of transgenic Bt corn technologies across the mid-southern region of the United States.

Conclusion

The F_2 and F_1 screening methods are being used to detect and monitoring rare Bt resistance alleles in field populations of sugarcane borer. No significant shift in Bt resistance frequency has been observed from 2004 to 2007. The resistance allele frequency in Louisiana populations of sugarcane borer to Cry1Ab in Bt corn varieties is still relatively low after nine years of using this technology. The current estimates appear to support the rare resistance allele requirement of the "high dose/refuge" resistance management strategy for Bt corn. Annual surveys to monitor Bt resistance in sugarcane borer is necessary for the sustainability of Bt corn technology for corn borer management across the mid-southern region of the United States.

References

Andow, D.A., and Alstad, D.N. (1998) F_2 screen for rare resistance alleles. *Journal of Economic Entomology, 91*, 572–578.

Andow, D.A., and Alstad, D.N. (1999) Credibility interval for rare resistance allele frequencies. *Journal of Economic Entomology, 92*, 755–758.

Andow, D.A., Alstad, D.N, Pang, Y.H, Bolin, P.C., and Hutchison, W.D. (1998) Using an F_2 screen to search for resistance alleles to *Bacillus thuringiensis* toxin in European corn borer (Lepidoptera: Crambidae). *Journal of Economic Entomology, 91*, 579–584.

Andow, D.A., Olson, D.M., Hellmich, R.L., Alstad, D.N., and Hutchison, W.D. (2000) Frequency of resistance to *Bacillus thuringiensis* toxin Cry1Ab in an Iowa population of

European corn borer (Lepidoptera: Crambidae). *Journal of Economic Entomology, 93*, 26–30.

Baute, T. (2004) A grower's handbook; controlling corn insect pests with Bt corn technology. Second edition. Canadian Corn Coalition. Ridgetown, Ontario. pp. 1-24.

Bourguet, D., Chaufaux, J., Séguin, M., Buisson, C., Hinton, J.L., Stodola, T.J., Porter, P., Cronholm, G., Buschman, L.L., and Andow, D.A. (2003) Frequency of alleles conferring resistance to Bt maize in French and US corn belt populations of the European corn borer, *Ostrinia nubilalis*. *Theoretical and Applied Genetics, 106*, 1225–1233.

Castro, B. A., Riley, T.J., Leonard, B.R., and Baldwin, J. (2004a) Borers galore: Emerging pests in Louisiana corn, grain sorghum and rice. *Louisiana Agriculture. 47*, 4-6.

Castro, B.A., Leonard, B.R., and Riley, T.J. (2004b) Evaluation of *Bacillus thuringiensis* transgenic field corn for management of southwestern corn borer and sugarcane borer (Lepidoptera: Crambidae). *Journal of Economic Entomology, 97*, 2106–2116.

Gould, F. (1998) Sustainability of transgenic insecticidal cultivars: integrating pest genetics and ecology. *Annual Review of Entomology, 43,* 701–726.

Gould F., Andenson, A., Jones, A., Sumerford, D., Heckel, D.G, Lopez, L., Micinski, S., Leonard, R., and Laster, M. (1997) Initial frequency of alleles for resistance to *Bacillus thuringiensis* toxins in field populations of *Heliothis virescens*. *Proceedings of National Academy of Sciences, USA, 94*, 3519–3523.

Huang, F. (2006) Detection and monitoring of insect resistance to Bt crops. *Insect Science 13*, 73-90.

Huang, F., Leonard, B.R., and Baldwin, J. (2006a). Corn borers and transgenic Bt corn technology. *Louisiana Agriculture 49*: 25-26.

Huang, F., B. R. Leonard, B.R., and Gable, R.H. (2006b) Comparative susceptibility of European corn borer, southwestern corn borer, and sugarcane borer (Lepidoptera: Crambidae) to Cry1ab protein in a commercial Bt-corn hybrid. *Journal of Economic Entomology, 99*,194-202.

Huang F., Leonard B.R., and Andow D.A. (2007a) Sugarcane borer resistance to transgenic *Bacillus thuringiensis*-maize. *Journal of Economic Entomology, 100*,164-171.

Huang, F., Leonard, B. R., and Andow D.A. (2007b) F_2 screen for resistance to a *Bacillus thuringiensis*-maize hybrid in the sugarcane borer (Lepidoptera: Crambidae). *Bulletin of Entomological Research, 97*, 437-444.

Huang, F., Leonard B.R., Cook, D.R, Lee, D.R., Andow, D.A., Baldwin, J.L., Tindall, K.V., and Wu, X. (2007c) Frequency of alleles conferring resistance to *Bacillus thuringiensis* maize in Louisiana populations of southwestern corn borer (Lepidoptera: Crambidae). *Entomologia Experimentalis et Applicata, 122*, 53 - 58.

Marçon, P.C.R.G., Siegfried, B.D., Spencer, T., and Hutchison, W.D. (2000) Development of diagnostic concentrations for monitoring *Bacillus thuringiensis* resistance in European corn borer (Lepidoptera: Crambidae). *Journal of Economic Entomology, 93*, 925–930.

McAllister, C.D., Gravois, K., Schexnayder, K.P., and Reagan, T.E. (2004) Transgenic Bt-corn affects sugarcane borer in Louisiana. *Southwestern Entomologist, 29*, 263-269.

NASS (National Agricultural Statistics Service). (2007) Acreage. USDA, Washington, DC http://usda.mannlib.cornell.edu/usda/current/Acre/Acre-06-29-2007.pdf. [Last accessed 10 January 2008].

Ostlie, K.R., Hutchison, W.D. and Hellmich, R.L. (1997) Bt-corn and European corn borer, Long-term success through resistance management. *North Central Region Extension Publication 602.* University of Minnesota, St. Paul, MN, USA.

Porter, P., Troxclair, N., Schuster, G., Porter, D.O., Cronholm, G., Bynum, E. Jr., Patrick, C., and Davis, S.G. (2005) Texas corn production: Emphasizing pest management and irrigation. Texas Cooperative Extension, *The Texas AandM University System B-6177*. pp 71.

Reagan, T.E. (2001) Integrated pest management in sugarcane. *Louisiana Agriculture. 44,* 16-18.

Roush, R.T., and Miller, G.L. (1986) Considerations for design of insecticide resistance monitoring programs. *Journal of Economic Entomology*, 79, 293–298.

Sankula, S., and Blumentehal, E. (2004) Impacts on US agriculture of biotechnology-derived crops planted in 2003-An update of eleven case studies. *National Center for Food and Agriculture Policy, 2004 report*. http://www.ncfap. org/whatwedo/ pdf/2004 finalreport. pdf. [Last accessed 10 January 2008].

Stodola, T.J., and Andow, D.A. (2004) F_2 screen variations and associated statistics. *Journal of Economic Entomology*, 97, 1756–1764.

Stodola, T. J., Andow, D.A., Hyden, A.R., Hinton, J.L., Roark, J.J., Buschman, L.L., Porter, P., and Cronholm, G. B. (2006) Frequency of resistance to *Bacillus thuringiensis* toxin Cry1Ab in Southern US Corn Belt populations of European corn borer (Lepidoptera: Crambidae). *Journal of Economic Entomology*, 99, 502-507.

Tabashnik, B.E., Patin, A.L., Dennehy, T.J., Liu, Y.B., Carrière, Y., Sims, M.A., and Antilla, L. (2000) Frequency of resistance to *Bacillus thuringiensis* in field populations of pink bollworm. *Proceedings of National Academy Sciences, USA*, 97, 12980–12984.

US EPA (United States Environmental Protection Agency). (2001) Biopesticides registration action document: *Bacillus thuringiensis* plant-incorporated protectants. http://www. epa.gov/oppbppd1/biopesticides/ pips/*Bt*_brad2/ 5-benefits.pdf. [Last accessed 24 April 2007].

US EPA (United States Environmental Protection Agency). (2005a) *Bacillus thuringiensis* Cry3Bb1 protein and the genetic material necessary for its production (Vector ZMIR13L) in event MON 863 corn and *Bacillus thuringiensis* Cry1Ab delta-endotoxin and the genetic material necessary for its production in corn (006430, 006484) fact sheet. http://www.epa.gov/pesticides/biopesticides/ingredients/factsheets/factsheet_006430-006484.htm. [Last accessed 10 January 2008].

US EPA (United States Environmental Protection Agency). (2005b) *Bacillus thuringiensis* Cry34Ab1 and Cry35Ab1 proteins and the genetic material necessary for their production (plasmid insert PHP 17662) in Event DAS-59122-7 corn and *Bacillus thuringiensis* Cry1F protein and the genetic material necessary for its production (plasmid insert PHI8999) in Event TC1507 corn Fact sheet. http://www.epa.gov/pesticides /biopesticides /ingredients/factsheets/factsheet_006430-006484.htm. [Last accessed 10 January 2008].

Wenes A.L., Bourguet, D., Andow, A.D., Courtin, C., Varre, G., Lorme, P., Sanchez, L., and Augustin, S. (2006). Frequency and fitness cost of resistance to *Bacillus thuringiensis* in *Chrysomeia tremulae* (Coleoptera: Chrysomelidae). *Heredity. 97*, 127-136.

This paper was reviewed by Carlos Blanco (USDA-ARS), Mike Stout, Jack Baldwin, Linxin Mao, and Shuanglin Dong. This article is published with the approval of the Director of the Louisiana Agricultural Experiment Station as manuscript No. 2008-234-1390.

In: Genetically Modified Plants
Editors: Timm Wolf and Jonas Koch

ISBN 978-1-60456-696-3

Chapter 2

PLANT-BASED ANTIBODIES AND VIRUS-LIKE PARTICLES: A LEAP TOWARDS NEW THERAPEUTIC DEVELOPMENT

Patricia Obregón, Alicia Fernandez-San Millan and Jon Veramendi
Instituto de Agrobiotecnología, Universidad Pública de Navarra-CSIC-Gobierno de Navarra, Campus Arrosadía, 31006-Pamplona, Spain

ABSTRACT

The generation of therapeutic antibodies and fusion proteins for medical application is one of the fastest growing areas of the pharmaceutical industry with more than 150 therapeutic antibodies and fusion proteins currently either in clinical trial or use. At the same time, the use of virus-like particles has become an interesting tool in the fight against viral infections. Thus, some devastating high-incidence diseases such as HIV or cancer are currently chosen as clear targets for this type of therapeutical strategy. However, the high production cost of the current manufacturing systems of these molecules is a latent hurdle to overcome. With the advent of biotechnology, transgenic plants have emerged as a more economical new strategy for recombinant protein production. Antibodies and virus-like particles have been demonstrated to be well expressed in plants. In addition, the achieved protein expression level of most of them in the plant system has been reported to be compatible with that established for commercial viability. These facts make the use of plants for the generation of these types of recombinant molecules a very promising strategy to the development of lower cost biopharmaceuticals. In consequence, it could lead to exert important economical and medical implications as being affordable for developing countries where the incidence of infectious diseases is the highest. The development and production of these therapeutic molecules in plants is reviewed in this chapter, and the medical implications, advantages and limitations of both the plant-system and plant-derived molecules for practical use are discussed.

1. Introduction

Since the advent of plant biotechnology in the 1980´s, the use of transgenic plants has proved an effective strategy for recombinant protein production. The use of plants as a surrogate expression system for generation of therapeutics provides an important number of advantages compared to other standard production systems: (i) Plants are eukaryotic organism and their endomembrane system is organized in identical manner to that in mammalians, (ii) recombinant protein production and storage can be made near the site of use, (iii) since plant-derived therapeutics could be heat stable formulations there would be no need for refrigeration, (iv) with the adoption of oral administration the need for needles could be reduced and so would the need for skilled medical practitioners, (v) the risk of human pathogens contamination is significantly reduced, (vi) importantly, the plant system offers high potential for scale-up (virtually limitless amounts of recombinant protein could be achieved depending only on the planting area) and lower costs than current production systems. In contrast, a hurdle to overcome is still the low expression levels reported for most pharmaceuticals, particularly antigens, (0,0005%-0.3% of total soluble protein, TSP) (Arntzen et al., 2005) and investigations in the area continue to be one of the most important driving factors in the development and optimization of this expression system. In fact, several molecular strategies have been approached and significant biotechnological improvements have already been achieved (Gil et al., 2001; Yu and Langridge, 2001) (Richter et al., 2000); (Walmsley et al., 2003); (Nugent and Joyce, 2005); (Satub et al., 2000). Notwithstanding this, high expression level of antibody molecules (>1% TSP) in transgenic plants, accepted as compatible with commercial viability, has been reported in most cases. Thus, it is a clear example of therapeutic molecules which can be favoured by the plant-based production system. In a similar manner, the generation and expression levels of virus-like particles (VLPs) indicate the plant system could be a promising alternative for production of these molecules.

Developing countries are the areas more affected by some of the most important diseases currently affecting the world, such as the HIV/AIDS epidemic, and other infectious diseases like hepatitis B, tuberculosis, rabies and malaria, for examples. Little economical and medical resources are two main reasons for the high prevalence and rapid propagation of many of these diseases in this part of the world. Studies on the field have demonstrated that plant-production systems could become a better option than the current fermentation or bioreactors systems to meet the production and availability of therapeutics demanded in the less favourable regions. In this context, the study and development of plant-based antibodies has transformed both the therapeutic options and application field of these molecules. In addition, the generation of highly structured particles such as VLPs in transgenic plants has led to expanding the window into the study of production of important therapeutics in plants.

In this chapter, the generation and production of plant-based antibodies and VLPs for human therapeutic use are reviewed.

2. Antibodies

2.1. Immunotherapy and Monoclonal Antibodies

Antibodies are protein molecules which form part of the immune system of vertebrates. They play a crucial role during immune response by neutralizing the antigen (foreign molecule presented in the organism as an isolated form or as part of an invading organism) and target it for elimination from the cell. There are five main types of serum antibodies or immunoglobulins (Ig), IgG, IgA, IgM, IgD and IgE. An antibody structure basically consists of two pairs of identical polypeptides. One pair is formed by those termed as heavy chains and another one by those polypeptides termed as light chain. The two heavy chains are joined to each other in a specific area of the molecules, the Fc region. The Fc region is also involved in elicit particular immune effector functions (Metzger and Kinet, 1988). Due to the conformation of each heavy chain, this bound structure results in a Y-shaped molecule. One light chain is connected to each arm of the Y molecule forming a complete antibody (Box 1, serum type IgG represents the most common type of antibody formed in the cells). The end of each chain, heavy and light, carries a variable region (Fv) involved in specific recognition of a particular antigen (Kiyoi and Naoe, 2001). Thus, heavy and light chains are formed by two regions, the constant and the variable region. The constant region of a heavy chain is constituted by three or four domains, or modules (abbreviated as Ch1, Ch2, Ch3, Ch4) depending on the Ig class.

The secretory antibodies (mainly sIgA, and less abundant pentameric sIgM, as specialized antibodies of the immune response at mucosal areas) are molecules which present a more complex structure. sIgA consist of two serum-type antibodies attached to each other by a small peptide, the joining chain (J-chain) (Box 1). In addition, they carry a fourth polypeptide known as the secretory component (SC). The SC protects the antibodies from proteases and adverse agents at the mucosal environment. Serum-type antibodies are formed by the specific immune system B cells developed into plasma cells. Secretory antibodies however require two types of cells for its generation, B/plasma cells and epithelial cells.

Monoclonal antibodies (MAbs) are antibodies that are identical because they are produced by one type of immune cell which is a clone of a unique single-parent cell. Thus, their specificity is the feature that makes the monoclonal technology so valuable. As a result, a definitive advantage of these molecules is their diversity of applications in, at least, medical, environmental protection and industrial areas. We will focus only on the human medical field.

After MAbs were developed for first time in 1975, the scientific community has sought to investigate their use in diagnostic and therapeutic application for different diseases (Kohler and Milstein, 1975). In the 1980´s, studies in humans had been already started and at least 150 antibodies in clinical trials and 19 already approved by the Food and Drug Administration (FDA) in clinical use can be currently listed. MAbs are proven as therapeutic molecules to protect against diseases but also to help in the diagnosis of a wide variety of illnesses and detection of any potential abnormal substance in the blood. Thus, the MAbs market is a continually increasing market, which has become one of the fastest growing areas of the pharmaceutical industry with an estimated value from 20.6 billion dollars in 2006 to at least 30.3 billion in 2010, and represents the single largest group of biotechnology-engineered recombinant molecules in clinical trials.

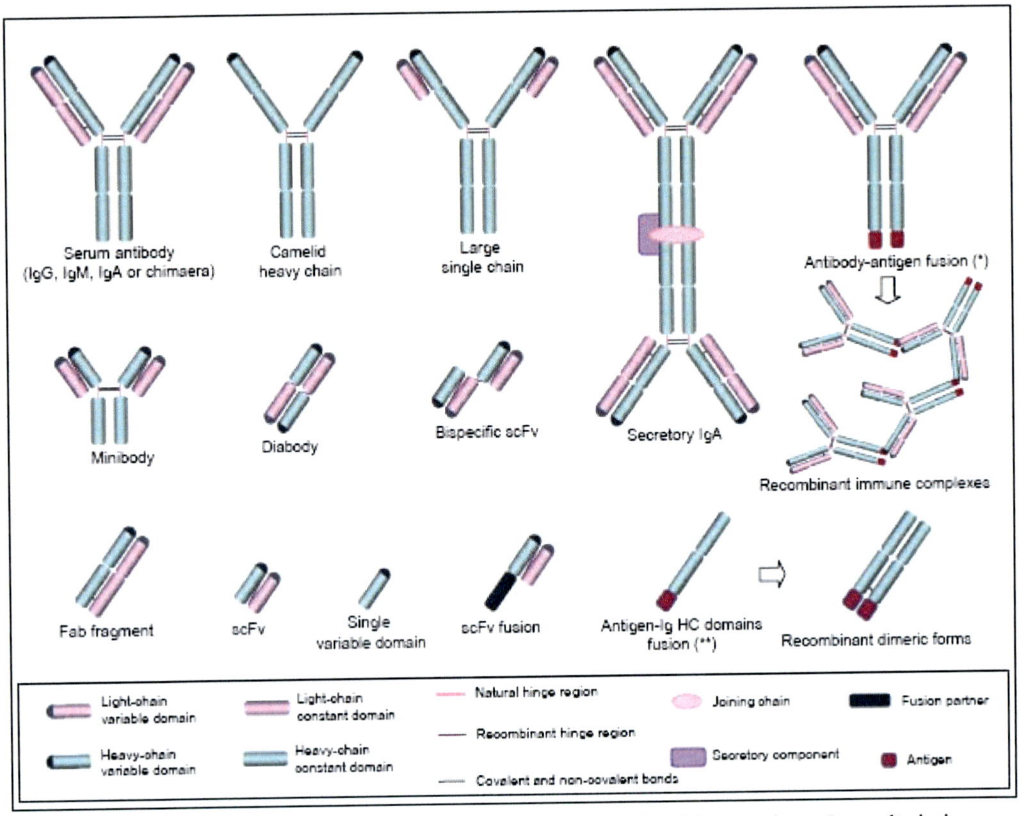

(*): IgG-TTFC fusion molecule. Fusion of a full size IgG antibody with an antigen. It results in immune complexes formation (RICs).

(**): p24/a2-a3 fusion molecule. An antigen is fused to two domains of an IgA heavy chain. It results in dimeric forms.

Box 1. Schematic representation of antibodies and antibody-derived forms expressed in plants.

At present the applicability of monoclonal antibodies has leaded the oncology products to dominate the market, however, other diseases-related products are expected to grow strongly and almost account for half of the manufactured goods by that time (2010). In this context, one of the most important medical targets is the development of efficient treatments for control of important infectious diseases like Tuberculosis, Hepatitis or HIV/AIDS.

Initially, late in the 1970´s no efficient technology was still developed to generate fully human antibodies. The first MAb approved for clinical use was an IgG2 antibody of murinic origin, the named as OKT3, or muromonab (1986) and since other murine antibodies have been generated. However, after several studies in humans one of the most important concerns for immunologists is the problem related to the immunogenicity, side effects (Human anti-mouse antibody response or hAMA) and hampered therapeutic efficacy of the vast majority of murine-derived antibodies generated. In consequence, in the 1980´s it was first reported the development of chimeric antibodies followed by studies on humanization of murine MAb (Jones et al., 1986; Morrison et al., 1984). An important feature that makes antibodies attractive drug candidates is also their organization into distinct structural and functional domains. Accordingly, MAbs are also a useful skill to the design and development of new recombinant molecules with different modules and bifunctional activities. This fact along with the advance of molecular engineering has contributed to the continuing success of the antibody use for clinical application in recent years. In addition, medical challenges increasingly target to the development of molecules with specific structural and functional characteristics necessary for particular medical applications. Thus, the pharmaceutical industry approach is since progressively switching from murine to humanized antibodies, and although the most scheduled objective is the generation of fully human antibodies, the generation of derived-antibody molecules is a crucial fact to the study and development of new therapeutic molecules with improved specific molecular, immunogenic and functional characteristics (Adams and Weiner, 2005; Wu and Senter, 2005);.

Several strategies have been utilized as a mean for the production of MAbs. In a first strategy immortalized human B cells were generated by infection with the Epstein-Bar virus (EBV) and later on the use of cellular engineering (hybridomas or molecules inducer of long-term B cell growth) has been other approach (Banchereau et al., 1989; Borrebaeck and Moller, 1986). However, while in the first case it has allowed the generation and cloning of B cells producing a number of approved and marketed MAbs (antiRheus D, antitetanus toxoid and anti-HIV MAbs) this technique is limited by important problems such as cell line stability and yield of antibody production. Similarly, the cellular engineering initially resulted in a poorly efficient and difficult technique to set up. More recently, however, the co-application of both EBV transformation and proliferation stimulation factors of B cells have lead the cellular engineering to resurge as a technique to produce large quantities of high-affinity specific Mabs (Karpas et al., 2001; Traggiai et al., 2004). Additionally, the use of immunocompromised mice for production of human antibodies from injected human cord blood cells has emerged as a possible new promising strategy. Notwithstanding this, the emergence of new generations of antibodies and antibodies-derived fusion molecules will represent an important subject for the biotech and big pharmaceutical companies to overcome in terms of both scale up and cost of production.

2.2. Plant-Expressed Antibodies

The expression of MAbs in transgenic plants is a well established technology. Hiatt et al, (Hiatt et al., 1989), reported the expression of the first antibody in transgenic plants (IgG 6D4) and the combination of two important growing technologies, the immunology and plant genetic engineering, commenced. Later on (During et al., 1990) reported the expression of another type of immunoglobulin (IgM) in tobacco plants. Since then a significant variety of plant-derived antibodies (also known as "plantibodies") have been generated in a vast range of vascular plants and more recently in algae (Fletcher et al., 2007; Mayfield et al., 2003; Smith and Glick, 2000; Stoger et al., 2002; Timko and Cahoon, 1999). It is interesting to recall that although full antibodies (e.g. IgG) can be generated by a unique type of cells (B cells of the immune system) in native conditions, the generation of secretory antibodies (e.g. sIgA) is a more complex process requiring two different types of cells (B/plasma cells and epithelial cells). Notwithstanding this, both antibody types, full IgG (Hiatt et al., 1989; Mayfield et al., 2003) and sIgA have been reported to be expressed and correctly processed in transgenic plants (Ma et al., 1995; Wieland et al., 2006). Additionally, the expression of plant-derived MAbs in tobacco plants is reported to be at higher levels (IgG 1% - IgA 5% TSP) equal or above the 1% TSP established to be compatible with purification technology and commercial applications (Hiatt, 1990; Hood et al., 2002; Kusnadi et al., 1998; Ma et al., 1995). This fact, along with the plant capability to produce different types of Igs, has lead to investigate the generation of new recombinant full Ig and Ig-derived therapeutic molecules in transgenic plants for clinical application by both routes systemic and mucosal. A schematic representation of the antibodies and antibody-based molecules generated in plant is shown in Box 1.

At least seven different plant species have been explored for antibody production (Hiatt et al., 1989; Hiatt and Ma, 1992; Khoudi et al., 1999; Mayfield et al., 2003; Stoger et al., 2002; Zeitlin et al., 1998) including tobacco, alfalfa, corn, rice, wheat and *Arabidopsis thaliana*. In the latter case the expression of a human anti-rhesus D IgG1 antibody (Bouquin et al., 2002) and, more recently, the expression of an anti-Hataan virus mouse/human chimeric antibody, 3G1MH (A virus that affects rodents and humans in contact with it producing Hemorrhagic fever and death) (Zhou et al., 2008) have been reported.

To date different plant transformation strategies have been used for plant-based antibody production. Two major transformation methods have been used for generation of full size antibodies: one is the nuclear transformation of the heavy and light chain genes separately into individual plants. This system requires the following cross-pollination of those plants expressing one of the two transgenes to yield a new T1 generation of plants carrying both the heavy and light genes and thus to derive into a full assembled antibody (Hein et al., 1991; Hiatt et al., 1989; Hiatt and Ma, 1992). A second alternative is the co-transformation of both heavy and light chain genes on a single plant expression cassette (De Neve et al., 1993; During et al., 1990). Although the first option is the most involved strategy, in both cases it is possible to generate a broad range of novel full antibodies, chimeric antibodies and antibody-derived molecules by combining different sets of modifications in any of the heavy or single antibody chains. In addition, the use of *Agrobacterium*-mediated transient expression (Agroinfiltration) and plant-virus-based vector systems have been also strategies successfully applied (Vaquero et al., 1999; Verch et al., 1998). Higher protein expression levels are achieved by these two last systems, however, due to its transient expression nature the use of these technological alternatives are thought to be possibly aimed for both particular production schemes and biotechnological requirements. Besides, though mainly non-glycosilated proteins are limited to this expression system, the chloroplast transformation has become a popular strategy in the effort to improve recombinant protein accumulation levels and more recently it has been also reported for the expression of specific antibodies (Mayfield et al., 2003).

Protein expression and stability of plant-derived Igs have been also investigated (Conrad and Fiedler, 1998; Fiedler et al., 1998). Plants are very efficient at producing Igs probably because the endomembrane system of plant and mammalian cells are organized in identical manner. In addition, plant chaperones homologous to those within the plant endoplasmic reticulum (ER) in mammalian cells and their interaction with Ig chains have been described. This chaperone-recombinant Ig interaction in plants has been suggested to determine the efficiency of protein folding and assembly (Nuttall et al., 2002). Moreover, studies on the plant expression of Ig heavy and light chain indicate that the assembly status of these molecules is a determinant of protein stability (Hiatt et al., 1989).

Full antibodies and antibody-derived molecules which contain glycosylation sites should be targeted to the plant endomembrane system for glycosylation, disulfide bonding and correct protein assembly. Most proteins are more stable in subcellular compartment or apoplastic space than in the cytosol. Likewise, antibody targeted to the plant secretory system usually accumulates at significantly higher levels than those of antibodies expressed in the cytosol. In several cases it has been reported that the highest accumulation of active full length antibodies was achieved by targeting to the apoplastic space (Conrad and Fiedler, 1998). However, the stability of antibodies is lower in the apoplast than in the lumen of ER where the antibody accumulation is also higher (Schouten et al., 1996). IgA antibodies accumulate predominantly within the plant endomembrane system and, in part, are targeted to vacuoles (Frigerio et al., 2000). (Hadlington et al., 2003) also described higher accumulation levels of the same hybrid immunoglobulin Ig A/G heavy chain expressed in tobacco plants. In this case a C-terminus cryptic sorting signal was identified in the C3□ tail (or Ch3 domain) of the IgA heavy chain as responsible for the vacuolar targeting. However, this subcellular targeting seems not to be disadvantageous for protein accumulation inside the cell as indeed the expression of IgA has been reported to be at higher levels than that of the IgG antibody secreted to the extracellular space. Accordingly, high expression levels of an HIV-1 antigen were achieved by fusion to human IgA sequences, which also contained the same sorting signal, and the recombinant protein was found to be retained inside the cell (Obregon et al., 2006)

Studies on post-translational protein processing have shown differences in protein glycosylation patterns between plants and animal cells. Although the base and frequency of glycans are the same in both systems there are specific differences in the processing of complex N-glycans in plants compared to that in mammalians. Glycan modifications could lead to shorter half life of recombinant antibodies alongside some limitations for its plant-based production and therapeutic use. Moreover, another reason for considering the N-glycan structure of plant-based immunoglobulin important is the potential for inducing immunogenicity/allergenicity reactions. People are continuously exposed to plant N-glycans contained in food and no adverse effects are normally developed. Thus, it would seem unlikely that those epitopes contained in plant-antibody glycans produce a different reaction.

It has been reported that no antibody response was detected against both the protein and the glycan part of a plant recombinant antibody in mice subcutaneously immunized with a plant-derived antibody and aluminum as adjuvant (Chargelegue et al., 2000). On the contrary, a minor but significant immune response to two plant glyco-epitopes was observed in C57BL/6 mice and a stronger response in rats was also described (Bardor et al., 2003). Additionally, an anti-plant glycan immune response could vary especially when there is a nasal or parenteral administration. In consequence, studies on generation of genetic modified plants to provide a glycosylation pattern similar to that from mammalians are being investigated (Bakker et al., 2001; Misaki et al., 2003; Palacpac et al., 1999) ; (Lerouge et al., 2000) ;(Bardor et al., 2006) (Gomord et al., 2005) and production of humanized antibodies expressed in plants has been reported (Schahs et al., 2007). Besides, a private company is also involved in pursuing this strategy (www.greenovation.com).

Table 1. Important therapeutic full length antibodies produced in plants

Pathology	Antibody type	Target and purpose	Production system	Highest expresión level	Relevance and comments	Reference
–	IgG1	Phosphonate ester; Catalitic antibody	Tobacco	1.3 % TSP	First antibody expressed in plants	(Hiatt et al., 1989)
–	B 1-8 antibody (IgM)	Neuropeptide hapten	Tobacco	–	First IgM expressed in plants. First antibody accumulated in plant chloroplast	(During et al., 1990)
Tooth decay	IgA/G (Carox Rx)	*Streptocuccus mutans*; Therapeutic	Tobacco	500 µg/g FW	First secretory antibody expressed in plants	(Ma et al., 1998) and (1995)
Herpes simplex (HSV)	Humanizad Anti-HSV-2 (IgG)	(HSV)-2 gly B; Therapeutic	Soybean	–	First pharmaceutical produced in soybean	(Zeitlin et al., 1998)
Detection of incomplete blood group antibodies	C5-1 (IgG)	Anti-human IgG (AHG); Diagnosis	Alfalfa	1.0% TSP	First example of molecular farming in perennial plant	(Khoudi et al., 1999)
Epithelial cancer	cT84.66 (IgG)	(CEA); Diagnosis and therapeutic	Tobacco Transient (Agrobac.)	1.0 µg/g FW	Functional T84.66 for cancer imaging and therapy	(Vaquero et al., 1999)
Colorectal cancer	Avicidin (IgG)	(EpCAM); Therapeutic	Corn	–	Withdrawn from phase II trials by developers companies	
Hepatitis B	CL4MAb (IgG)	Hepatitis B virus; Therapeutic	Tobacco cell culture	2 % TSP	Feasibility of producing clinically used-anti-hepatitis B monoclonal antibodies in plants	(Yano et al., 2004)
Rabies	SO57 (IgG)	Rabies virus; Therapeutic	Tobacco	0.07 % TSP	First IgG expressed in plants showing therapeutic activity by systemic injection	(Ko et al., 2003)
Tumor carcinomas	BR55-2 (IgG2a)	Lewis Y oligosacaride *; Therapeutic	Tobacco	30 mg/Kg FW	Suitable for efficient cancer immunotherapy	(Brodzik et al., 2006)
HIV-1/AIDS	2G12 (IgG)	HIV-1 gp120 env; Therapeutic	Maize (seeds)	38.8 µg/g DW	First full length anti-HIV neutralizing antibody expressed in plants	(Rademacher et al., 2008)

AHG: Monoclonal anti-human globulin used in anti-human globulin test; DW: Dry weight, FW: Fresh weight, HIV/AIDS: Human Immunodeficiency virus, TSP: Total soluble protein, -: Not reported, (*): Tumor-associated non-protein antigen, IgA/G: Chimeric full length antibody, (HSV)-2: HSV-2 glycoprotein B antigen, CEA: Carcinoembrionyc antigen, EpCAM: Epithelian cellular adhesion molecule, HIV-1 gp120 env: Major HIV-1 envelope antigen

Plants have proven effective as production system for human antibodies manufacture. In summary, they are correctly expressed, folded and assembled, and a yield in excess of 1% TSP has been reported in most cases. Despite differences in glycan pattern generation of plant-derived biological active antibodies has been also demonstrated (Brodzik et al., 2006; Ko et al., 2005; Koprowski, 2005). As consequence, production of antibodies in transgenic plants has moved from theory to Phase II clinical trials, and some biotechnology companies with molecular farming portfolios are currently involved in the scale up production of plant-derived therapeutic proteins for commercial purposes. Some examples are the Large Scale Biology Corp. (http://www.Isbc.com/), Planet Biotechnology (http://www.planetbiotechnology.com/), Medicago (http://www.medicago.com), Epicyte Pharmaceutical (http://www.epicyte.com) and Sigma Aldrich http://www.sigmaaldrich.com/). Other relevant companies involved in the development of plant-based pharmaceuticals are Monsanto (http://www.monsanto.com/), and Neo Rx Corp. (http://www.neorx.com/).

At least 11 full size antibodies have been expressed in plants to target some of the most relevant diseases affecting human health such as cancer, HIV/AIDS or hepatitis B. Important antibodies produced in plant along with detailed information are shown in Table 1. Plant-based antibodies in comparison with those parental antibodies isolated from hybridomas are proven to show conserved biological functions suitable for efficient immunotherapy of cancer and other diseases. Furthermore, although the efficiency of some antibodies depends mainly on binding to virus antigens (neutralizing Abs), both antibody activity binding to antigens and interaction with Fc receptors to exert antibody-dependant cell citotoxicity (ADCC) effectors functions have been confirmed from plant-derived antibodies. These are particularly important antibody functions for cancer therapy and development of an efficient HIV/AIDS vaccine. Some of the most relevant plant-derived antibodies are following described.

Cancer-Immunotherapy-Related Antibodies

One of the first anticancer monoclonal antibodies produced in plants is the counterpart of the *T84.66* mAb. T84.66 recognizes the carcinoembryonic antigen (CEA). This is a well characterized marker of epithelial cancers detected in practically all colon cancers, and at least 50% of breast cancer. It has been successfully tested for both in vivo cancer imaging and diagnosis of colorectal carcinoma (Neumaier et al., 1990). Thus, a functional full length chimeric mouse/human T84.66 antibody (cT84.66) was expressed in tobacco (Vaquero et al., 1999). Chimeric heavy and light chain genes were engineered and the mouse light and heavy chain constant domain sequences were exchanged with their human counterparts. They were cloned into two independent plant expression vectors and transformed into *Agrobacterium tumefaciens*. Tobacco plant leaves were then vacuum infiltrated with both populations of recombinant *Agrobacterium*. T84.66-derivatives molecules have been also engineered and produced in a variety of plant systems for the study of expression efficiency and quality of product of recombinant molecules in transgenic plants (see section “smaller antibody-derivatives”), (Stoger et al., 2002; Vaquero et al., 1999).

Avicidin is a full size IgG that recognizes a specific antigen of colorectal cancer (anti-epithelial cellular-adhesion molecule EpCAM). This antibody has been expressed in corn and was the first one in being administered to humans and show anticancer activity as its counterpart produced in animal cells. It was tested to treat resistant tumors to standard therapies and reduction of tumor size was achieved. However, this is a mouse-derived molecule and limitation in the number of doses to be administered had to be determined due to the hAMA response generated in most patients. Despite this molecule has been withdrawn of the market by the developers (NeoRx and Monsanto) during its test in phase II clinical trials those side effects in patients were not identified as specific to the antibody produced in plants (See review Fischer and Emans, 2000; For the expression of anti-EpCAM antibodies by plant-based virus vectors see Verch et al., 2004; Verch et al., 1998). More recently a murine anti-EpCAM GA733-2 antigen antibody (CO17-1A, an IGg2a) has been produced in tobacco and reported to be efficient in binding SW948 colorectal cells. Furthermore, it has been also demonstrated to inhibit tumor growth of xenotransplanted tumor cells in nude mice. The antibody expression levels achieved were reported to be low (0.02% TSP). However, in this case the antibody was not targeted to the plant ER (Ko et al., 2005).

BR55-2 (IgG2a) is a plant-expressed MAb more recently reported for its application in this medical field. It recognizes a non-protein tumor-associated antigen, Lewis Y oligosaccharide, overexpressed in human carcinomas, particularly in breast and colorectal cancer. It bounds specifically to both SK-BR3 breast cancer and SW948 colorectal cancer cells. In addition, the Fc domain of the plant-derived MAb shows similar Fc□RI receptor (CD64) binding to that of its counterpart from mammalian and comparables levels of citotoxicity against SK-BR3 cells were also shown for both MAbs in antibody-dependent cell-mediated citotoxicity assay. Furthermore, the plant-derived BR55-2 efficiently inhibited SW948 tumor growth xenografted in nude mice.

Alkaloids are toxin compounds that have represented a cause of concern in regard to the purity level of purified recombinant biopharmaceutical from tobacco plants. BR55-2 production in a low-alkaloid tobacco line (var. LAMD609) has first demonstrated that biological activity of antibodies produced in these plants is suitable for efficient cancer immunotherapy and that plants modified to reduce the content of alkaloids are efficient as manufacturer of recombinant antibodies (Brodzik et al., 2006).

Anti-hCG antibody (PIPP) is a full size IgG that recognizes the human corionic gonadotropin (hCG). hCG is a pituitary hormone synthesized and secreted soon after fertilization (Fishel at al., 1984). The use of the PIPP has been proposed as passive immunization strategy for contraceptive purposes (Talwar et al., 1994) and other applications in diagnosis as well as prognosis and immunotherapy of cancer (Okamoto et al., 2001). Scientists, (Kathuria et al., 2002), have investigated the expression of a chimeric full length IgG antibody (and derivative molecules, scFv and diabody; see section 2.3 and Box 1) from the original anti-hCG murine MAb (PIPP) in transiently transformed tobacco plants. The plant-derived antibody conserved its specificity for the B-subunit of hCG secreted by the most aggressive hCG-producing tumors (B-hCG) (Butler et al., 2000). In addition, the hCG induced testosterone production by Leydig cells was inhibited by the plantibody in vitro and in vivo studies demonstrated that the use of plant-derived PIPP antibodies block the hCG induced-increase in mouse uterine weight. The antibody studies on efficacy demonstrated that the full length antibody was up to 1000 fold more active than its derivative molecules. These results indicate that passive immunization with the hCG-specific plantibody may have clinical utility in diagnosis and therapy for hCG-expressing cancers and as contraceptive measures.

Non-Cancer Disease Immunotherapy-Related Antibodies

Other clinical areas have also been explored for the expression of antibodies in transgenic plants.

The termed *Carox Rx* antibody is a first example. It recognizes the antigen from *Streptococcus mutans*, the pathogen bacteria responsible of tooth decay in humans, and has been shown to prevent oral colonisation by those bacteria in phase II clinical trials in humans (Larrick and Thomas, 2001; Ma et al., 1998). This molecule is a chimeric secretory antibody (IgA/G) produced in transgenic tobacco plants by crossing of four independent transformed plants with a single transgene each of them. By this study it was demonstrated that plants have the ability to accumulate transgenes to express more complex multimeric proteins as these types of antibodies. Carox Rx was shown to have similar antigen affinity as its parental murine IgG antibody, however, its half life was three fold higher that that of its murine parent molecule due to its dimeric structure (see secretory Ig in Box 1). Planet Biotechnology Inc. completed US FDA-approved phase I and II clinical trials in 2000. However, further results have not been reported.

The Anti-HSV-2 is a humanized full length IgG antibody under development by The Epicyte Pharmaceutical biotechnology Company. It recognizes the Herpes simplex virus (HSV)-2 glycoprotein B antigen and has been expressed in transgenic soybean plants. It has been shown to prevent vaginal HSV-2 transmission as its counterpart produced in CHO cells (Chinese ovarian cells) in mice after topical application. Its stability in human semen and cervical mucus as well as its ability to disperse in cervical mucus was also investigated. Thus, the use of plant-derived antibodies would represent a good cost-effective preventative option for sexually transmitted diseases if its efficiency is proven in humans (Zeitlin et al., 1998).

SO57 is a human anti-rabies virus neutralizing MAb. Rabies is a major zoonotic disease and thus has a significant impact in global health. A vast majority of human rabies death is sited in developing countries, particularly in India where more than 50% of the worldwide rabies death are reported. In addition, limitations inherent to the combined administration of Igs (RIG) used currently as vaccine have leaded to look into the use of monoclonal antibodies as a safer and efficacious alternative. (Ko et al., 2003) have reported the expression of SO57 in tobacco plants. This MAb has as effective neutralizing activity of the rabies virus as the mammalian-derived antibody or human rabies Igs formulation (HRIG). In addition, although it shows a shorter half-life than mammalians MAbs it is as efficient as (HRIG) for post-exposure prophylaxis against rabies virus in hamsters (Ko et al., 2005; Ko et al., 2003).

CL4MAb is an anti-hepatitis B antigen (HBsAg) monoclonal antibody. Chronic Hepatitis B infection occurs worldwide and more than the 10% of population in developing countries is infected. The current immunization methods are based on the use of serum from antibody positive donor's and the product safety has to bc assured by strict product standards. However, both the use of donors and the standard production controls are expensive for its application in developing countries. Human CL4MAb has been expressed in tobacco suspension cultures. Parallely, monoclonal antibodies derived from parental CL4MAb (B294 and B303) were selected and produced in plant. B294 and B303 MAbs have shown anti-HBs relative affinity similar to that of the parental CL4MAb. In addition, B303 MAb showed to induce complement-dependent cytotoxicity in a manner that was similar to the clinically used anti-HBs human Igs (HBIg). These results have pointed out the production of HBsAg monoclonal antibodies in plants as a feasible alternative to HBIg (Yano et al., 2004)

2G12 is the most recent IgG produced in plants. The parental 2G12 is a potent and broadly active human anti-HIV-1 neutralizing antibody. It recognizes oligomannose-type- N-glycans on the major HIV-1 gp120 envelop antigen. In vivo studies have demonstrated that it is efficient in prevention and transmission of HIV infection (Hofmann-Lehmann et al., 2001;

Mascola, 2002; Mascola et al., 1999; Veazey et al., 2003). HIV infection is one of the worst epidemic currently affecting the global health. In addition, approximately 95% of the total HIV infected-individuals live in developing countries. As example Sub-Saharan Africa which has been particularly hit by the HIV epidemic counting with almost 70% of all HIV-infected people living there. In most of those countries, poverty and lack of resources are a constant that makes the necessary caring for HIV-infected people and effective prevention efforts practically impossible. This fact has worsened the situation by collaborating to a major dissemination of the infection. Therefore, an efficient and cost-effective HIV vaccine is urgently needed for its application worldwide, particularly in the less favourable regions.

Plant-derived 2G12 has been produced in maize seed endosperm and studies demonstrate that it presents antigen-binding properties and neutralizing activity in vitro nearly identical to its counterpart produced in CHO cells. Although differences in the glycosylation patter have been identified, the complete functionality of this corn-derived antibody has been confirmed.

Additionally, the use of the *Discosoma* red fluorescent protein (DsRed) as a visual marker has been reported in this investigation. Thus, an easy strategy to identify genetically modified maize and facilitate plant breeding optimization to selection of the best transgenic lines has been also proposed (Rademacher et al., 2008).

2.3. Antibody-Based Molecules Expressed in Plants

An alternative to the engineering of humanized full size monoclonal antibodies is the generation of antibody-derived molecules and fusion proteins (FPs) (Huston et al., 1991). In the first instance the generation of smaller antibody-derivatives includes the engineering of one or more antibody modules tailored to suit a particular therapeutic use. An example is the development of bispecific antibodies (BsAbs) which may present different antigen affinity patterns and crosslink different functions (e.g. effector and target cells) (Weiner, 1999). One of the limitations of using large molecular size Igs is the undesirable increased time in circulation of these molecules when used in clinical applications. More to the point, although in some cases it may be required, the effectors functions mediated by the Ig Fc region are another undesirable aspect in some applications. Thus, initial attempts to generate smaller antibody fragments were based on the digestion of intact IgGs by proteolytic enzymes such as pepsin and papain by which antigen binding fragments (Fab´or $F(ab')_2$) were obtained. The main advantages of these fragments compared to complete Igs are that they conserve their specific antigen binding activity, are usually cleared from the blood stream faster that the complete Igs but acquire sufficient half life to make useful clinical applications (important aspect to determine the efficiency of these molecules in the medical field) and show a higher penetration of target tissues. As a consequence, they are frequently more effective as drugs than full size Igs.

FPs are artificial antibody-non-antibody protein fusion molecules which are able to retain the functionality of both fusion partners. The generation of FPs was initially achieved by covalent link of antibodies to other non-Ig molecules such as haptens, peptides or toxins. Later on, in the 1980,s, the introduction of genetic engineering represented a leap forward in this area of research. The tools necessary to develop gene fusion procedures in which a protein or peptide could be fused to another one at DNA level had been achieved. The main advantageous characteristic of FPs lays on the capability of these engineered molecules to exhibit different origin-protein modules with different functional activities. Thus, to help prevent misfolding of FPs the introduction of a small DNA sequence, which uses to be only a few amino acids, between the different fusion partner domains is a common feature in their molecular design. At present there are at least two FPs (Etanercept, Abatacept) produced by standard production systems in clinical trials or market, and the design of new more complex FPs is increasing as well as it has opened the study to new therapy approaches (Moreland et al., 2002; Weinblatt et al., 1999).

A wide range of antibody-based molecules for clinical application have been generated by standard systems and tested in clinical trials with initial success (Ward, 1993); (Kipriyanov and Gall, 2004). However, clinical application has been hampered by high costs of development and production in bulk by those procedures. As alternative, the generation and production of these molecules in transgenic plants have been investigated and at present the plant expression of most of those Ig-derived molecules has been demonstrated. Furthermore, the absence of glycosylation and disulfide bond requirement in some cases represents an advantage for the expression of these molecules in the plant-system. A schematic representation of the antibody-based molecules produced in plants is shown in Box 1.

Smaller Antibody-Derivatives

The expression of this type of molecules in plant include single variable region and single chain Fv molecules (scFv) as univalent molecules; Bispecific scFv which recognizes two different antigens; and minibodies, diabodies and large single chains (LSC) as dimeric molecules (Benvenuto et al., 1991; Conrad and Fiedler, 1998; Schillberg et al., 2003; Stoger et al., 2002). Additionally, expression of camelid antibodies in transgenic plants has been also reported (Ismaili et al., 2007; Jobling et al., 2003). Camels and Llamas produce a unique class of antibody termed heavy-chain antibodies. This antibody is formed by a dimmer of a unique heavy chain and lacks of light chains. Hence, it favours the expression of the complete antibody molecule by a single transgene. It comprises intrinsic features of wide interest in antibody engineering and biotechnology fields. First, their variable domain (V_HH or nanobody) is the smallest domain (15 kDa) capable of independently binding an antigen. Additionally, an important fact is that the V_HH have longer hipervariable or complementarity determining regions (CDR) than those of the conventional Ig variable domains (V_H) which enable it to improve penetration on hidden epitopes on the target antigens. Secondly, V_HHs display higher thermal and intracellular stability, solubility and affinities compared with the V_H (De Genst et al., 2006; Nguyen et al., 2001), and an enthusiasm on its use in genetic engineering and research to medical applications has been generated. Some of the most important antibody-derivatives molecules expressed in plants are shown in Table 2.

ScFv are small (~30 KD) monovalent molecules with a useful short serum half life for medical applications and diagnosis. This type of molecules are formed by the variable regions of both the antibody heavy and light chains linked by a small flexible peptide. They have been produced in bacteria, yeast, insect cell and mammalian cell culture (Jost et al., 1994; Ridder et al., 1995; zu Putlitz et al., 1990). However, the requirement of difficult purification procedures, low yield of recombinant protein in some cases and high production costs are some limitations intrinsic to these expression systems. Besides, their uncomplicated assembly and lack of processing through the ER make the scFv molecules an excellent candidate for production in transgenic plants. Hence, scFvs have become the group of antibody-derived molecules most commonly expressed in the plant system. In a first molecular approach scFv were targeted to plant cell apoplast and cytosol but only low protein expression levels were achieved (< 0.5% of total soluble protein, TSP). However, studies on the expression of these molecules in different plant species (tobacco, corn, rice), plant organs (leaves, seeds, fruits) and subcellular compartments (e.g. ER retention by fusion of the KDEL sequence to the transgene) have led to obtain a significant increase in recombinant protein accumulation levels (up to 3% TSP in seeds and 6.8% TSP when the transgene is retained in ER), (Conrad and Fiedler, 1998; Fiedler et al., 1997; Fischer et al., 1999). Additionally, high expression levels of a LSC (Box 1) in algae chloroplasts has been reported (Mayfield et al., 2003). As a result, the major limiting factor of recombinant molecule accumulation was determined to be the protein stability; this effect however could be partially controlled by subcellular targeting. Thus, the plant production system has also turned out in a promising strategy for the development of at least a number of these chimeric molecules. Relevant plant-based smaller antibody-derivatives are following described.

scFvT84.66 is a derivative molecule of the IgG1 T84.66 anti-cancer antibody (table 2). For its expression in rice and wheat it was targeted to the cell apoplast and endoplasmic reticulum (ER). It represents the first step in the transition of antibody production from model plants as tobacco and *Arabidopsis* to widely cultivated cereals crops. Thus, it has been demonstrated the viability of molecular farming in cereals for important pharmaceutical production and their potential to development of edible plant-derived therapeutics for oral administration. Other T84.66-derived molecules expressed in plants are the scFv-CH3 and T84.66/GS8. The scFv-CH3 molecule was formed by an antibody fragment engineered to contain the sequence VL-VH-CH3 of IgG antibody modules and thus a minibody generated (Hu et al., 1996). Differently, the T84.66/GS8 is a diabody or bivalent molecule (Box 1). All of T84.66 antibody-derivatives recognize the carcinoembryonic antigen (CEA) and have been reported to conserve antigen specificity when expressed in plants.

PIPP scFv and PIPP diabody are recombinant anti-Human Chorionic Gonadothropin (hCG) antibody-derivatives. See (Fishel et al., 1984) about hCG fisiological function. As mention before, measurement of hCG levels are widely used in pregnancy testing, HCG neutralizing antibodies as contraceptive measure (Bagshawe, 1983; Talwar et al., 1994) and also the use of anti-hCG antibodies are used in diagnosis and immunotherapy of a number of hCG producing cancers (carcinoma of bladder, pancreatic carcinoma, cervix carcinoma, carcinoma of the nervous system) (Acevedo et al., 1997; Hameed et al., 1999; Okamoto et al., 2001); (Butler et al., 2000). Β-hCG subunit molecule can be recognized by monoclonal antibodies, and passive immunization with antibodies prevents tumor growth.

The scFv PIPP molecule and PIPP diabody have been transiently expressed in tobacco leaves. The antigen specificity for the Β-hCG of the antibodies was confirmed. Plant-ScFv was shown to neutralize hCG activity in vitro almost in an identical manner to that exerted by the plant-diabody. This result has been proposed as the effect of the formation of scFv dimeric molecules. Thus, the efficiency of the recombinant PIPP antibody forms was confirmed. Since any of the engineered molecules retained its efficacy when investigated in vitro, any of them could potentially be used in pregnancy detection, diagnosis and/or treatment of hCG producing tumors (Kathuria et al., 2002).

Table 2. Relevant antibody-derived molecules for clinical use produced in plants

Pathology	Antibody type	Target and purpose	Production system	Highest expression level	Relevance and comments	Reference
CD40-non-Hodgkin's lymphoma	ntBD1-G28-5 sFv (scFv)	CD40 expressing cancer cells; Therapeutic	Tobacco cell culture	–	First pharmaceutical ScFv produced in plants.	(Francisco et al., 1997)
38C13 B cell non-Hodgkin's lymphoma	Murine 38C13 (scFv)	Idiotype vaccine; Therapeutic	*N. benthamiana* Transient	30.2 µg/g leaves 100 µg/ml IF	Efficient anti-specific malignant idiotypes response in mice and activity on human B-cell tumor. Completed Phase I trials	(McCormick et al., 1999); (McCormick et al., 2003)
Epithelial cancer	T84.66 (scFv) (a), (b) and (d), T84.66/GS8 (diabody) (c)	(CEA); Therapeutic and diagnosis	(a): rice and wheat; (b):Tobacco Transient; (c): Tobacco; (d): Pea	(a): Up to 30 µg/g in leaves and seeds of both species, (b): 5 mg/kg leaves; (c): 5 mg/ kg in apoplast and 12 mg/Kg in ER; (d) 9 µg/g seed	Expression of a clinically efficient anti-cancer antibody in widely cultivated cereals crops	(Stoger et al., 2000) ; (Vaquero et al., 1999) ; (Vaquero et al., 2002) ; (Perrin et al., 2000)
Pregnancy test and hCG producing-tumors	PIPP (IgG) (scFv) and (diabody)	(hCG) ; Therapeutic and diagnosis	Tobacco Transient	20-40 mg of pure protein / kg FW	For use in contraceptives, standard pregnancy test and treatment/ detection of hCG- expressing tumor	(Kathuria et al., 2002)
Herpes simplex (HSV)	LSC	(HSV)-Glycoprotein D; Therapeutic	*C. reinhardtii* Chloroplast transformation	> 1% TSP (deduced)	First example of molecular farming in algae and antibody expression in algae chloroplasts	(Mayfield et al., 2003)

Table 2. (Continued)

Pathology	Antibody type	*Target and purpose*	Production system	Highest expression level	Relevance and comments	Reference
Tetanus	IgG-TTFC (FP) (IgG2a)	*Clostridium tetanii* (TTFC); Therapeutic and vaccine development	Tobacco	> 0.8% TSP	First example of functional recombinant immune complexes. Candidate as a new vaccine model	(Chargelegue et al., 2005)
HIV/AIDS	p24/□2-□3 (FP) (Antigen-antibody)	HIV; Diagnosis and vaccine development	Tobacco	1.4% TSP	First functional HIV-1 Ag-Ig FP in plants. Strategy to new vaccine model design	(Obregon et al., 2006)
Eye diseases	Anti-CD4/CD28 (scFv)	CD4/CD28 molecules; Therapeutic and diagnosis	Wheat (seeds)	50-180 μg/g FW	Plant system for eye drops development. Reduced risk of endotoxin contamination	(Brereton et al., 2007)

DW: Dry weight, FW: Fresh weight, IF: Crude plant interstitial fluid (viral vector-based protein secretion system); resultant fraction after vacuum infiltration, centrifugation and sterile filtration from plant leaves, ER: Endoplasmic reticulum, HIV/AIDS: Human Immunodeficiency virus, HIV-1: Human immunodeficiency virus type 1, LSC: Large single chain, TSP: Total soluble protein, CD40: Cellular molecule; ntBD1-G28-5 sFv: scFv (CD 40)-bryodin 1 immunotoxin, CEA: Carcinoembrionyc antigen; hCG: Human chorionic gonadotropin; HSV: Herpes simplex virus, TTFC: Tetanus toxin fragment C, CD4 and CD28: molecules involved in specific T-cell stimulation during cellular-mediated immune response, LSC: Large single chain, IgG-TTFC (FP): IgG antibody-tetanus toxin C Fragment fusion molecule, HIV-1 p24-IgA (FP): HIV-1 p24 antigen-Immunoglobulin A heavy chain α2-α3 domains fusion molecules. Ag: Antigen.

38C13 is a scFv molecule. It is based on the malignant idioptype of B lymphocytes from the mouse lymphoma cell line 38C13. This antibody was produced in plant by using a virus-based expression system. A tobamoviral vector was engineered to code the idiotype-specific scFv fragment of the Ig and a rapid technique for production of tumor-specific vaccines in transgenic plants reported (McCormick et al., 1999). High expression levels of scFv molecule secreted to the extracellular compartment were also achieved. In addition, mice immunized with plant-derived 38C13 were protected against a lethal dose of the syngeneic 38C13 tumor in a similar manner to that of the native 38C13 IgM-keyhole limpet hemocyanin conjugate vaccine. Therefore, it has been reported that this technique could provide a viable strategy for production of therapeutic molecules for non-Hodgkin's lymphoma treatment. Although scFv 38C13 was not expressed in stable transgenic plants the rapid production of plant-derived recombinant molecules achieved by this system could suit to lower scale-up productions for the development of specific personalised treatments. Studies of the plant 38C13 Ig molecule on phase I clinical trials have been completed by Large Scale Biology Corp. (LSBC). LSBC´s further clinical developments have not been reported.

Anti-CD4/CD28 scFv. It has been also reported that scFv antibody fragments are promising candidates as topical drug for ophthalmic use. In contrast to full size antibodies the scFv molecules can penetrate well into the eye when applied as eye drops. Bacteria culture is the expression system most used for production of scFv and antibody fragments. However, the compulsory endotoxin removal during purification is a difficult and high cost process. The endotoxin cause eye inflammation and thus elimination of these molecules is a crucial requisite in preparation of ocular therapeutic formulations. Transgenic plants provide a system low in endotoxin containment and free of animal pathogens for protein pharmaceuticals production.

The CD4 and CD28 are molecules involved in stimulation of T-cells during a cell-mediated immune response. In some cases, this can produce undesirable effects against donor tissue from recipients of corneal implants. An anti-CD4/CD28 scFv molecule has been recently generated in a commercial variety of wheat (Westonia) (Brereton et al., 2007). High recombinant protein expression levels similar to those achieved by bacterial system for commercial purposes were obtained and anti-CD4/CD28 specificity was confirmed in a rat model. The use of the plant-derived anti-CD4/CD28 scFv in corneal transplants and treatment of eye diseases is proposed.

Antibody-Based Fusion Molecules

There are two main types of FPs. A first group is represented by molecules which retain the antibody specificity. They are formed by either a complete antibody or antibody fragments (e.g. scFv, Fab fragments) fused to molecules other than antibodies. This strategy results in the generation of bifunctional molecules and has been particularly used for oncology-therapeutic molecules design (Berger M et al., 2002). A second group is formed by FPs which retain the Fc portion of an immunoglobulin fused to other types of molecules such as receptors or antigens. The Fc region can provide different advantages to these types of molecules. It facilitates formation of dimeric forms and thus increases the avidity to bind the target molecule of the otherwise monovalent forms. It also increases the serum half life and tissue diffusion of the molecules which contain it and eventually facilitates the clearance of immune complexes. In addition, the Fc region can interact with Fc receptors expressed by immune cells which can result in stabilization of the fusion molecules interaction and trigger effectors functions (antigen uptake and presentation and ADCC).

Generation and expression of antibody-based fusion molecules in transgenic plants have been also investigated (Box 1). Plant-derived fusion proteins have been designed to meet different objectives. Some of them are involved in facilitating recombinant protein immunochemical detection and purification (Kuhnel et al., 2003; Schouten et al., 1996; Semeniuk et al., 2003), increase of recombinant protein production in transgenic plants (Obregon et al., 2006; Scheller et al., 2006) or the design, study and production of bifunctional chimeric molecules for therapeutic use (Chargelegue et al., 2005; Obregon et al., 2006; Vaquero et al., 2002) (Tabla 2). Two antibody-based fusion pharmaceuticals which have been successfully expressed in tobacco plants are following described.

IgG-TTFC

It is one of the most complex functional recombinant molecules obtained in the field of plant-based antibody engineering. The production of recombinant immune complexes (RICs) is described for first time and plant-derived functional RICs are proposed as a new vaccine model.

The co-administration of antigen and antibody has been determined to form immune complexes and enhance the immunogenicity of the antigen. Thus, the use of immune complexes preparations is a relevant option for the design of vaccines with the advantage that not adjuvant may be needed. However, the optimum antigen-antibody ratio necessary in this type of formulation is not easy to achieve and so either the use of immune complexes for vaccine development.

In this work a murine anti-tetanus toxin fragment C monoclonal antibody (murine Mab 278.02, IgG2a) fused with the Tetanus toxin fragment C antigen was engineered and expressed in tobacco plants. The 3´end of the IgG heavy chain DNA sequence was fused to a DNA sequence encoding the Tetanus toxin fragment C (TTFC) and a mouse IgG heavy chain-Tetanus toxin fragment C fusion molecule generated (Heavy chain-TTFC). The Heavy chain-TTFC genetic fusion was generated to ensure an optimal expression ratio between both fusion partners. Plants stably transformed with this chimeric Heavy chain-TTFC genetic construct were crossed with transgenic tobacco plants expressing the counterpart mouse IgG light chain and the full length IgG antibody-Tetanus toxin Fragment C fusion molecule obtained (IgG-TTFC) (Box 1). The recombinant IgG antibody binding region (Fab) specific to recognize the Tetanus toxin was reported to maintain its affinity and functionality and the formation of plant-derived IgG-TTFC antibody-antigen immune complexes demonstrated (Box 1). In addition, it was reported that RICs elicited antibody titers that were at least 10,000 fold higher than those achieved in mice immunized with the Tetanus toxin antigen alone, and adjuvant administration was not required. Furthermore, full protection of IgG-TTFC subcutaneously immunized mice was demonstrated with lethal Tetanus toxin challenge.

It has been demonstrated that adequate fusion between antibody and antigen results in an ensured optimal ratio between both moieties and entirely functional recombinant immune complexes for the development of a new vaccine model (Chargelegue et al., 2005).

p24/α2-α3

A chimeric HIV-1 antigen-human antibody (IgA) fusion protein was expressed in tobacco plants. The enhancing effect of antibody-derived sequences on the plant-expression of non-Ig proteins was demonstrated for first time.

The HIV-1 p24 antigen is a highly conserved core protein (It belongs to the termed Gag proteins or group-specific antigens) and it is likely to be an important part of any multicomponent HIV vaccine. Studies have demonstrated anti-Gag cross-clade antibody response in HIV infected individuals, the absence of anti-Gag humoral response has been correlated to disease progression and p24 antigen has been identified as target of T-cell immune response in infected individuals, which is probably the most important protective mechanism against HIV (McMichael and Rowland-Jones, 2001). HIV/AIDS has particularly hit the developing countries and the current anti-HIV therapeutic products are too expensive for these regions of the world. An efficient and cost-effective system for production of HIV pharmaceuticals/vaccines is crucially needed. The p24 antigen has been expressed in different production systems (*Escherichia coli*, yeast and baculovirus) and preliminary studies on the expression of this antigen in transgenic plants had been done (Zhang et al., 2002).

Now, a new strategy for the high expression of the p24 antigen in plants was investigated. A DNA sequence encoding the heavy chain Ch2 and Ch3 (α2-α3) domains of a human IgA antibody was fused to the 3´end of the HIV-1 p24 antigen DNA sequence. As a result, an HIV-1 p24/IgA heavy chain α2-α3 domains antigen-antibody fusion molecule was generated (p24/α2-α3) (Box 1). Transgenic tobacco plants expressing p24/α2-α3 fusion molecule were generated. Alternatively, tobacco plants expressing the p24 antigen alone were generated as control of the antigen expression level. In the context of the p24/α2-α3 antigen-antibody fusion molecule the p24 antigen expression levels were up to 13 fold higher than those of the antigen expressed alone. In addition, the formation of p24/α2-α3 dimeric forms by a functional IgA heavy chain α2-α3 domains fusion partner was demonstrated (Box 1). In addition, the plant-derived HIV-1 p24 antigen expressed as p24/α2-α3 fusion molecule conserves its protein conformation and antigenicity. Furthermore, anti-HIV-1 p24 antibody response and stimulation of T-cell proliferation was demonstrated in mice subcutaneously immunized with purified p24/α2-α3 fusion molecule. The relevance of these findings for either, the protein expression, and stability, of non-Ig proteins in plants and applicability of antigen-antibody domains fusion molecules in the medical field is reported.

The use of Ig fusion partner is proposed as a generic technology for both increasing the production of high value-proteins in transgenic plants and design of antigen-antibody fusion protein based-vaccines with specific immunological characteristics (Obregon et al., 2006)

3. Virus-Like Particles

3.1 Virus-Like Particles Produced in Plants: Expression, Particle Assembly and Immunogenicity

Virus-like particles (VLPs) consist of highly repetitive and ordered structures that self-assemble structural proteins derived from many viruses. VLPs lack viral nucleic acid and hence they are non-infectious. One or more chimeric capsid proteins can be expressed and self-assemble.

Up to date, the major application field of VLPs is their use for vaccine development, especially as a safer alternative to attenuated live or inactivated killed virus-based vaccines, which include non-cultivable viruses (Santi et al., 2006). Since a single particle presents thousands of epitopes, VLPs are commonly more immunogenic than subunit vaccines or recombinant protein immunogens, and are able to stimulate both the humoral and cellular arms of the immune system. VLPs can be targeted to antigen presenting cells (APC) indirectly by endosomal uptake or directly displaying or incorporating specific molecules that interact with dendritic cells. Contact with APC induces their maturation and triggers a lymphoproliferative response together with production of inflammatory cytokines (Santi et al., 2006). In addition, VLPs are capable of inducing strong cellular and humoral responses as direct immunogens without the use of adjuvants. So far, most VLPs derived from different viruses have proven to be strong stimulators of the immune system and several studies have been performed. As a result, the first recombinant vaccine against hepatitis B, consisting on yeast-produced VLPs, was marketed in 1982. During the past decade, subviral particles of different origins have been considered as potential vaccines for cognate virus infections including HIV, Human papillomavirus (HPV), Norwalk virus (NV), rotavirus or parvovirus (Ludwig and Wagner, 2007). While HPV VLPs produced in yeast or insect cells have recently made successful vaccines, VLPs from pathogens that directly affect immune cells and those that successfully evade the immune system, such as HIV and hepatitis C virus have proven to be more challenging.

Commercial vaccines composed of VLPs are administered by injection. However, VLPs could also be used as oral vaccines since the compact, highly ordered structures very likely provide resistance to degradative enzymes in the gut (e.g. NV VLPs are acid and protease resistant). The particulate nature of VLPs allow them to be efficiently sampled by the "M" cells of the gut epithelium that overlie the gut-associated lymphoid tissue and be efficiently transported across the mucosal barrier and presented to dendritic cells. Moreover, the presence of a structure that mimics the authentic viral particle may present a "danger signal" that can overcome the perception of gut antigens as benign and thus provoke potent immune responses (Huang et al., 2005b).

Expression of viral genes and VLP assembly has been performed in different heterologous systems such as *E. coli* and other bacteria, yeast (*Saccharomyces cerevisiae* and *Pichia pastoris*), baculovirus/insect cell system and mammalian cell lines. Yeast system has some advantages as easy of expression and ability to scale-up. However, considerations such as codon optimisation, correct assembly and protein glycosylation patterns make necessary alternative production systems. *E. coli* neither allows for glycosylation and presents difficulties to in vivo folding of VLPs. Baculovirus-driven expression of VLPs overcomes the majority of these problems and mammalian cell culture systems are favoured for appropriate modifications and authentic assembly. However these are high cost demanding systems (Grgacic and Anderson, 2006).

Expression of antigens in plants has been proposed as a novel way of generating inexpensive subunit vaccines. As mentioned before, production of protein antigens in plants is substantially cheaper than in other systems and offers significant advantages over other vaccine development strategies in cost, manufacturing, packaging, storage, transportation, and most important, in the ease of administration and safety of the vaccine. Vaccine production in plants offers the possibility of parenteral administration after convenient purification steps or oral administration by expression of proteins in edible organs. In the later case, plant cell wall would potentially protect the antigen in the stomach and intestine (Tacket and Mason, 1999). The production of a subunit vaccine in transgenic plants was first reported by Mason et al. (1992). A DNA fragment encoding the hepatitis B surface antigen (HBsAg) was inserted into the genome of tobacco plants using *Agrobacterium*-mediated transformation. In this system, the expressed HBsAg retains the capacity to self-assemble into highly immunogenic VLPs (Tacket and Mason, 1999).

Besides to their use as prophylactic vaccines, VLPs can also be exploited for the presentation of foreign epitopes and/or targeting molecules on chimeric VLPs, such as small therapeutics. Of special interest is their use as viral vectors carriers for gene therapy and DNA vaccines. Hepatitis B virus (HBV) envelope L particles display a peptide that is indispensable for liver-specific infection by HBV in humans. Nucleic acids were introduced in yeast-derived L particles and specific targeting was demonstrated in human hepatocytes cultures and in a mouse xenograft model (Yamada et al., 2003). Chimeric HBV VLPs including a synthetic DNA fragment encoding a poliovirus epitope induced neutralizing antibodies against poliovirus (Delpeyroux et al., 1986). In addition, taking advantage of their structural stability and tolerance towards manipulation, VLPs can also be used as building blocks for novel nanomaterials (Ludwig and Wagner, 2007). The tobacco mosaic virus-derived capsid protein has proven to be a suitable scaffold for chemical modification allowing assembly of nanobiopolymers (reviewed in (Gleba et al., 2007).

Correctly assembled VLPs have been produced in plants by three different methods. First, plant virus vectors can be used to achieve transient expression of the foreign protein. This approach usually results in rapid and high yields of protein. Second, transgenic plants can be engineered to express VLPs through the stable integration into the plant nucleus genome, but protein yields achieved by this approach are generally low. The third approach is to insert the desired gene into the chloroplast genome. Despite some publications on VLP assembly by infection with viral vectors, most of the reported articles use nuclear transformation. More recently, it has been reported assembly of VLPs in chloroplasts by plastid transformation (Fernandez-San Millan et al., 2008). This system has a number of advantages over the insertion of genes into the nuclear genome (Daniell et al., 2002; Maliga, 2002; Maliga, 2003), but one of the most interesting for VLPs expression is the achieved high expression level of the recombinant proteins (up to 46% of total soluble protein, TSP, (De Cosa et al., 2001).

One of the main advantages of plant-derived vaccines is the possibility to practice the oral administration. The adoption of oral vaccines will help to provide an effective vaccination strategy with major benefits for generation of both mucosal and systemic immunity and ease of administration.

Because of the ease of transformation and *in vitro* culture, tobacco and potato plants have been used as a proof-of-concept of plant-derived vaccines. However, tobacco leaves are not edible and the high levels of alkaloids, make protein extraction and purification a necessary step prior to oral delivery. Raw potatoes are not eaten, but cooking would denature the proteins and greatly reduce or eliminate the immunogenicity. Alternative edible plants (e.g. banana, tomato, maize, lettuce and alfalfa) are being used for studies on plant-derived vaccine

Table 3. Virus-like particles produced in plants

					Immune response		
Pathogen	Antigen	Production system	Highest expression level	VLP proof	IP[1]	Oral[2]	Reference
HBV	HBsAg	Tobacco	66 ng/mg TSP	Gradients, EM	-	-	(Tacket and Mason, 1999)
	HBsAg	Potato tubers	8.5 μg/g FW	EM	-	+	(Kong et al., 2001)
	HBsAg	Tobacco cell culture	2 μg/g FW	Gradients	-	-	(Kumar et al., 2003)
	HBsAg	Tobacco cell culture	5.8 ng/μl 226 ng/mg TSP	Gradients	+	-	(Sojikul et al., 2003)
	HBsAg M protein	*N. benthamiana*	394 ng/mg TSP	Gradients	+	-	(Huang et al. 2005b)
	HBsAg	Rice seeds	31.5 ng/g DW	Gradients, EM	+	-	(Qian et al., 2007)
	GFP-HBsAg	*N. benthamiana* Transient (*Agrobac.*)	24.7 ng/mg TSP	Gradients, EM FM	-	-	(Huang and Mason, 2004)
	HBcAg	Tobacco	24 μg/g FW	EM	-	-	(Tsuda et al., 1998)
	HBcAg	*N. benthamiana* Transient (MagnICON)	2.38 mg/g FW	Gradients, EM	+	+	(Huang et al., 2006)
	HBsAg	*N. benthamiana* Transient (MagnICON)	295 μg/g FW	Gradients, EM	+	-	(Huang et al., 2008)
	HBcAg	*N.benthamiana* Transient (PVX)	100 μg/g FW	EM	-	-	(Mechtcheriakova et al., 2006)
	HBcAg	*Vigna angustifolia* Transient (CPMV)	10 μg/g FW	EM	-	-	(Mechtcheriakova et al., 2006)
	HBsAg	Banana	38 ng/g FW	Gradients	-	-	(Kumar et al., 2005)
HBV-HIV	HIV-HBsAg	Tobacco *A. thaliana*	257.8 ng/mg TSP	Conformational antibody	-	-	(Greco et al., 2007)
HPV	HPV16-L1	Tobacco Potato tubers	0.5% TSP 0.2% TSP	Gradients, EM	+	+	(Biemelt et al., 2003)
	HPV16-L1	Tobacco	4 μg/kg FW	EM	+	-	(Varsani et al., 2003)
	HPV16-L1	Tobacco *N. benthamiana*	887 μg/g FW 533 μg/g FW	Gradients, EM	+	-	(Maclean et al., 2007)
	HPV16-L1	*N. benthamiana* Transient (TMV)	37 μg/g FW	EM	+	-	(Varsani et al., 2006)
	HPV11-L1	Tobacco cell culture Potato tubers	- 23 ng/g FW	EM	-	+	(Warzecha et al., 2003)
	HPV11-L1	*A. thaliana* Tobacco	12 μg/g FW 2 μg/g FW	EM	+	-	(Kohler and Milstein, 1975)
	HPV16-L1	Tobacco Chloroplast transformation	3 mg/g FW	Gradients, EM	+	-	(Fernandez-San Millan et al., 2008)

Pathogen	Antigen	Production system	Highest expression level	VLP proof	Immune response IP[1]	Immune response Oral[2]	Reference
Norwalk virus	NVCP	Tobacco	0.23% TSP	Gradients, EM	-	+	(Tacket and Mason, 1999)
	NVCP	Potato tubers	34 μg/g FW	Gradients	-	+	(Tacket and Mason, 1999)
	NVCP	Tomato fruits	20 μg/g FW 30 μg/g DW	Gradients	-	+	(Huang et al. 2005b)
	Plant optimized NVCP	Tomato fruits	8% TSP	Gradients, EM	-	+	(Zhang et al., 2006)
	Plant optimized NVCP	Potato tubers	0.4% TSP	Gradients, EM	-	+	(Zhang et al., 2006)
	Plant optimized NVCP	*N. benthamiana* Transient (MagnICON)	0.86 mg/g FW	Gradients, EM	-	+	(Santi et al., 2008)
Rotavirus	PVX-VP6 VP6	*N. benthamiana* Transient (PVX)	1 μg/g FW 50 μg/g FW	EM	-	-	(O'Brien et al., 2000)
	VP2-VP6	Tomato fruits	1% TSP	EM	+	-	(Saldaña et al., 2006)

1, IP: Intraperitoneal or subcutaneous administration,
2, Oral: Oral or intranasal administration,
+: specific immune response after vaccination
CPMV: Cowpea mosaic virus, DW: Dry weight, EM: Electron microscopy, FM: Fluorescence microscopy, FW: Fresh weight, HBV: Hepatitis B virus, HIV: Human Immunodeficiency virus, HPV: Human papillomavirus, PVX: Potato virus X, TMV: Tobacco mosaic virus, TSP: Total soluble protein.

production and oral administration (Streatfield et al., 2003; Tacket and Mason, 1999; Wigdorovitz et al., 1999). In addition to their use as oral vaccines, extraction and purification of VLPs will also be viable if high expression levels are achieved. This requirement can be fulfilled with viral vectors for transient expression and chloroplasts transformation. In fact, these two systems can compete with microbial fermentation or baculovirus-insect cell systems for production of VLPs.

In the following sections plant-produced VLPs against important human infection diseases are described in detail. Table 3 summarizes the different plant systems used for the production of VLPs and target diseases.

3.2. Hepatitis B Virus (HBV)

Infant immunization is the most effective preventive measure against HBV infection. As HBV replicates in the hepatocytes of humans but does not grow in artificial cell cultures, it was difficult to develop a vaccine. The first-generation vaccine consisted on VLPs purified from the plasma of persons with chronic HBV infection. The hepatitis B surface antigen (HBsAg) is a lipoprotein of the viral envelope that is produced in excess and circulates in the blood as spherical particles 22 nm in size. The production of HBsAg in recombinant yeast resulted in homogeneous particles that, after extraction and purification, constituted the second-generation vaccine (McAleer et al., 1984). This was the first example of a vaccine produced from recombinant cells which is effective against a human viral infection. The recombinant hepatitis B vaccine has been available since 1982. However, in 2000, there were an estimated 5.7 million cases of acute hepatitis B infection and more than 500 000 deaths from hepatitis B-related disease worldwide. Clearly, vaccine costs preclude the use of the recombinant vaccine in developing countries, where they are needed most. A plant-derived vaccine could solve the cost problem.

The first attempt to develop a plant-derived vaccine was the expression of the HBsAg in tobacco (Tacket and Mason, 1999). The foreign gene was correctly transcribed and translated, yielding a maximum of 66 ng/mg of TSP from leaf. Purified recombinant HBsAg self-assembled into spherical particles with an average diameter of 22 nm, equivalent to the subviral particles detected in serum from infected individuals. HBsAg has also been expressed in tobacco cell suspension cultures (Kumar et al., 2003; Sojikul et al., 2003). To facilitate oral delivery, HBsAg was expressed in potato tubers at levels up to 16 μg/g tuber (Richter et al., 2000) by optimizing 5' and 3' flanking sequences and by targeting the antigen to the endoplasmic reticulum. The expression of HBsAg in an edible fruit has been reported for banana (Kumar et al., 2005). HBsAg was detected in leaves, but no data of HBsAg expression in fruits was reported.

Partially purified HBsAg from tobacco was delivered by intraperitoneal (i.p.) injection to mice. The antibody and T cell responses were qualitatively similar to that obtained with the yeast-derived commercial vaccine (Richter et al., 2000). Oral immunogenicity was achieved in mice after feeding raw potato tuber with cholera toxin as adjuvant. Mice produced HBsAg-specific serum antibodies and, on parenteral boosting, generated a strong long-lasting secondary antibody response (Kong et al., 2001). In addition, the effectiveness of oral delivery by using a prime-oral boost immunization schedule was demonstrated. For humans, the ability of plant-derived HBsAg to elicit specific antibody responses after oral administration was demonstrated in two clinical trials with lettuce leaves (Kapusta et al., 1999) and uncooked potato tubers (Richter et al., 2000).

Chimeric proteins of HBsAg with N- or C-terminal fusions are able to self-assemble into VLPs without altering significantly their antigenic properties. HBsAg fusion with the green fluorescence protein (GFP) at the N-terminus (but not at the C-terminus), transiently expressed in *Nicotiana benthamiana*, formed VLPs similar to yeast-derived vaccine (Huang and Mason, 2004). Interestingly, co-expression of HBsAg and GFP:HBsAg resulted in heterodimer formation. This would be particularly useful for the development of multicomponent vaccines. Other example is a fusion protein consisting of amino acids 21-47 of the hepatocyte receptor-binding presurface 1 region fused to the truncated C-terminus of HBsAg (Qian et al., 2007). This chimeric protein was specifically expressed in rice seeds. It was observed the formation of VLPs and the induction of immune responses against both the HBsAg and the presurface 1 region. This strategy of fusion proteins was practically developed by the successful expression of novel recombinant HIV-1/HBV VLPs in tobacco and *Arabidopsis thaliana* (Greco et al., 2007). Both plants yielded comparable level of recombinant VLPs. As both virus spread via similar transmission pathways, a bivalent vaccine would be very useful, especially for developing countries where the two diseases pose major public health problems.

In order to increase expression levels, HBsAg was targeted to the endoplasmic reticulum by the signal peptide from soybean vegetative storage protein vspA (Sojikul et al., 2003). Despite the signal peptide was not cleaved, the product was more stable and accumulated up to 226 ng/mg of TSP. The signal peptide was not an obstacle for VLP formation, and these chimeric VLPs stimulated higher levels of serum IgG than native HBsAg when injected into mice. Recently, a high-yield rapid production of HBsAg using a novel viral transient expression system based on TMV-derived vectors has been described (Huang et al., 2008). *N. benthamiana* leaves infiltrated with the MagnICON viral modular vectors (Gleba et al., 2007) produced HBsAg at high levels (~295 μg/g FW). HBsAg assembled into VLPs and elicited specific antibodies in mice immunized with partially purified extracts.

Despite most reports have focused on the major surface antigen (S-protein), some groups have studied the expression of HBsAg middle protein (M-protein) in plant systems. M-protein has an additional pre-S2 sequence at the N-terminus of the S-protein and confers higher and earlier immunological response. Huang et al. (2005b) expressed the HBsAg M protein in suspension cultures of *N. benthamiana* and demonstrated VLP formation. It was observed that i.p. administration of *N. benthamiana*-derived HBsAg M protein provoked stronger serum antibody responses in mice than did S protein. However, expression levels of M protein in *N. benthamiana* (Huang et al., 2005b) and in potato (Ehsani et al., 1997) are much lower than those of S protein.

In addition to HBsAg, HBV core antigen (HBcAg) has been expressed in plants. During HBV infection, HBcAg self-assembles into subviral particles. Specific antibodies developed against HBcAg do not provide protection against HBV infection. However, HBcAg has an immunoenhancing effect on co-delivered HBsAg (Lobaina et al., 2006; Lobaina et al., 2005). The formulation containing the core and surface antigens could be a promising alternative for the design of new therapeutic and preventive vaccines against HBV infection. HBcAg was expressed in tobacco plants and self-assembled into spherical particles of 25-30 nm (Tsuda et al., 1998). HBcAg was also transiently expressed in plants using two viral vectors, potato virus X and cowpea mosaic virus for *N. benthamiana* and cowpea infection, respectively (Mechtcheriakova et al., 2006). Assembly into VLPs was demonstrated, but the immunogenicity of these VLPs was not tested. By using the MagnICON system (Gleba et al., 2007), HBcAg was transiently expressed in *N. benthamiana*. The antigen was efficiently produced in infected leaves with high expression levels (7% of TSP equivalent to 2 mg/g FW) (Huang et al., 2006). Partially purified HBcAg stimulated strong serum IgG responses. Furthermore, mice immunized mucosally in the absence of adjuvants also developed specific serum IgG as well as intestinal IgA. This result suggests that HBcAg could be an excellent carrier for epitope presentation and mucosal delivery. Indeed, the group of H. Mason produced a fusion between HBcAg and the neutralizing epitope of the human papillomavirus 16 L2 protein in plant leaves, and found that the fusion formed chimeric VLPs (unpublished result). In the same way, the VP21 epitope of the VP1 protein from the foot-and-mouth disease virus (affecting cloven-hoofed animals) was fused to the internal region of HBcAg and used for the transformation of tobacco plants. The formation of regular 27-nm particles in the crude extracts from transgenic tobacco leaves was demonstrated (Huang et al., 2005a). Mice, immunized i.p. with a soluble crude extract of transgenic tobacco leaves, were found to produce specific antibody responses to both HBcAg and VP1. Importantly, a virus challenge demonstrated that the immunized mice were highly protected against virulent foot-and-mouth disease virus.

3.3. Human Papillomavirus (HPV)

Cervical cancer is the second most prevalent cancer among women worldwide and the most common cancer in developing countries. Every year, approximately 500 000 women develop cervical cancer and the disease causes about 200 000 deaths. Most of these cancers are associated with infection by high-risk HPVs. HPV-16 and HPV-18 types account together for around 70% of cervical cancers. HPVs are small, non-enveloped viruses with two structural proteins, L1 and L2, that form the viral capsid. The major capsid L1 protein produced in yeast or insect cells self-assembles into VLPs, induces high titers of antibodies and prevents the development of lesions induced by authentic virions (Suzich et al., 1995). Recently, two subunit vaccines against HPV based on intramuscular injection have been marketed (Cervarix® from GlaxoSmithKline produced in insect cells and Gardasil® from Merck produced in yeast). In both cases, the selected antigen is the L1 protein and the expression systems are yeast or insect cells (Lowy and Schiller, 2006). Gardasil® was recently approved in some developed countries as a prophylactic cervical cancer vaccine. The market for HPV treatments is expected to increase from 135 million US dollars in 2002 to 1.4 billion in 2012, with prophylactic vaccines driving this growth (http://www.drugresearch.com). An ideal HPV vaccine should be inexpensive to manufacture, protect against all oncogenic HPV types and act both therapeutically and prophylactically (Schiller and Nardelli-Haefliger, 2006). None of the marketed vaccines meet this ideal. In addition, the high cost of the current vaccines will prohibit a widespread use in developing countries where they are needed most. Therefore, there is need for the development of a second generation vaccine.

HPV-L1 protein has been produced in different plants, via stable or transient expression. HPV-16 L1 was synthesized in transgenic tobacco and potato (Biemelt et al., 2003; Maclean et al., 2007; Varsani et al., 2003) and in *N. benthamiana* by TMV infection (Varsani et al., 2006). Biemelt et al. (2003) found that either the original L1 gene or an L1 gene with a codon usage optimized for expression in plants failed. Surprisingly, L1 protein was detected with a L1 gene optimized for expression in human cells. HPV-11 L1 protein was expressed in transgenic potato, tobacco and *A. thaliana* (Kohler and Milstein, 1975; Warzecha et al., 2003). Both groups observed that HPV-11 L1 expression was enhanced by removal of the carboxy-terminal nuclear localization signal sequence. In general, yields were modest, independent of the plant species or expression system, ranging from 4 ng to 12 μg/g of leaf FW. Maclean et al. (2007) analysed different HPV-16 L1 gene variants and different subcellular localization. It was found that a humanized gene expressed better than the native or plant-optimized genes. Moreover, chloroplast localization by a transit peptide allowed significantly higher levels of accumulation of L1 protein (up to 887 μg/g FW) than did cytoplasmic or endoplasmic reticulum localization (Maclean et al., 2007). Recently, Fernandez-San Millan et al. (2008) expressed the HPV-16 L1 in tobacco chloroplasts by plastid transformation. A very high yield production was achieved in mature plants (3 mg/g FW). A single mature plant synthesized 240 mg of L1 protein.

Despite the great differences in L1 expression, in all cases the plant-produced L1 assembled into VLPs of ~55 nm as evidenced by electron microscopy (EM). It was shown that VLPs produced in *N. benthamiana* had a similar size to those derived from insect-cells (Maclean et al., 2007), while VLPs of lower size were also detected in the case of potato cytoplasm and tobacco chloroplasts (Biemelt et al., 2003; Fernandez-San Millan et al., 2008).

Plant-derived HPV VLPs were highly immunogenic in mice after i.p. or subcutaneous injection and elicited neutralizing antibodies (Fernandez-San Millan et al., 2008; Maclean et al., 2007). Oral immunization of mice with transgenic potatoes did not evoke detectable antibody responses; the addition of cholera toxin subunit or CpG plasmid DNA as adjuvants to the raw material developed weak and transient responses in some animals (Biemelt et al., 2003; Warzecha et al., 2003). Both groups observed that a subimmunogenic dose of insect-cell derived VLPs administered at the end of the vaccination schedule significantly boosted the L1-specific responses. This indicates the development of a specific immune memory induced by transgenic potato tuber feeding. Further research using alternative adjuvants is needed, since oral delivery is the most attractive advantage of plant vaccines for large-scale human vaccination. For example, the *E. coli* heat-labile enterotoxin significantly improved anti-VLP humoral and mucosal responses when co-administered orally with insect-cell derived VLPs (Gerber et al., 2001).

3.4. Norwalk virus (NV)

Noroviruses are enteric viruses of the *Caliciviridae* family. NV is the prototype of human noroviruses, the major cause of nonbacterial gastroenteritis in developed and developing countries. Its viral morphology is generated by the icosahedral arrangement of the 58 kDa NV capsid protein (NVCP) that assembles into a non-enveloped capsid. Expression of rNVCP in baculovirus-infected insect cells has shown that this protein alone can self-assemble into VLPs of 38 nm antigenically and morphologically similar to the native NV particles (Jiang et al., 1992). Naked NV-VLPs are immunogenic when administered parenterally and intranasally in experimental animal models and when given orally to volunteers (Ball et al., 1998; Guerrero et al., 2001; Tacket and Mason, 1999). Since there is no commercial vaccine to prevent norovirus infections, VLPs are a potential vaccine candidate.

The first report of NVCP expression in plants as VLPs was done in tobacco and potato (Tacket and Mason, 1999). They obtained rNV expression levels of up to 0.23% of TSP in tobacco leaves and up to 0.37% in potato tubers (34 μg/g of tuber weight). The presence of correctly assembled 38 nm VLPs purified from transgenic tobacco leaves was demonstrated. They showed that either partially purified tobacco NVCP given orally or potato tubers expression NVCP fed directly to mice, with or without the use of cholera toxin as mucosal adjuvant, stimulated the production of humoral and mucosal antibody responses. However, the immune response was lower than that obtained by gavage with a similar dose, perhaps because the tuber material was consumed more slowly and was less pure. Afterwards, human volunteers that received 2 or 3 doses of transgenic potato also developed immunogenic response (Tacket and Mason, 1999). However, low levels of expression and the fact that the degree of assembly of NVCP subunits in potato was relatively poor (25-50%), probably limited its immunogenicity.

In an attempt to improve palatability and expression levels of the oral inmunogen to avoid that volunteers eat large bolus of plant sample, NVCP was later produced in tomato at up to 20 μg/g fruit fresh weight, mainly in the form of the 23 nm VLPs. They also showed that freeze-drying of transgenic tomato fruit yielded stable preparations that stimulated good humoral and mucosal responses when fed to mice (Huang et al., 2005b). (Zhang et al., 2006) studied the expression of a NVCP plant-optimized gene in tomato and potato. Expression of assembled VLPs was increased fourfold in tomato and potato plants compared to the native gene (8% and 0.4% of TSP respectively). VLPs of 23 and 38 nm were formed averaged about 41 and 22% respectively in fresh tomato fruit and about 24 and 40% in fresh potato tubers. Authors found that air dried tomato fruit can induce more robust immune responses than freeze-dried powder, indicating that a very convenient stabilization process has utility for tomato-derived vaccine production. They saw better immunological results by feeding dried tomato powder than dried potato tubers or than purified NVCP. Their hypothesis is that plant cell encapsulation is more substantially retained in dried tomato fruit than in dried potato tubers.

Recentlly, (Santi et al., 2008) utilized the magnICON expression system to reach higher levels of expression using leaves of *N. benthamiana*. They tested different cellular targeting sequences to determine which plant cell compartment allowed maximal accumulation levels (endomembrane, chloroplast or cytosol). Cytoplasmic accumulation of the antigen was found to clearly give the highest expression. They reported a high level of rNV expression of 0.86 mg/g fresh weight at 12 days post infection. They also showed proper assembly of VLPs. Oral immunization of mice by gavage with a partially purified rNV enriched preparation, with or without cholera toxin adjuvant, elicited systemic and mucosal immune responses. In conclusion, this system has validated the transient technology for rapid production of high yields of properly assembled VLPs.

3.5. Rotavirus

Rotavirus belongs to the *Reoviridae* family and produce infections that are the most important cause of severe infantile gastroenteritis, accounting for more than 125 million cases of diarrhea and an estimated 600 000 deaths per year, mainly in developing countries (O'Ryan et al., 2005). Rotavirus-related deaths represent approximately 5% of all deaths in children younger than 5 years of age and causes 39% of childhood diarrhea hospitalizations worldwide (Dennehy, 2008). Rotavirus also constitutes an important source of economic losses in animal production related to death, treatment costs and reduction in weight gain of affected animals (Molinari et al., 2008).

The need of an effective vaccine is an international priority and initial efforts have been focused on the development of oral, live attenuated human-animal vaccines. Several whole attenuated rotavirus virions have been tested as candidate vaccines, showing a very variable efficacy in clinical trials. For example, a oral rhesus-human rotavirus vaccine had been licensed in USA, but it lasted for just over 10 months before its use was suspended because of a series of intussusception cases in small children. A new live-attenuated vaccine (Rotarix®) has recently been licensed for use in Mexico and the Dominican Republic and a pentavalent vaccine (RotaTeq®) has been submitted for FDA approval in the USA (Dennehy, 2008).

An alternative to live rotavirus vaccines are recombinant VLPs produced by co-expression of a different combination of rotavirus structural proteins. The rotavirus virion is a nonenveloped icosahedral particle consisting of four main proteins. VP6 and VP2 represent 50% and 15% of total virion mass, respectively. The rotavirus outer coat is composed of the glycoprotein VP7 and spikes of VP4. In general, the use of rotavirus VLPs produced in cell culture containing VP2/VP6, VP2/VP6/VP7 or VP2/VP6/VP7P4 for protective immunization has yielded the best immune response in comparison to immunization with individual rotavirus proteins. For example, VLPs produced by baculovirus infection of insect cells composed of VP2 and VP6 capsid proteins administered parenterally, mucosally or intrarectally provided protection against oral challenge in mice (Agnello et al., 2006; Madore et al., 1999).

There are several studies about rotavirus capsid proteins VP6, VP7 and VP2 expression in plants. However, only two of them have been able to demonstrate VLPs formation. (O'Brien et al., 2000) produced VP6 (1-50 □g/g FW) by transient expression from a potato virus x vector (PVX) in leaves of *N. benthamiana*. VP6 was expressed either independently or as a fusion with the PVX coat protein separated by a catalytic peptide. When expressing VP6 independently, they were able to see assembly of VP6 into VLPs when combined with VP2 produced in baculovirus-infected cells. Expression as a fusion protein yielded PVX rods that presented an external overcoat of VP6. But what is more interesting, VP6 released by self-cleavage also was able to assemble into VLPs, although at a low efficiency. The assembly of viral proteins into VLPs suggests that prior display of VP6 on the flexuous PVX rod facilitates the subsequent assembly of VP6 into stable icosahedral particles. This was the first report of the intracellular self-assembly of VP6 into VLPs. In a later work, Birch-Machin et al. (2004) were not able to detect VP6-VLPs formation inside tobacco chloroplasts although the amount of protein accumulated was six-fold higher. This finding suggests the importance of the redox environment and/or the PVX presence.

Afterwards, Saldaña et al. (2006) based their work on co-expression of VP2 and VP6 coat proteins in tomato plants. The result indicated that only a small proportion of VP2/VP6 assembled into VLPs in tomato fruits. The tomato-derived VLPs were slightly smaller than purified virus, probably due to the fact that they consist of only two proteins as opposed to the intact particle. When they inoculated mice i.p. with tomato-derived VLPs with a strong adjuvant, a significant humoral immune response was obtained. This was similar to what had been obtained previously using VP2/VP6 VLPs produced in baculovirus-infected insect cells.

In summary, higher levels of expression and production of efficiently and properly assembled VLPs are needed to investigate the development of a plant-based rotavirus vaccine.

4. Conclusion

Plants offer significant advantages over other expression systems for the production of antibodies and vaccines. Thus, a promising prospect appears to sustain the optimized development and practical use of this system in the future.

Expression of recombinant proteins in different plant species has progressed from single functional molecules such as antigens to more complex biological active molecules such as immune complexes and VLPs. Three main encouraging aspects of the plant expression system could be emphasized: (i) The capability of this system to scale-up, (ii) the real potential to the establishment of a cost-effective production system, and (iii) the availability of a safer production method by the practically inexistent contamination with animal derived pathogens, including virions or prions.

On the contrary, specific issues still must be conquered to overcome specific limitations currently affecting the advance to the use of plants as certified production technology. Thus, although a significant number of pharmaceuticals has entered clinical trial stages approved by the FDA, no therapeutics, including both antibodies and VLPs, have been commercialised so far. Three main limitations could also be listed. They are (i) the low expression levels of most recombinant protein, (ii) differences in post-translational protein processing mainly pointing out the modifications in the glycosylation pattern between species and (iii) absence of clear regulatory guidelines related to the choice of right plant species, harvesting and post-harvesting processing procedures and biosafety and bioethical issues.

It has been widely assumed that a minimum level of recombinant protein expression equivalent to 1% TSP is required for commercialization of plant-derived therapeutics. In contrast, figures lower than 1% are commonly found in the literature and continuous efforts are being made to set up efficient strategies to overcome this setback. In general terms, high expression levels have been reported to be achieved mainly by two principal means: transient expression of foreign proteins by viral-based vectors or chloroplast transformation. In fact, expression levels higher than 20% TSP are commonly achieved by these systems. Notwithstanding this, in the particular case of antibody production, nuclear transformation and adequate subcellular targeting have demonstrated to reach protein expression levels estimated feasible for commercial exploitation. In addition, conservation of the transgene expression through plant generations and the possibility of providing a natural protein storage system in seeds makes nuclear transformation an easy and attractive strategy with views to develop a long-term stable production platform, cold-chain free protein storage and undemanding product delivery schemes. Furthermore, the expression level of antibodies in transgenic plants also can be increased by the use of appropriate regulatory elements in the expression construct, optimizing codon usage of the recombinant transgene, and enhancing the stability of the antibody. Similarly, the expression of VLPs has been demonstrated to render up to 24 % TSP when expressed in chloroplasts.

An important factor affecting the acceptance and use of transgenic plants for pharmaceutical production, particularly in the case of antibodies, is the foreign glycan structures that these molecules have been shown to contain compared to those of their counterparts from mammalian cells (Rudd et al., 2001; Wright and Morrison, 1997). There are at least two concerns about modifications in the glycosylation pattern of plant-derived antibodies. Firstly, it could change their longevity or biological activity. Secondly, it could potentially provoke allergic responses of immunized individuals against plant glycan epitopes. However, studies indicate that modified glycosylation does not affect the activity of the antibodies and allergic responses have not been observed, if only in a few particular cases, in the animal models utilized (Chargelegue et al., 2000; Bardor et al., 2003). Equally, the EPIcyte company reported that a plant-derived mouse antibody did not elicit an immune response in mice (Morrow, 2002). Furthermore, several strategies to avoid these concerns including humanization of plant glycans have been approached (Bakker et al., 2001) and particular strategies used in other expression systems for this objective are now applied to plants. A guideline of those plant genes that could need to be expressed, removed or regulated to match the glycosylation pattern of recombinant molecules as much as possible to their native structure has been reported (Warner, 2000).

For plants to be feasible as production systems one must determine how to optimize the manufacturing procedures from plant cultivation to protein purification and commercialization. However, although initial regulatory guidance for plant-based therapeutic production has already been addressed, Good Manufacturing Practice guidelines specific for purification and processing of these molecules has to be clearly defined. Validation of specific facilities, protocols and analytical methods could be mentioned as a main purpose. For example, purification of plant-antibodies has been a major concern because phenolic compounds contained in the plant material are both toxic and can irreversibly alter the native properties of the protein. In this regard, several strategies to enhance protein purification levels have been investigated such as improved immunoprecipitation/affinity chromatography methods (Ko et al., 2004), oleosin-fusion technology (Kuhnel et al., 2003) and rhizosecretion through the transgenic plant roots (Drake et al., 2003).

Another important aspect is the selection of adequate plant species to achieve high yields of recombinant protein. Many different plant species have been investigated and described for production of high value proteins. These include edible and non-edible leafy crops, fruits and vegetables, cereals, seeds, cell cultures, algae, hydroponic systems and aquatic plants (Fischer et al., 2004; Twyman et al., 2003). In order to facilitate post-harvesting processing, tissue specific promoters can be used to direct transgene expression in storage organs such as seeds and tubers. An additional benefit of this approach is the higher protein stability and accumulation in such organs. The use of seeds for large-scale production of recombinant molecules has been demonstrated to provide many advantages compared to other plant-based technologies and thus seeds, particularly maize, have become leading candidates. However, several studies provide significant diversity of results between different crops and thus further investigations are required. In this context, factors influencing protein processing, compartmentalization and accumulation have a clear impact on the final protein yield. Therefore, specific features inherent to particular species affecting the expression, targeting and subcellular location of foreign proteins need to be identified. At the same time, the choice of the right plant species for production of biopharmaceuticals is also a controversial issue. The use of edible crops represents a real risk of contamination of the food chain with biopharmaceuticals if transgenic and conventional crops coexistence assurance programs and controls fail. Similarly, cross-pollination of non-edible transgenic crops with wild relatives could not be tolerated. The importance placed on this issue by the regulatory agencies is highlighted by the case of Prodigene Inc. This company accepted in 2002 to pay the US Department of Agriculture more than $3 million in penalties for contaminating soybean crops with genetically modified maize that contained an animal vaccine. Therefore, as done with other production systems precise monitoring methods and required infrastructure for adequate control of therapeutics production in plants need to be strictly established. Thus, independent of the plant species used, all of them will require the following criteria to be met: isolation from other crops if out-crossing is possible, identity preservation throughout crop development to final product formulation, compliance with authorities' regulations and quality control with standard operating procedures to ensure final product quality (Nikolov and Hammes, 2002). The next steps to product development will have to involve the optimization to homogeneity of batches, product presentation and stability. However, as exposed in this text, in spite of continuing advances, important issues need to be answered before this new strategy can be put into practice. Furthermore, public acceptance of transgenic plants and plant-based products, and definition of real final costs of plant-derived pharmaceuticals are critical aspects under consideration.

In conclusion, achievement of functional plant-derived molecules and progressing studies on plant biotechnology are leading to an increasing advance to the practical field. Antibodies and VLPs are prime candidates for immunotherapy and vaccine development. Plant antibodies seem to appear as the first generation of plant therapeutics that could be released. Pharmaceutical crops will need to satisfy the strict regulations that apply to genetically modified plants as well as regulations related to the production of pharmaceuticals. The harmonization of human health, agriculture and environmental regulations is one of the most relevant challenges of the plant-derived pharmaceuticals production.

REFERENCES

Acevedo, H.F., Hartsock, R.J. and Maroon, J.C. (1997) Detection of membrane-associated human chorionic gonadotropin and its subunits on human cultured cancer cells of the nervous system. *Cancer Detect. Prev.*, 21, 295-303.

Adams, G.P. and Weiner, L.M. (2005) Monoclonal antibody therapy of cancer. *Nat. Biotechnol.*, 23, 1147-1157.

Agnello, D., Herve, C.A., Lavaux, A., Darniot, M., Guillon, P., Charpilienne, A. and Pothier, P. (2006) Intrarectal immunization with rotavirus 2/6 virus-like particles induces an antirotavirus immune response localized in the intestinal mucosa and protects against rotavirus infection in mice. *J. Virol.*, 80, 3823-3832.

Arntzen, C., Plotkin, S. and Dodet, B. (2005) Plant-derived vaccines and antibodies: potential and limitations. *Vaccine*, 23, 1753-1756.

Bagshawe, K.D. (1983) Human chorionic gonadotropin as a model for a fetal antigen. *Ciba Found Symp.*, 96, 146-159.

Bakker, H., Bardor, M., Molthoff, J.W., Gomord, V., Elbers, I., Stevens, L.H., Jordi, W., Lommen, A., Faye, L., Lerouge, P. and Bosch, D. (2001) Galactose-extended glycans of antibodies produced by transgenic plants. *Proc. Natl. Acad. Sci. U S A*, 98, 2899-2904.

Ball, J.M., Hardy, M.E., Atmar, R.L., Conner, M.E. and Estes, M.K. (1998) Oral immunization with recombinant Norwalk virus-like particles induces a systemic and mucosal immune response in mice. *J. Virol.*, 72, 1345-1353.

Banchereau, J., De Paoli, P., Valle, A., Garcia, E. and Rousset, F. (1989) Long-term human B cell lines dependent on interleukin-4 and antibody to CD40. *Science*, 251, 70-72.

Bardor, M., Cabrera, G., Rudd, P.M., Dwek, R.A., Cremata, J.A. and Lerouge, P. (2006) Analytical strategies to investigate plant N-glycan profiles in the context of plant-made pharmaceuticals. *Curr. Opin. Struct. Biol.*, 16, 576-583.

Bardor, M., Faveeuw, C., Fitchette, A.C., Gilbert, D., Galas, L., Trottein, F., Faye, L. and Lerouge, P. (2003) Immunoreactivity in mammals of two typical plant glyco-epitopes, core alpha(1,3)-fucose and core xylose. *Glycobiology*, 13, 427-434.

Benvenuto, E., Ordas, R.J., Tavazza, R., Ancora, G., Biocca, S., Cattaneo, A. and Galeffi, P. (1991) 'Phytoantibodies': a general vector for the expression of immunoglobulin domains in transgenic plants. *Plant Mol. Biol.*, 17, 865-874.

Biemelt, S., Sonnewald, U., Galmbacher, P., Willmitzer, L. and Muller, M. (2003) Production of human papillomavirus type 16 virus-like particles in transgenic plants. *J. Virol.*, 77, 9211-9220.

Birch-Machin, I., Newell, C.A., Hibberd, J.M. and Gray, J.C. (2004) Accumulation of rotavirus VP6 protein in chloroplasts of transplastomic tobacco is limited by protein stability. *Plant. Biotechnol. J.*, 2, 261-270.

Borrebaeck, C.A. and Moller, S.A. (1986) In vitro immunization. Effect of growth and differentiation factors on antigen-specific B cell activation and production of monoclonal antibodies to autologous antigens and weak immunogens. *J. Immunol.*, 136, 3710-3715.

Bouquin, T., Thomsen, M., Nielsen, L.K., Green, T.H., Mundy, J. and Dziegiel, M.H. (2002) Human anti-rhesus D IgG1 antibody produced in transgenic plants. *Transgenic. Res.*, 11, 115-122.

Brereton, H.M., Chamberlain, D., Yang, R., Tea, M., McNeil, S., Coster, D.J. and Williams, K.A. (2007) Single chain antibody fragments for ocular use produced at high levels in a commercial wheat variety. *J. Biotechnol.*, 129, 539-546.

Brodzik, R., Glogowska, M., Bandurska, K., Okulicz, M., Deka, D., Ko, K., van der Linden, J., Leusen, J.H., Pogrebnyak, N., Golovkin, M., Steplewski, Z. and Koprowski, H. (2006) Plant-derived anti-Lewis Y mAb exhibits biological activities for efficient immunotherapy against human cancer cells. *Proc. Natl. Acad. Sci. U S A*, 103, 8804-8809.

Butler, S.A., Ikram, M.S., Mathieu, S. and Iles, R.K. (2000) The increase in bladder carcinoma cell population induced by the free beta subunit of human chorionic gonadotrophin is a result of an anti-apoptosis effect and not cell proliferation. *Br. J. Cancer*, 82, 1553-1556.

Conrad, U. and Fiedler, U. (1998) Compartment-specific accumulation of recombinant immunoglobulins in plant cells: an essential tool for antibody production and immunomodulation of physiological functions and pathogen activity. *Plant. Mol. Biol.*, 38, 101-109.

Chargelegue, D., Drake, P.M., Obregon, P., Prada, A., Fairweather, N. and Ma, J.K. (2005) Highly immunogenic and protective recombinant vaccine candidate expressed in transgenic plants. *Infect. Immun.*, 73, 5915-5922.

Chargelegue, D., Vine, N.D., van Dolleweerd, C.J., Drake, P.M. and Ma, J.K. (2000) A murine monoclonal antibody produced in transgenic plants with plant-specific glycans is not immunogenic in mice. *Transgenic Res.*, 9, 187-194.

Daniell, H., Khan, M.S. and Allison, L. (2002) Milestones in chloroplast genetic engineering: an environmentally friendly era in biotechnology. *Trends Plant Sci.*, 7, 84-91.

De Cosa, B., Moar, W., Lee, S.B., Miller, M. and Daniell, H. (2001) Overexpression of the Bt cry2Aa2 operon in chloroplasts leads to formation of insecticidal crystals. *Nat. Biotechnol.*, 19, 71-74.

De Genst, E., Silence, K., Decanniere, K., Conrath, K., Loris, R., Kinne, J., Muyldermans, S. and Wyns, L. (2006) Molecular basis for the preferential cleft recognition by dromedary heavy-chain antibodies. *Proc. Natl. Acad. Sci. U S A*, 103, 4586-4591.

De Neve, M., De Loose, M., Jacobs, A., Van Houdt, H., Kaluza, B., Weidle, U., Van Montagu, M. and Depicker, A. (1993) Assembly of an antibody and its derived antibody fragment in Nicotiana and Arabidopsis. *Transgenic. Res.*, 2, 227-237.

Delpeyroux, F., Chenciner, N., Lim, A., Malpiece, Y., Blondel, B., Crainic, R., van der Werf, S. and Streeck, R.E. (1986) A poliovirus neutralization epitope expressed on hybrid hepatitis B surface antigen particles. *Science*, 233, 472-475.

Dennehy, P.H. (2008) Rotavirus Vaccines: an Overview. *Clinical Microbiology Reviews*, 198-208.

Drake, P.M., Chargelegue, D.M., Vine, N.D., van Dolleweerd, C.J., Obregon, P. and Ma, J.K. (2003) Rhizosecretion of a monoclonal antibody protein complex from transgenic tobacco roots. *Plant Mol. Biol.*, 52, 233-241.

During, K., Hippe, S., Kreuzaler, F. and Schell, J. (1990) Synthesis and self-assembly of a functional monoclonal antibody in transgenic Nicotiana tabacum. *Plant Mol. Biol.*, 15, 281-293.

Ehsani, P., Khabiri, A. and Domansky, N.N. (1997) Polypeptides of hepatitis B surface antigen produced in transgenic potato. *Gene*, 190, 107-111.

Fernandez-San Millan, A., Ortigosa, S.M., Hervás-Stubbs, S., Corral-Martínez, P., Seguí-Simarro, J.M., Gaétan, J., Coursaget, P. and Veramendi, J. (2008) Human papillomavirus L1 protein expressed in tobacco chloroplasts self-assembles into virus-like particles that are highly immunogenic. *Plant Biotechnol. J.*, In press.

Fiedler, U., Phillips, J., Artsaenko, O. and Conrad, U. (1997) Optimization of scFv antibody production in transgenic plants. *Immunotechnology*, 3, 205-216.

Fiedler, W., Borchers, C., Macht, M., Deininger, S.O. and Przybylski, M. (1998) Molecular characterization of a conformational epitope of hen egg white lysozyme by differential chemical modification of immune complexes and mass spectrometric peptide mapping. *Bioconjug. Chem.*, 9, 236-241.

Fischer, R. and Emans, N. (2000) Molecular farming of pharmaceutical proteins. *Transgenic Res.*, 9, 279-299; discussion 277.

Fischer, R., Schumann, D., Zimmermann, S., Drossard, J., Sack, M. and Schillberg, S. (1999) Expression and characterization of bispecific single-chain Fv fragments produced in transgenic plants. *Eur. J. Biochem.*, 262, 810-816.

Fischer, R., Stoger, E., Schillberg, S., Christou, P. and Twyman, R.M. (2004) Plant-based production of biopharmaceuticals. *Curr. Opin. Plant Biol.*, 7, 152-158.

Fishel, S.B., Edwards, R.G. and Evans, C.J. (1984) Human chorionic gonadotropin secreted by preimplantation embryos cultured in vitro. *Science*, 223, 816-818.

Fletcher, S.P., Muto, M. and Mayfield, S.P. (2007) Optimization of recombinant protein expression in the chloroplasts of green algae. *Adv. Exp. Med. Biol.*, 616, 90-98.

Francisco, J.A., Gawlak, S.L., Miller, M., Bathe, J., Russell, D., Chace, D., Mixan, B., Zhao, L., Fell, H.P. and Siegall, C.B. (1997) Expression and characterization of bryodin 1 and a bryodin 1-based single-chain immunotoxin from tobacco cell culture. *Bioconjug. Chem.*, 8, 708-713.

Frigerio, L., Vine, N.D., Pedrazzini, E., Hein, M.B., Wang, F., Ma, J.K. and Vitale, A. (2000) Assembly, secretion, and vacuolar delivery of a hybrid immunoglobulin in plants. *Plant Physiol.*, 123, 1483-1494.

Gerber, S., Lane, C., Brown, D.M., Lord, E., DiLorenzo, M., Clements, J.D., Rybicki, E., Williamson, A.L. and Rose, R.C. (2001) Human papillomavirus virus-like particles are efficient oral immunogens when coadministered with Escherichia coli heat-labile enterotoxin mutant R192G or CpG DNA. *J. Virol.*, 75, 4752-4760.

Gil, C., Chaib-Oukadour, I., Blasi, J. and Aguilera, J. (2001) HC fragment (C-terminal portion of the heavy chain) of tetanus toxin activates protein kinase C isoforms and phosphoproteins involved in signal transduction. *Biochem. J.*, 356, 97-103.

Gleba, Y., Klimyuk, V. and Marillonnet, S. (2007) Viral vectors for the expression of proteins in plants. *Curr. Opin. Biotechnol.*, 18, 134-141.

Gomord, V., Chamberlain, P., Jefferis, R. and Faye, L. (2005) Biopharmaceutical production in plants: problems, solutions and opportunities. *Trends Biotechnol.*, 23, 559-565.

Greco, R., Michel, M., Guetard, D., Cervantes-Gonzalez, M., Pelucchi, N., Wain-Hobson, S., Sala, F. and Sala, M. (2007) Production of recombinant HIV-1/HBV virus-like particles in Nicotiana tabacum and Arabidopsis thaliana plants for a bivalent plant-based vaccine. *Vaccine*, 25, 8228-8240.

Grgacic, E.V. and Anderson, D.A. (2006) Virus-like particles: passport to immune recognition. *Methods*, 40, 60-65.

Guerrero, R.A., Ball, J.M., Krater, S.S., Pacheco, S.E., Clements, J.D. and Estes, M.K. (2001) Recombinant Norwalk virus-like particles administered intranasally to mice induce systemic and mucosal (fecal and vaginal) immune responses. *J. Virol.*, 75, 9713-9722.

Hadlington, J.L., Santoro, A., Nuttall, J., Denecke, J., Ma, J.K., Vitale, A. and Frigerio, L. (2003) The C-terminal extension of a hybrid immunoglobulin A/G heavy chain is responsible for its Golgi-mediated sorting to the vacuole. *Mol. Biol. Cell*, 14, 2592-2602.

Hameed, A., Miller, D.S., Muller, C.Y., Coleman, R.L. and Albores-Saavedra, J. (1999) Frequent expression of beta-human chorionic gonadotropin (beta-hCG) in squamous cell carcinoma of the cervix. *Int. J. Gynecol. Pathol.*, 18, 381-386.

Hein, M.B., Tang, Y., McLeod, D.A., Janda, K.D. and Hiatt, A. (1991) Evaluation of immunoglobulins from plant cells. *Biotechnol. Prog.*, 7, 455-461.

Hiatt, A. (1990) Antibodies produced in plants. *Nature*, 344, 469-470.

Hiatt, A., Cafferkey, R. and Bowdish, K. (1989) Production of antibodies in transgenic plants. *Nature*, 342, 76-78.

Hiatt, A. and Ma, J.K. (1992) Monoclonal antibody engineering in plants. *FEBS Lett.*, 307, 71-75.

Hofmann-Lehmann, R., Vlasak, J., Rasmussen, R.A., Smith, B.A., Baba, T.W., Liska, V., Ferrantelli, F., Montefiori, D.C., McClure, H.M., Anderson, D.C., Bernacky, B.J., Rizvi, T.A., Schmidt, R., Hill, L.R., Keeling, M.E., Katinger, H., Stiegler, G., Cavacini, L.A., Posner, M.R., Chou, T.C., Andersen, J. and Ruprecht, R.M. (2001) Postnatal passive immunization of neonatal macaques with a triple combination of human monoclonal antibodies against oral simian-human immunodeficiency virus challenge. *J. Virol*, 75, 7470-7480.

Hood, E.E., Woodard, S.L. and Horn, M.E. (2002) Monoclonal antibody manufacturing in transgenic plants--myths and realities. *Curr. Opin. Biotechnol.*, 13, 630-635.

Hu, S., Shively, L., Raubitschek, A., Sherman, M., Williams, L.E., Wong, J.Y., Shively, J.E. and Wu, A.M. (1996) Minibody: A novel engineered anti-carcinoembryonic antigen antibody fragment (single-chain Fv-CH3) which exhibits rapid, high-level targeting of xenografts. *Cancer Res.*, 56, 3055-3061.

Huang, Y., Liang, W., Wang, Y., Zhou, Z., Pan, A., Yang, X., Huang, C., Chen, J. and Zhang, D. (2005a) Immunogenicity of the epitope of the foot-and-mouth disease virus fused with a hepatitis B core protein as expressed in transgenic tobacco. *Viral Immunol.*, 18, 668-677.

Huang, Z., Elkin, G., Maloney, B.J., Beuhner, N., Arntzen, C.J., Thanavala, Y. and Mason, H.S. (2005b) Virus-like particle expression and assembly in plants: hepatitis B and Norwalk viruses. *Vaccine*, 23, 1851-1858.

Huang, Z., Lepore, K., Elkin, G., Thanavala, Y. and Mason, H.S. (2008) High-yield rapid production of hepatitis B surface antigen in plant leaf by a viral expression system. *Plant Biotechnol. J.*, 6, 202-209.

Huang, Z. and Mason, H.S. (2004) Conformational analysis of hepatitis B surface antigen fusions in an Agrobacterium-mediated transient expression system. *Plant Biotechnol. J.*, 2, 241-249.

Huang, Z., Santi, L., LePore, K., Kilbourne, J., Arntzen, C.J. and Mason, H.S. (2006) Rapid, high-level production of hepatitis B core antigen in plant leaf and its immunogenicity in mice. *Vaccine*, 24, 2506-2513.

Huston, J.S., Mudgett-Hunter, M., Tai, M.S., McCartney, J., Warren, F., Haber, E. and Oppermann, H. (1991) Protein engineering of single-chain Fv analogs and fusion proteins. *Methods Enzymol.*, 203, 46-88.

Ismaili, A., Jalali-Javaran, M., Rasaee, M.J., Rahbarizadeh, F., Forouzandeh-Moghadam, M. and Memari, H.R. (2007) Production and characterization of anti-(mucin MUC1) single-domain antibody in tobacco (Nicotiana tabacum cultivar Xanthi). *Biotechnol. Appl. Biochem.*, 47, 11-19.

Jiang, X., Wang, M., Graham, D.Y. and Estes, M.K. (1992) Expression, self-assembly, and antigenicity of the Norwalk virus capsid protein. *J. Virol.*, 66, 6527-6532.

Jobling, S.A., Jarman, C., Teh, M.M., Holmberg, N., Blake, C. and Verhoeyen, M.E. (2003) Immunomodulation of enzyme function in plants by single-domain antibody fragments. *Nat Biotechnol*, 21, 77-80.

Jones, P.T., Dear, P.H., Foote, J., Neuberger, M.S. and Winter, G. (1986) Replacing the complementarity-determining regions in a human antibody with those from a mouse. *Nature*, 321, 522-525.

Jost, C.R., Kurucz, I., Jacobus, C.M., Titus, J.A., George, A.J. and Segal, D.M. (1994) Mammalian expression and secretion of functional single-chain Fv molecules. *J. Biol. Chem.*, 269, 26267-26273.

Kapusta, J., Modelska, A., Figlerowicz, M., Pniewski, T., Letellier, M., Lisowa, O., Yusibov, V., Koprowski, H., Plucienniczak, A. and Legocki, A.B. (1999) A plant-derived edible vaccine against hepatitis B virus. *Faseb J*, 13, 1796-1799.

Karpas, A., Dremucheva, A. and Czepulkowski, B.H. (2001) A human myeloma cell line suitable for the generation of human monoclonal antibodies. *Proc. Natl. Acad. Sci. U S A*, 98, 1799-1804.

Kathuria, S., Sriraman, R., Nath, R., Sack, M., Pal, R., Artsaenko, O., Talwar, G.P., Fischer, R. and Finnern, R. (2002) Efficacy of plant-produced recombinant antibodies against HCG. *Hum. Reprod.*, 17, 2054-2061.

Khoudi, H., Laberge, S., Ferullo, J.M., Bazin, R., Darveau, A., Castonguay, Y., Allard, G., Lemieux, R. and Vezina, L.P. (1999) Production of a diagnostic monoclonal antibody in perennial alfalfa plants. *Biotechnol. Bioeng.*, 64, 135-143.

Kipriyanov, S. and Gall, F.L. (2004) Generation and production of engineered antibodies. *Mol. Biotechnol..*, 26, 39-60.

Kiyoi, H. and Naoe, T. (2001) Immunoglobulin variable region structure and B-cell malignancies. *Int. J. Hematol.*, 73, 47-53.

Ko, K., Steplewski, Z., Glogowska, M. and Koprowski, H. (2005) Inhibition of tumor growth by plant-derived mAb. *Proc. Natl .Acad. Sci. U S A*, 102, 7026-7030.

Ko, K., Tekoah, Y., Rudd, P.M., Harvey, D.J., Dwek, R.A., Spitsin, S., Hanlon, C.A., Rupprecht, C., Dietzschold, B., Golovkin, M. and Koprowski, H. (2003) Function and glycosylation of plant-derived antiviral monoclonal antibody. *Proc. Natl. Acad. Sci. U S A*, 100, 8013-8018.

Ko, K., Wei, X., Crooks, P.A. and Koprowski, H. (2004) Elimination of alkaloids from plant-derived human monoclonal antibody. *J. Immunol. Methods*, 286, 79-85.

Kohler, G. and Milstein, C. (1975) Continuous cultures of fused cells secreting antibody of predefined specificity. *Nature*, 256, 495-497.

Kong, Q., Richter, L., Yang, Y.F., Arntzen, C.J., Mason, H.S. and Thanavala, Y. (2001) Oral immunization with hepatitis B surface antigen expressed in transgenic plants. *Proc. Natl. Acad .Sci. U S A*, 98, 11539-11544.

Koprowski, H. (2005) Vaccines and sera through plant biotechnology. *Vaccine*, 23, 1757-1763.

Kuhnel, B., Alcantara, J., Boothe, J., van Rooijen, G. and Moloney, M. (2003) Precise and efficient cleavage of recombinant fusion proteins using mammalian aspartic proteases. *Protein. Eng.*, 16, 777-783.

Kumar, G.B.S., Ganapathi, T.R., Revathi, C.J., Prasad, K.S. and Bapat, V.A. (2003) Expression of hepatitis B surface antigen in tobacco cell suspension cultures. *Protein. Expr. Purif.*, 32, 10-17.

Kumar, G.B.S., Ganapathi, T.R., Revathi, C.J., Srinivas, L. and Bapat, V.A. (2005) Expression of hepatitis B surface antigen in transgenic banana plants. *Planta*, 222, 484-493.

Kusnadi, A.R., Evangelista, R.L., Hood, E.E., Howard, J.A. and Nikolov, Z.L. (1998) Processing of transgenic corn seed and its effect on the recovery of recombinant beta-glucuronidase. *Biotechnol. Bioeng.*, 60, 44-52.

Larrick, J.W. and Thomas, D.W. (2001) Producing proteins in transgenic plants and animals. *Curr .Opin. Biotechnol.*, 12, 411-418.

Lerouge, P., Bardor, M., Pagny, S., Gomord, V. and L, F. (2000) N-glycosylation of recombinant pharmaceutical glycoproteins produced in transgenic plants: towards an humanisation of plant N-glycans. *Curr. Phar. Biotechnol.*, 1, 347-354.

Lobaina, Y., Palenzuela, D., Garcia, D., Rodriguez, D., Pichardo, D., Muzio, V. and Aguilar, J.C. (2006) Comparative study of the immunogenicity and immunoenhancing effects of two hepatitis B core antigen variants in mice by nasal administration. *Vaccine*, 24 Suppl 2, S2-58-59.

Lobaina, Y., Palenzuela, D., Pichardo, D., Muzio, V., Guillen, G. and Aguilar, J.C. (2005) Immunological characterization of two hepatitis B core antigen variants and their immunoenhancing effect on co-delivered hepatitis B surface antigen. *Mol. Immunol.*, 42, 289-294.

Lowy, D.R. and Schiller, J.T. (2006) Prophylactic human papillomavirus vaccines. *J. Clin. Invest.*, 116, 1167-1173.

Ludwig, C. and Wagner, R. (2007) Virus-like particles-universal molecular toolboxes. *Curr. Opin. Biotechno.l*, 18, 537-545.

Ma, J.K., Hiatt, A., Hein, M., Vine, N.D., Wang, F., Stabila, P., van Dolleweerd, C., Mostov, K. and Lehner, T. (1995) Generation and assembly of secretory antibodies in plants. *Science*, 268, 716-719.

Ma, J.K., Hikmat, B.Y., Wycoff, K., Vine, N.D., Chargelegue, D., Yu, L., Hein, M.B. and Lehner, T. (1998) Characterization of a recombinant plant monoclonal secretory antibody and preventive immunotherapy in humans. *Nat. Med.*, 4, 601-606.

Maclean, J., Koekemoer, M., Olivier, A.J., Stewart, D., Hitzeroth, II, Rademacher, T., Fischer, R., Williamson, A.L. and Rybicki, E.P. (2007) Optimization of human papillomavirus type 16 (HPV-16) L1 expression in plants: comparison of the suitability of different HPV-16 L1 gene variants and different cell-compartment localization. *J. Gen. Virol.*, 88, 1460-1469.

Madore, H.P., Estes, M.K., Zarley, C.D., Hu, B., Parsons, S., Digravio, D., Greiner, S., Smith, R., Jiang, B., Corsaro, B., Barniak, V., Crawford, S. and Conner, M.E. (1999) Biochemical and immunologic comparison of virus-like particles for a rotavirus subunit vaccine. *Vaccine*, 17, 2461-2471.

Maliga, P. (2002) Engineering the plastid genome of higher plants. *Curr. Opin. Plant. Biol.*, 5, 164-172.

Maliga, P. (2003) Progress towards commercialization of plastid transformation technology. *Trends Biotechnol.*, 21, 20-28.

Mascola, J.R. (2002) Passive transfer studies to elucidate the role of antibody-mediated protection against HIV-1. *Vaccine*, 20, 1922-1925.

Mascola, J.R., Lewis, M.G., Stiegler, G., Harris, D., VanCott, T.C., Hayes, D., Louder, M.K., Brown, C.R., Sapan, C.V., Frankel, S.S., Lu, Y., Robb, M.L., Katinger, H. and Birx, D.L. (1999) Protection of Macaques against pathogenic simian/human immunodeficiency virus 89.6PD by passive transfer of neutralizing antibodies. *J Virol*, 73, 4009-4018.

Mayfield, S.P., Franklin, S.E. and Lerner, R.A. (2003) Expression and assembly of a fully active antibody in algae. *Proc .Natl. Acad. Sci. U S A*, 100, 438-442.

McAleer, W.J., Markus, H.Z., Wampler, D.E., Buynak, E.B., Miller, W.J., Weibel, R.E., McLean, A.A. and Hilleman, M.R. (1984) Vaccine against human hepatitis B virus prepared from antigen derived from human hepatoma cells in culture. *Proc. Soc. Exp. Biol. Med*,. 175, 314-319.

McCormick, A.A., Kumagai, M.H., Hanley, K., Turpen, T.H., Hakim, I., Grill, L.K., Tuse, D., Levy, S. and Levy, R. (1999) Rapid production of specific vaccines for lymphoma by expression of the tumor-derived single-chain Fv epitopes in tobacco plants. *Proc. Natl. Acad. Sci. U S A*, 96, 703-708.

McCormick, A.A., Reinl, S.J., Cameron, T.I., Vojdani, F., Fronefield, M., Levy, R. and Tuse, D. (2003) Individualized human scFv vaccines produced in plants: humoral anti-idiotype responses in vaccinated mice confirm relevance to the tumor Ig. *J. Immunol. Methods*, 278, 95-104.

McMichael, A.J. and Rowland-Jones, S.L. (2001) Cellular immune responses to HIV. *Nature*, 410, 980-987.

Mechtcheriakova, I.A., Eldarov, M.A., Nicholson, L., Shanks, M., Skryabin, K.G. and Lomonossoff, G.P. (2006) The use of viral vectors to produce hepatitis B virus core particles in plants. *J. Virol. Methods*, 131, 10-15.

Metzger, H. and Kinet, J. (1988) How antibodies work: focus on Fc receptors. *FASEB J.*, 2, 3-11.

Misaki, R., Kimura, Y., Palacpac, N.Q., Yoshida, S., Fujiyama, K. and Seki, T. (2003) Plant cultured cells expressing human beta1,4-galactosyltransferase secrete glycoproteins with galactose-extended N-linked glycans. *Glycobiology*, 13, 199-205.

Molinari, P., Peralta, A. and Taboga, O. (2008) Production of rotavirus-like particles in Spodoptera frugiperda larvae. *Journal of Virological Methods*, 147, 364-367.

Moreland, L.W., Bucy, R.P., Weinblatt, M.E., Mohler, K.M., Spencer-Green, G.T. and Chatham, W.W. (2002) Immune function in patients with rheumatoid arthritis treated with etanercept. *Clin. Immunol.*, 103, 13-21.

Morrison, S.L., Johnson, M.J., Herzenberg, L.A. and Oi, V.T. (1984) Chimeric human antibody molecules: mouse antigen-binding domains with human constant region domains. *Proc. Natl. Acad. Sci. U S A*, 81, 6851-6855.

Morrow, J. (2002) Economics in antibody production. *Genet. Eng. News*, 22, 1-39.

Neumaier, M., Shively, L., Chen, F., Gaida, F., Ilgen, C., Paxton, R., Shively, J.E. and Riggs, A. (1990) Cloning of the genes for T84.66, an antibody that has a high specificity and affinity for carcinoembryonic antigen, and expression of chimeric human/mouse T84.66 genes in myeloma and Chinese hamster ovary cells. . *Cancer Res.*, 2128-2134.

Nguyen, V.K., Desmyter, A. and Muyldermans, S. (2001) Functional heavy-chain antibodies in Camelidae. *Adv. Immunol.*, 79, 261-296.

Nikolov, Z.L. and Hammes, D. (2002) Production of recombinant proteins from transgenic crops. In Plants as Factories for protein Production, Hood EE and Howard JA, eds (Kluwer Academic Publishers, The Netherlands.

Nugent, J. and Joyce, S. (2005) Producing human therapeutic proteins in plastids. *Curr. Pharm. Des.*, 11, 2459-2470.

Nuttall, J., Vine, N., Hadlington, J.L., Drake, P., Frigerio, L. and Ma, J.K. (2002) ER-resident chaperone interactions with recombinant antibodies in transgenic plants. *Eur. J. Biochem.*, 269, 6042-6051.

O'Brien, G.J., Bryant, C.J., Voogd, C., Greenberg, H.B., Gardner, R.C. and A.R., B. (2000) Rotavirus VP6 Expressed by PVX Vectors in Nicotiana benthamiana Coats PVX Rods and Also Assembles into Viruslike particles. *Virology*, 270, 444-453.

O'Ryan, M., Prado, V. and Pickering, L.K. (2005) A millennium update on pediatric diarrheal illness in the developing world. *Semin Pediatr Infect Dis*, 16, 125-136.

Obregon, P., Chargelegue, D., Drake, P.M., Prada, A., Nuttall, J., Frigerio, L. and Ma, J.K. (2006) HIV-1 p24-immunoglobulin fusion molecule: a new strategy for plant-based protein production. *Plant Biotechnol. J.*, 4, 195-207.

Okamoto, T., Matsuo, K., Niu, R., Osawa, M. and Suzuki, H. (2001) Human chorionic gonadotropin (hCG) beta-core fragment is produced by degradation of hCG or free hCG beta in gestational trophoblastic tumors: a possible marker for early detection of persistent postmolar gestational trophoblastic disease. *J. Endocrinol.*, 171, 435-443.

Palacpac, N.Q., Yoshida, S., Sakai, H., Kimura, Y., Fujiyama, K., Yoshida, T. and Seki, T. (1999) Stable expression of human beta1,4-galactosyltransferase in plant cells modifies N-linked glycosylation patterns. *Proc. Natl. Acad .Sci. U S A*, 96, 4692-4697.

Perrin, Y., Vaquero, C., Gerrard, I., Sack, M., Drossard, J. and Stoger, E. (2000) Transgenic pea seeds as bioreactors for the production of a single-chain Fv fragment (scFV) antibody used in cancer diagnosis and therapy. *Mol. Breeding*, 345-352.

Qian, B., Shen, H., Liang, W., Guo, X., Zhang, C., Wang, Y., Li, G., Wu, A., Cao, K. and Zhang, D. (2007) Immunogenicity of recombinant hepatitis B virus surface antigen fused with preS1 epitopes expressed in rice seeds. *Transgenic Res.*

Rademacher, T., Sack, M., Arcalis, E., Stadlmann, J., Balzer, S., Altmann, F., Quendler, H., Stiegler, G., Kunert, R., Fischer, R. and Stoger, E. (2008) Recombinant antibody 2G12 produced in maize endosperm efficiently neutralizes HIV-1 and contains predominantly single-GlcNAc N-glycans. *Plant Biotechnol. J.*, 6, 189-201.

Richter, J., Thanavala, Y., Arntzen, C. and Mason, H. (2000) Production of hepatitis B surface antigen in transgenic plants for oral immunisation. *Nat. Biotechnol.*, 1167-1171.

Ridder, R., Schmitz, R., Legay, F. and Gram, H. (1995) Generation of rabbit monoclonal antibody fragments from a combinatorial phage display library and their production in the yeast Pichia pastoris. *Biotechnology (N Y)*, 13, 255-260.

Rudd, P.M., Elliott, T., Cresswell, P., Wilson, I.A. and Dwek, R.A. (2001) Glycosylation and the immune system. *Science*, 291, 2370-2376.

Saldaña, S., Esquivel Guadarrama, F., Olivera Flores Tde, J., Arias, N., López, S., Arias, C., Ruiz-Medrano, R., Mason, H., Mor, T., Richter, L., Arntzen, C. and Gómez Lim, M.A. (2006) Production of Rotavirus-Like Particles in Tomato (Lycopersicon esculentum L.) Fruit by Expression of Capsid Proteins VP2 and VP6 and Immunological Studies. *Viral Immunol.*, 19, 42-53.

Santi, L., Batchelor, L., Huang, Z., Hjelm, B., Kilbourne, J., Arntzen, C., Chen, Q. and Mason, H. (2008) An efficient plant viral expression system generating orally immunogenic Norwalk virus-like particles. *Vaccine*, In press.

Santi, L., Huang, Z. and Mason, H. (2006) Virus-like particles production in green plants. *Methods*, 40, 66-76.

Satub, J., B Garcia, Graves, J., Hajdukiewicz, P., Hunter, P., Nehra, N., Paradkar, V., Schlitter, M., Carroll, J., Spatola, L., Ward, D., Yeand, G. and DARussell. (2000) High-yield production of a human therapeutic protein in tobacco chloroplasts. *Nat. Biotechnol*, 333-338.

Schahs, M., Strasser, R., Stadlmann, J., Kunert, R., Rademacher, T. and Steinkellner, H. (2007) Production of a monoclonal antibody in plants with a humanized N-glycosylation pattern. *Plant Biotechnol J*, 5, 657-663.

Scheller, J., Leps, M. and Conrad, U. (2006) Forcing single-chain variable fragment production in tobacco seeds by fusion to elastin-like polypeptides. *Plant Biotechnol. J.*, 4, 243-249.

Schillberg, S., Fischer, R. and Emans, N. (2003) Molecular farming of recombinant antibodies in plants. *Cell Mol Life Sci*, 60, 433-445.

Schiller, J.T. and Nardelli-Haefliger, D. (2006) Chapter 17: Second generation HPV vaccines to prevent cervical cancer. *Vaccine*, 24 Suppl 3, S147-153.

Schouten, A., Roosien, J., van Engelen, F.A., de Jong, G.A., Borst-Vrenssen, A.W., Zilverentant, J.F., Bosch, D., Stiekema, W.J., Gommers, F.J., Schots, A. and Bakker, J. (1996) The C-terminal KDEL sequence increases the expression level of a single-chain antibody designed to be targeted to both the cytosol and the secretory pathway in transgenic tobacco. *Plant Mol. Biol.*, 30, 781-793.

Semeniuk, E.G., Stremovskii, O.A., Orlova, I.V., Baladin, T.G., Nosov, A.M., Bur'ianov Ia, I., Deev, S.M. and Petrov, R.V. (2003) [Biosynthesis of mini-antibodies to human ferritin in plant and bacterial producers]. *Mol. Biol. (Mosk)*, 37, 916-923.

Smith, M.D. and Glick, B.R. (2000) The production of antibodies in plants: an idea whose time has come? *Biotechnol .Adv.*, 18, 85-89.

Sojikul, P., Buehner, N. and Mason, H.S. (2003) A plant signal peptide-hepatitis B surface antigen fusion protein with enhanced stability and immunogenicity expressed in plant cells. *Proc. Natl. Acad. Sci. U S A*, 100, 2209-2214.

Stoger, E., Sack, M., Fischer, R. and Christou, P. (2002) Plantibodies: applications, advantages and bottlenecks. *Curr. Opin. Biotechnol.*, 13, 161-166.

Stoger, E., Vaquero, C., Torres, E., Sack, M., Nicholson, L., Drossard, J., Williams, S., Keen, D., Perrin, Y., Christou, P. and Fischer, R. (2000) Cereal crops as viable production and storage systems for pharmaceutical scFv antibodies. *Plant Mol. Biol.*, 42, 583-590.

Streatfield, S.J., Lane, J.R., Brooks, C.A., Barker, D.K., Poage, M.L., Mayor, J.M., Lamphear, B.J., Drees, C.F., Jilka, J.M., Hood, E.E. and Howard, J.A. (2003) Corn as a production system for human and animal vaccines. *Vaccine*, 21, 812-815.

Suzich, J.A., Ghim, S.J., Palmer-Hill, F.J., White, W.I., Tamura, J.K., Bell, J.A., Newsome, J.A., Jenson, A.B. and Schlegel, R. (1995) Systemic immunization with papillomavirus L1 protein completely prevents the development of viral mucosal papillomas. *Proc. Natl. Acad. Sci. U S A*, 92, 11553-11557.

Tacket, C.O. and Mason, H.S. (1999) A review of oral vaccination with transgenic vegetables. *Microbes Infect.*, 1, 777-783.

Talwar, G.P., Singh, O., Pal, R., Chatterjee, N., Sahai, P., Dhall, K., Kaur, J., Das, S.K., Suri, S., Buckshee, K. and et al. (1994) A vaccine that prevents pregnancy in women. *Proc. Natl. Acad. Sci. U S A*, 91, 8532-8536.

Timko, M.P. and Cahoon, A.B. (1999) Transgenic plants for the production of human therapeutics. In: Chopra VL, Malik VS and Bhat SR (eds). *Applied Plant Biotechnology. Science Publishers Inc. New Hampshire, USA.*, 155-180.

Traggiai, E., Becker, S., Subbarao, K., Kolesnikova, L., Uematsu, Y., Gismondo, M.R., Murphy, B.R., Rappuoli, R. and Lanzavecchia, A. (2004) An efficient method to make human monoclonal antibodies from memory B cells: potent neutralization of SARS coronavirus. *Nat. Med.*, 10, 871-875.

Tsuda, S., Yoshioka, K., Tanaka, T., Iwata, A., Yoshikawa, A., Watanabe, Y. and Okada, Y. (1998) Application of the human hepatitis B virus core antigen from transgenic tobacco plants for serological diagnosis. *Vox Sang.*, 74, 148-155.

Twyman, R.M., Stoger, E., Schillberg, S., Christou, P. and Fischer, R. (2003) Molecular farming in plants: host systems and expression technology. *Trends. Biotechnol.*, 21, 570-578.

Vaquero, C., Sack, M., Chandler, J., Drossard, J., Schuster, F., Monecke, M., Schillberg, S. and Fischer, R. (1999) Transient expression of a tumor-specific single-chain fragment and a chimeric antibody in tobacco leaves. *Proc Natl Acad Sci U S A*, 96, 11128-11133.

Vaquero, C., Sack, M., Schuster, F., Finnern, R., Drossard, J., Schumann, D., Reimann, A. and Fischer, R. (2002) A carcinoembryonic antigen-specific diabody produced in tobacco. *Faseb. J.*, 16, 408-410.

Varsani, A., Williamson, A.L., Rose, R.C., Jaffer, M. and Rybicki, E.P. (2003) Expression of Human papillomavirus type 16 major capsid protein in transgenic Nicotiana tabacum cv. Xanthi. *Arch. Virol.*, 148, 1771-1786.

Varsani, A., Williamson, A.L., Stewart, D. and Rybicki, E.P. (2006) Transient expression of Human papillomavirus type 16 L1 protein in Nicotiana benthamiana using an infectious tobamovirus vector. *Virus Res.*, 120, 91-96.

Veazey, R.S., Shattock, R.J., Pope, M., Kirijan, J.C., Jones, J., Hu, Q., Ketas, T., Marx, P.A., Klasse, P.J., Burton, D.R. and Moore, J.P. (2003) Prevention of virus transmission to macaque monkeys by a vaginally applied monoclonal antibody to HIV-1 gp120. *Nat. Med.*, 9, 343-346.

Verch, T., Hooper, D.C., Kiyatkin, A., Steplewski, Z. and Koprowski, H. (2004) Immunization with a plant-produced colorectal cancer antigen. *Cancer Immunol. Immunother.*, 53, 92-99.

Verch, T., Yusibov, V. and Koprowski, H. (1998) Expression and assembly of a full-length monoclonal antibody in plants using a plant virus vector. *J. Immunol. Methods*, 220, 69-75.

Walmsley, A.M., Alvarez, M.L., Jin, Y., Kirk, D.D., Lee, S.M., Pinkhasov, J., Rigano, M.M., Arntzen, C.J. and Mason, H.S. (2003) Expression of the B subunit of Escherichia coli heat-labile enterotoxin as a fusion protein in transgenic tomato. *Plant Cell Rep.*, 21, 1020-1026.

Ward, E.S. (1993) Antibody engineering using Escherichia coli as host. *Adv. Pharmacol.*, 24, 1-20.

Warner, T. (2000) Metabolic engineering glycosylation: biotechnology´s challenge to the glycobiologist in the new millennium. *In Carbohydrates in Chemistry and Biology. Edited by Ernst B, hart GW, Sanay P., (New York: Wiley-VCH)*, 1043-1064.

Warzecha, H., Mason, H.S., Lane, C., Tryggvesson, A., Rybicki, E., Williamson, A.L., Clements, J.D. and Rose, R.C. (2003) Oral immunogenicity of human papillomavirus-like particles expressed in potato. *J. Virol.*, 77, 8702-8711.

Weinblatt, M.E., Kremer, J.M., Bankhurst, A.D., Bulpitt, K.J., Fleischmann, R.M., Fox, R.I., Jackson, C.G., Lange, M. and Burge, D.J. (1999) A trial of etanercept, a recombinant tumor necrosis factor receptor:Fc fusion protein, in patients with rheumatoid arthritis receiving methotrexate. *N. Engl. J. Med.*, 340, 253-259.

Weiner, L.M. (1999) An overview of monoclonal antibody therapy of cancer. *Semin Oncol*, 26, 41-50.

Wieland, W.H., Lammers, A., Schots, A. and Orzaez, D.V. (2006) Plant expression of chicken secretory antibodies derived from combinatorial libraries. *J. Biotechnol.*, 122, 382-391.

Wigdorovitz, A., Carrillo, C., Dus Santos, M.J., Trono, K., Peralta, A., Gomez, M.C., Rios, R.D., Franzone, P.M., Sadir, A.M., Escribano, J.M. and Borca, M.V. (1999) Induction of a protective antibody response to foot and mouth disease virus in mice following oral or parenteral immunization with alfalfa transgenic plants expressing the viral structural protein VP1. *Virology*, 255, 347-353.

Wright, A. and Morrison, S.L. (1997) Effect of glycosylation on antibody function: implications for genetic engineering. *Trends Biotechnol.*, 15, 26-32.

Wu, A.M. and Senter, P.D. (2005) Arming antibodies: prospects and challenges for immunoconjugates. *Nat. Biotechnol.*, 23, 1137-1146.

Yamada, T., Iwasaki, Y., Tada, H., Iwabuki, H., Chuah, M.K., VandenDriessche, T., Fukuda, H., Kondo, A., Ueda, M., Seno, M., Tanizawa, K. and Kuroda, S. (2003) Nanoparticles for the delivery of genes and drugs to human hepatocytes. *Nat .Biotechnol.*, 21, 885-890.

Yano, A., Maeda, F. and Takekoshi, M. (2004) Transgenic tobacco cells producing the human monoclonal antibody to hepatitis B virus surface antigen. *J Med Virol*, 73, 208-215.

Yu, J. and Langridge, W.H. (2001) A plant-based multicomponent vaccine protects mice from enteric diseases. *Nat. Biotechnol.*, 19, 548-552.

Zeitlin, L., Olmsted, S.S., Moench, T.R., Co, M.S., Martinell, B.J., Paradkar, V.M., Russell, D.R., Queen, C., Cone, R.A. and Whaley, K.J. (1998) A humanized monoclonal antibody produced in transgenic plants for immunoprotection of the vagina against genital herpes. *Nat. Biotechnol*, 16, 1361-1364.

Zhang, G.G., Rodrigues, L., Rovinski, B. and White, K.A. (2002) Production of HIV-1 p24 protein in transgenic tobacco plants. *Mol Biotechnol*, 20, 131-136.

Zhang, X., Buehner, N., Hutson, A.M., Estes, M.K. and Mason, H. (2006) Tomato is a highly effective vehicle for expression and oral immunization with Norwalk virus capsid protein. *Plant Biotechnol. J.*, 4, 419-432.

Zhou, W., Wu, X.A., Luo, W., Hu, G., Zhang, F.L., Bai, W.T., Zhang, L., Yu, L., Shi, M.Y. and Xu, Z.K. (2008) [Transformation of full-length gene of mouse/human chimeric antibody against Hantaan virus into Arabidopsis thaliana.]. *Xi Bao Yu Fen Zi Mian Yi Xue Za Zhi*, 24, 49-51.

zu Putlitz, J., Kubasek, W.L., Duchene, M., Marget, M., von Specht, B.U. and Domdey, H. (1990) Antibody production in baculovirus-infected insect cells. *Biotechnology (N Y)*, 8, 651-654.

In: Genetically Modified Plants
Editors: Timm Wolf and Jonas Koch

ISBN 978-1-60456-696-3

Chapter 3

GENETIC MODIFICATION IN *LOLIUM* AND *FESTUCA* SPECIES

Wataru Takahashi[*] and Tadashi Takamizo
National Institute of Livestock and Grassland Science, 768 Senbonmatsu, Nasushiobara, Tochigi 329-2793, Japan

ABSTRACT

Lolium and *Festuca* species are the most important forage and turf grasses, and are cultivated in temperate zones around the world. Since most of these species are outcrossing plants, genetic diversity is observed even within a single cultivar developed by means of conventional breeding. Thus, genetic engineering is expected to be an attractive and powerful tool for improving the sophistication of breeding programs and ensuring that even within a genetically diverse cultivar, it is possible to incorporate useful genes in all individuals; as a result, this approach has been experimentally adopted in both groups of grass species. In this review, we present the current state of genetic modification that has been achieved in *Lolium* and *Festuca* species, and discuss how we can produce practical transgenic plants of these species.

INTRODUCTION

Lolium and *Festuca* species are the most important grasses grown as forage and turf crops in temperate zones around the world. For instance, perennial ryegrass (*Lolium perenne* L.) and Italian ryegrass (*Lolium multiflorum* Lam.) show high yield potential, fast establishment, and the ability to establish on heavy and waterlogged soils, among other useful characteristics. Tall fescue (*Festuca arundinacea* Schreb.), the most important grass in the genus *Festuca*, is valued for its tolerance of a wide range of soil and climatic factors and its high yield potential. Since the majority of *Lolium* and *Festuca* species are self-infertile,

[*] Corresponding author: Wataru Takahashi, Tel.: +81-287-37-7694, Fax: +81-287-36-6629, E-mail: twataru@affrc.go.jp

progeny are necessarily produced by outcrossing, and breeding lines and cultivars are produced by crossing among several selected genotypes. Thus, genetic diversity and phenotypic differences may be high even within a cultivar, because precise control of an agronomically useful trait by conventional breeding methods is time-consuming and troublesome to achieve. Therefore, other sophisticated breeding techniques are required to develop new cultivars. In this context, genetic engineering is an attractive and powerful tool, and has been experimentally adopted in several *Lolium* and *Festuca* species. This review presents the current progress in genetic modification of these species to improve their tolerance of abiotic and biotic stresses, improve their nutritional value, control flowering, and permit the use of low-input stand management. Based on this review, we describe a perspective for the practical application of transgenic plants in these species.

Transformation Systems

Gene transfer can be generally carried out using polyethylene glycol, electroporation, particle bombardment, whiskers, and *Agrobacterium*-mediated transformation. During the last decade, systems for the transformation of both *Lolium* and *Festuca* species have been developed using each of these gene transfer methods. Polyethylene glycol–mediated transformation systems are often used to produce transgenic plants of *Lolium* and *Festuca* species (Z.Y. Wang et al. 1992; Spangenberg et al. 1994; Dalton et al. 1995; G.R. Wang et al. 1997; Kuai et al. 1999). Electroporation was used in *Festuca rubra* L. (Ha et al. 1992). These two methods, however, require delicate and complicated manipulation of protoplasts. Thus, in the late 1990s, particle bombardment– and whiskers-mediated gene transfer were developed as alternative transformation systems in *Lolium* and *Festuca* species, since these methods can use callus and suspension cells as targets without the need for troublesome protoplast cultures (Spangenberg et al. 1995; Ye et al. 1997; Dalton et al. 1998, 1999; Altpeter et al. 2000; Takahashi et al. 2002). Although whiskers-mediated gene transfer is the simplest, cheapest, and fastest direct gene transfer method, it has a lower transformation frequency and can treat only limited amounts of target tissue compared with particle bombardment (Dalton et al. 1998). For this reason, particle bombardment is currently the dominant gene transfer method. However, some research groups have devoted considerable effort to establishing an *Agrobacterium*-mediated transformation system, even though neither genus is a primary host of *Agrobacterium* (Bettany et al. 2003; Z.Y. Wang and Ge 2005; Bajaj et al. 2006; Cao et al. 2006; Sato and Takamizo 2006; Ge et al. 2007). Since *Agrobacterium*-mediated gene transfer seems to produce a low transgene copy number and results in stable gene expression in the transgenic plants, this method may prevail over the competing methods, along with particle bombardment, in the near future.

High-Throughput Transformation Systems

To incorporate genetic transformation into breeding programs, practical, high-throughput transformation systems are indispensable. A high-throughput transformation system combines high transformation efficiency with an efficient supply of a large amount of homogeneous regenerable target tissue. As the self-infertility of most *Lolium* and *Festuca* species necessitates outcrossing, many genotypes exist within a cultivar, and the regeneration potential depends strongly on each genotype. Further, since regenerable genotypes are not common in *Lolium* and *Festuca* species, screening to detect them is essential to obtaining large quantities of such tissue. These ideas have underlain the breeding of *Lolium* species. For example, Takahashi et al. (2004) screened 11 317 genotypes of Italian ryegrass (*L. multiflorum*) through callus culture of mature seeds of 12 cultivars, and aseptically preserved and propagated the regenerable genotypes *in vitro*. Highly regenerable calluses were routinely induced from meristems isolated from the screened genotypes and were incorporated in a high-throughput particle bombardment–mediated transformation system (Takahashi et al. 2002). Similarly, Bajaj et al. (2006) screened genotypes of perennial ryegrass with high embryogenic potential during tissue culture and maintained them as plants *in vitro*. In this work, embryogenic callus derived from meristematic regions of the vegetative tillers was subjected to *Agrobacterium*-mediated gene transfer, and a large number of transgenic plants were efficiently produced. In both transformation systems, the highly regenerative genotypes are maintained *in vitro* as plants. This permits the induction of a large amount of regenerable calluses derived from a single genotype at any time. This is an advantageous approach, since transgenic plants derived from a single genotype can be precisely evaluated by comparison with original nontransgenic plants preserved *in vitro* as a control. These transformation systems are effective in transforming outcrossing plants, especially in annual plant species such as Italian ryegrass.

Genetic Modification of Agronomically Useful Traits

Table 1 summarizes the recent attempts to genetically modify *Lolium* and *Festuca* species to improve their tolerance of biotic and abiotic stresses, improve their nutritional value, induce late flowering, reduce the allergenicity of their pollen, and permit the use of low-input turf management. Details of this research are described in the following sections.

Table 1. Transgenic *Lolium* and *Festuca* species that have been engineered to express various useful genes, and the results of the engineering

Traits Plant species	Transgenes (marker genes in brackets)	Outcome	Reference
Abiotic stress tolerance			
Lolium multiflorum	*sacB* , [*hpt*]	Stunted phenotype	Ye et al. (2001)
Lolium perenne	*ipt, uidA*[†], [*hpt*]	Stay green	Li et al. (2004)
Lolium perenne	*wft1, wft2,* [*bar*]	Freezing tolerance	Hisano et al. (2004)
Festuca arundinacea	*ipt,* [*bar*]	Freezing tolerance	Hu et al. (2005)
Festuca arundinacea	*CuZn-SOD, APX,* [*hpt*]	ROS* – stress tolerance	Lee et al. (2007)
Festuca arundinacea	*DREB1A/CBF3* , *uidA,* [*hpt*]	Drought tolerance	Zhao et al. (2007a)
Lolium perenne	*OsNHX1,* [*bar*]	Salt tolerance	Y.Y. Wu et al. (2005)
Festuca arundinacea	*AtNHX1,* [*nptII*]	Salt tolerance	Zhao et al. (2007b)
Festuca arundinacea	*AeNHX1,* [*hpt*]	Salt tolerance	Qiao et al. (2007)

Table 1. (Continued)

Traits	Plant species	Transgenes (marker genes in brackets)	Outcome
Festuca arundinacea	*CBF1, uidA,* [*hpt*]	Salt, freezing, heat, and drought tolerance	G.T. Wu et al. (2006)
Biotic stress tolerance			
Lolium perenne	*RgMV-CP,* [*nptII*]	Viral disease resistance	Xu et al. (2001)
Lolium multiflorum	*RCC2,* [*hpt*]	Fungal disease resistance	Takahashi et al. (2005)
Festuca arundinacea	*AGLU1, Pi9, dermaseptin SI,* [*hpt*]	Fungal disease resistance	Dong et al. (2007)
Festuca arundinacea	T4 *e,* [*hpt*]	Fungal disease resistance	Dong et al. (2008)
High nutritional value			
Festuca arundinacea	*sfa8,* [*hpt*]	Protein accumulation	Z.Y. Wang et al. (2001)
Festuca arundinacea	*CAD,* [*hpt*]	High digestibility	Chen et al. (2003)
Festuca arundinacea	*COMT,* [*hpt*]	High digestibility	Chen et al. (2004)
Lolium multiflorum	*FAE,* [*hpt*]	High digestibility	Buanafina et al. (2006)
Lolium perenne	*1-SST, 6G-FFT,* [*bar*], [*nptII*]	Fructan accumulation	Gadegaard et al. (2008)
Late flowering			
Festuca rubra	*LpTFL1,* [*bar*]	Late flowering	Jensen et al. (2004)
Lolium perenne	*ATH1,* [*hpt*]	Late flowering	Van der Valk et al. (2004)
Hypoallergenicity			
Lolium rigidum	*Lol p 5,* [*nptII*]	Hypoallergenicity of pollen	Bhalla et al. (1999)
Lolium perenne	*Lol p 1, Lol p 2,* [*hpt*]	Hypoallergenicity of pollen	Petrovska et al. (2004)
Low-input management			
Lolium perenne	*argE,* [*hpt*]	Selective mortality	Chen et al. (2005)

†*uidA*: *β*-glucuronidase gene (reporter gene).
*ROS: reactive oxygen species.

Tolerance of Abiotic Stress

Potential crop productivity is greatly limited by unfavorable environments (Boyer 1982). Therefore, conferring stress tolerance is a critical strategy for ensuring stable production of crops, and genetic engineering is expected to be a powerful tool for the production of stress-tolerant plants. The largest category of the studies in Table 1 were designed to produce transgenic plants capable of tolerating various forms of abiotic stress. Modification of fructan metabolism appears to be one of the most important approaches to developing stress-tolerant grass species, since fructan is a main component of the carbohydrates that accumulate in temperate C_3 grasses and is associated with winter hardiness and the ability to regrow immediately after defoliation; it also contributes to the nutritive value of feed, as discussed later in this chapter. Ye et al. (2001) produced transgenic Italian ryegrass that constitutively expressed a *Bacillus subtilis sacB* gene that could be used for functional analysis of fructan synthesis. The transgenic plants accumulated small amounts of bacterial levan, whereas levels of high-molecular-weight native fructan decreased, and the pattern of accumulation of oligosaccharides with a degree of polymerization ranging from 5 to 35 was distorted. Growth of the levan-accumulating *sacB*-transgenic ryegrass plants slowed down with onset of the reproductive stage, and flowering plants were stunted and had narrower leaves and a poorly developed root system. Such aberrant phenotypes may have resulted from the accumulation of bacterial-type levan in the plants. Unfortunately, the levan-accumulating *sacB*-transgenic plants were not evaluated against any stresses. Hisano et al. (2004) produced transgenic perennial ryegrass plants that overexpressed two wheat-derived fructan metabolism–related genes, *wft1* and *wft2*, which encode sucrose-fructan 6-fructosyltransferase (6-SFT) and sucrose-sucrose 1-fructosyltransferase (1-SST), respectively. Significant increases in fructan content were detected in transgenic plants expressing 6-SFT or 1-SST. Freezing tolerance of the transgenic plants was confirmed at the cellular level by means of a freezing test using the electrical conductivity method to monitor detached leaves. In contrast with transgenic plants that expressed *sacB*, no aberrant development was observed in transgenic plants that accumulated a greater amount of fructan than non-transgenic plants. Gadegaard et al. (2008) produced transgenic perennial ryegrass plants expressing onion fructosyltransferase genes *1-SST* and *6G-FFT* (fructan-fructan 6G-fructosyltransferase gene). Details of their study are provided in the section "*Nutritional value*", since the aim of their study was to improve the nutritional value of the crop through fructan accumulation.

Freezing-tolerant transgenic tall fescue (*F. arundinacea*) was also produced using an *Agrobacterium tumefaciens*–derived isopentenyl transferase (*ipt*) gene (Hu et al. 2005). A freezing test using detached leaves indicated increased freezing tolerance in the transgenic plants at the cellular level. Field trials of the transgenic plants showed higher tillering ability and a phenotype that stayed green longer than non-transgenic plants under lower temperatures. Li et al. (2004) also produced transgenic perennial ryegrass plants that expressed the *ipt* gene. The transgenic plants also displayed a phenotype that stayed green longer, although the stress tolerance of these plants was not evaluated.

To produce salt-tolerant plants, some researchers used the gene encoding a vacuolar Na^+/H^+ antiporter. The gene controls the process used to pump Na^+ from the cytoplasm into the vacuole to maintain a higher K^+/Na^+ ratio in the cytoplasm than in vacuoles. This mechanism for the compartmentalization of Na^+ protects the cell from sodium toxicity. The rice vacuolar Na^+/H^+ antiporter gene (*OsNHX1*) was transferred into perennial ryegrass to produce salt-tolerant transgenic plants (Y.Y. Wu et al. 2005). Transgenic plants that expressed the antiporter gene survived irrigation with a nutrient solution containing 350 mmol/L NaCl for 10 weeks, whereas non-transgenic plants died. Similarly, Zhao et al. (2007b) produced transgenic tall fescue plants expressing the *Arabidopsis*-derived vacuolar Na^+/H^+ antiporter gene (*AtNHX1*). A higher germination rate, greater biomass, and less-deleterious effects were observed in the transgenic plants than in nontransgenic plants under saline conditions: transgenic plants grown hydroponically were not affected by NaCl concentrations below 200 mM, whereas nontransgenic plants showed progressive chlorosis, reduced leaf size, and growth inhibition as the NaCl concentration increased. Qiao et al. (2007) also isolated a Na^+/H^+ antiporter gene (*AeNHX1*) from the root tissue of *Agropyron elongatum* L. that conferred strong tolerance of salt stress in tall fescue, and transgenic plants that expressed the *AeNHX1* gene showed normal growth after irrigation with 300 mM NaCl at

5-day intervals for 45 days and higher relative dry weight than nontransgenic control plants, which showed chlorosis and wilting.

Genes that encode transcription factors that govern the plant's response to abiotic stresses are promising tools for producing transgenic plants that are able to tolerate multiple abiotic stresses, since many genes related to stress tolerance may be simultaneously regulated by the transcription factor. For example, G.T. Wu et al. (2006) showed that transgenic turf-type tall fescue plants that strongly expressed the stress tolerance–related *CBF1* gene from *Arabidopsis* showed obviously higher performance under high salinity than non-transgenic plants, and their tolerance of multiple stresses (low and high temperatures, dehydration, and high salinity) were demonstrated by analysis of the electrical conductivity of detached leaves *in vitro*. Similarly, the *Arabidopsis DREB1A/CBF3* gene, which encodes a transcription factor, was introduced into tall fescue by Zhao et al. (2007a). The transgenic plants contained high levels of the compatible osmolyte proline, and showed higher drought survival than nontransgenic plants under drought.

In addition, transgenic tall fescue plants capable of tolerating stress related to reactive oxygen species (ROS) have been produced by Lee et al. (2007). Transgenic plants that co-expressed two foreign genes that encoded anti-oxidative enzymes, cassava CuZn-superoxide dismutase (CuZn-SOD) and pea ascorbate peroxidase (APX), under the control of the oxidative stress–inducible promoter sweet potato peroxidase anionic 2 (SWAP2) were produced by means of *Agrobacterium*-mediated transformation. Expression of the transgenes in transgenic plants was upregulated during ROS-related stress, and lower amounts of ROS were detected in transgenic plants exposed to the stresses than in nontransgenic plants, resulting in decreased chlorophyll degradation and decreased cellular damage. This finding indicates that the anti-oxidative defense in transgenic tall fescue was improved by co-overexpression of *CuZn-SOD* and *APX*, and these transgenic plants might mitigate the oxidative damage caused by abiotic and biotic stresses.

Tolerance of Biotic Stress

In addition to the tolerance of abiotic stress, research on transgenic tolerance of biotic stress has also been performed. For example, Xu et al. (2001) have produced transgenic perennial ryegrass plants that express an untranslatable ryegrass mosaic virus (RgMV) coat protein (CP) under the control of the rice *Act1* promoter. After inoculation with RgMV, the most resistant transgenic line showed no immunodetectable RgMV-CP in the primary transformant and its sexual progeny. Molecular analysis clarified that the resistance of the transgenic line to RgMV is triggered by targeted RNA degradation, resulting in post-transcriptional gene silencing.

Takahashi et al. (2005), on the other hand, targeted enhanced resistance to the fungal pathogen *Puccinia coronata* Corda, the causal agent of crown rust disease in Italian ryegrass, by introducing the rice chitinase (*RCC2*) gene. Chitinase hydrolyzes a structural component of the fungal cell wall, thereby weakening the fungus, and the resulting hydrolysate, *N*-acetyl-D-glucosamine, elicits the plant's defense system. Bioassay of detached leaves of transgenic plants expressing *RCC2* showed increased resistance to crown rust. The most resistant transgenic plants had approximately 8.7 times the chitinase activity of the nontransgenic plants. Dong et al. (2008) used another approach for the production of transgenic tall fescue plants resistant to fungal disease by introducing the T4 phage lysozyme gene *e* (T4 *e*). Lysozymes contain 1,4-β-*N*-acetylmuramidase, which hydrolyzes peptidoglycan in bacterial cell walls and exhibits membrane-disrupting antifungal activity. Resistance conferred against two fungal diseases, gray leaf spot (*Magnaporthe grisea*) and brown patch (*Rhizoctonia solani*), was clearly observed in the transgenic plants. This research group also produced transgenic tall fescue that expressed the gene (*AGLU1*) for alfalfa β-1,3-glucanase, which degrades β-1,3-glucan, a major structural component in fungal cell walls. They also included a truncated version of the frog *dermaseptin SI* gene that is cytolytic to bacteria, yeast, filamentous fungi, and protozoa, or the rice *Pi9* gene, which is an R gene that confers resistance to rice blast (Dong et al. 2007). Transgenic plants expressing the *AGLU1* gene or the dermaseptin SI gene showed high resistance to both *M. grisea* and *R. solani*. A transgenic plant containing the rice *Pi9* gene resisted turfgrass isolates of *M. grisea*.

Nutritional Value

The digestibility of dry matter is one of the most important traits of forage grasses. Slight improvements in forage digestibility can have significant impacts on animal performance, such as meat and milk production. A major factor limiting forage digestibility is the lignin in plant cell walls. Lignins are phenolic heteropolymers associated with cellulose and hemicellulose in the cell walls, and their presence limits the utilization of these cell wall polysaccharides and thereby reduces energy extraction from forage grasses by livestock. Therefore, altering the lignin content and thereby improving forage digestibility is a crucial issue for efficient production of livestock products. Chen et al. (2003) isolated a gene (*CAD*) that encodes the key lignin biosynthesis enzyme, cinnamyl alcohol dehydrogenase (CAD), from tall fescue, and introduced it back into tall fescue with a construct for either sense or antisense expression of *CAD*. Transgenic plants carrying sense *CAD* transgenes showed severely reduced transcription of *CAD*, and plants carrying antisense *CAD* transgenes showed significantly reduced endogenous CAD enzymatic activity. A significant reduction in the Klason lignin content (by 14%–15%) and a significant increase in *in vitro* dry matter digestibility (by 7.2%–9.5%) were found in the plants with downregulated *CAD* compared with the control plants.

Chen et al. (2004) subsequently isolated another lignin biosynthesis gene (*COMT*) that encodes caffeic acid *O*-methyltransferase (COMT) from tall fescue and produced transgenic tall fescue plants with either sense or antisense *COMT* constructs. Two transgenic plants with co-suppressed *COMT* were obtained with the sense *COMT* construct. These plants, with downregulated *COMT*, showed drastically reduced COMT enzymatic activity, a significant reduction (by 28%–29%) in the Klason lignin content, and an increase (by 9.8%–10.8%) in *in vitro* dry matter digestibility. In contrast, Buanafina et al. (2006) focused on the ferulic acid esters, which play a key role in crosslinking hemicellulose in plant cell walls. The crosslinks hinder degradation of the cell wall by the microbes that help ruminants digest plant matter. To address this problem, they produced transgenic Italian ryegrass that expressed a ferulic acid esterase that releases both monomeric and dimeric ferulic acids from hemicellulose. Production of this esterase in transgenic plants released monomeric and dimeric ferulic acids from cell walls and increased *in vitro* dry matter digestibility compared with non-transgenic plants.

In addition to approaches based on increasing forage digestibility, the accumulation of nutritionally useful proteins in the leaves of forage grasses is another way to improve forage quality. For instance, forage crops that contain the sulphur-containing amino acids methionine and cysteine are attractive in ruminant animal nutrition, since such forage improves the growth of beef animals, milk production in dairy cows, and wool growth in sheep. To achieve these improvements, Z.Y. Wang et al. (2001) produced transgenic tall fescue that expressed the gene for sulphur-rich sunflower albumin 8 (SFA8) protein under the control of the cauliflower mosaic virus 35S promoter or the light-regulated wheat *Cab* promoter. SFA8 accumulated to as much as 0.2% of the total soluble protein in the leaves of transgenic plants. However, a nutritionally useful level of SFA8 is estimated to be approximately 4% of the total leaf protein. Thus, these researchers are exploring strategies to improve the accumulation of SFA8 in transgenic plants, including the use of a stronger promoter and the development of a chloroplast transformation technique.

In addition to improving the stress tolerance of temperate grasses, fructans also contribute to their nutritive value as feed for ruminant animals (Chalmers et al. 2005). High carbohydrate levels in forage crops lead to efficient utilization of amino acids by micro-organisms in the rumen of these animals. This increased efficiency also inhibits the release of ammonia into the environment. In addition, a recent study (Miller et al. 2001) showed that a high sugar content in forage increased its nutritional value to support milk production by dairy cows. To improve the nutritional value of perennial ryegrass by controlling carbohydrate biosynthesis, Gadegaard et al. (2008) produced transgenic perennial ryegrass plants that expressed the onion fructosyltransferase genes *1-SST* and *6G-FFT*. The transgenic plants showed up to three times the fructan content of the nontransgenic plants, and the fructan content was higher and more stable than that of a high-sugar elite variety.

Flowering Control

The nutritional value and digestibility of forage grasses decline significantly as the plants mature. This is largely caused by the high amounts of low-digestibility compounds that accumulate in floral stems, such as lignins and cell wall compounds coupled to lignins. Late flowering as a result of an extended vegetative growth phase is therefore another option for improving the digestibility of forage grasses. Late flowering is also expected to result in high density and persistence of the turf and is thus an attractive feature for amenity use of grasses. Jensen et al. (2004) isolated the floral repressor gene *LpTFL1* from perennial ryegrass and used it to manipulate the transition between vegetative growth and flowering in red fescue (*F. rubra*). Transgenic plants that expressed *LpTFL1* at low to intermediate levels flowered approximately 2 weeks later than the control plants, whereas very high levels of *LpTFL1* expression produced plants that remained in a vegetative state after exposure to strong vernalization conditions in two successive years. Van der Valk et al. (2004) introduced the *Arabidopsis* TALE homeobox gene *ATH1* into perennial ryegrass. In *ATH1*-expressing transgenic plants, heading was delayed, and in many cases transgenic plants never flowered at all. These transgenic plants produced a leafy phenotype, and when they eventually underwent heading, they generally produced a reduced number of inflorescences.

Hypoallergenic Pollen

Hay fever and seasonal asthma are both caused by grass pollen. Ryegrass pollen is a major cause of these allergic diseases, and several allergenic proteins have been identified and characterized from ryegrass pollen. To downregulate the expression of these allergenic proteins, antisense strategies have been employed. Bhalla et al. (1999) targeted Lol p 5, the most important widespread grass pollen allergen, since it reacts with IgE antibodies of more than 90% of patients who are allergic to grass pollen, and transgenic *Lolium rigidum* L. plants that expressed antisense *Lol p 5* under the regulatory control of the rice pollen–specific promoter Ory s1 were produced. A dramatic reduction in the allergenicity of the transgenic pollen was confirmed by means of Western blot analysis and slot blot analysis using the serum of a patient with a grass allergy. Similarly, Petrovska et al. (2004) generated and analyzed transgenic ryegrass plants with antisense expression of two major pollen allergens, Lol p 1 and Lol p 2, under the control of the pollen-specific maize promoter *Zm13*, and molecular analysis revealed downregulation of the target gene expression, resulting in reduced accumulation of the target proteins in transgenic pollen.

Low-input Crop Management

A bialaphos resistance gene (*bar*) is often used as a selectable marker gene in the transformation of many plants, including *Lolium* and *Festuca* species (Z.Y. Wang et al. 1992; Spangenberg et al. 1994; Kuai et al. 1999; Cho et al. 2000; Hisano et al. 2004; Hu et al. 2005; Y.Y. Wu et al. 2005). The resulting transgenic plants have resistance to glufosinate, bialaphos, or phosphinothricin herbicides. This trait is useful for weed control in turf management. However, it might have negative aspects if the transgene escapes into the environment. Thus, herbicide resistance is viewed with ambivalence.

Aunique type of research was attempted with turf-type perennial ryegrass. Turf-type perennial ryegrass is often used as a turf for winter overseeding of warm-season turfgrass species in golf courses and athletic fields. In late spring to early summer, warm-season turfgrass starts growing and turns green. Therefore, at this time, perennial ryegrass should decline, since its persistence would disturb the growth of the warm-season turfgrass. Chen et al. (2005) proposed an ideal scenario in which managers would selectively remove the perennial ryegrass when it is no longer needed by using genetically engineered vulnerable ryegrass lines. To do so, they produced transgenic perennial ryegrass plants that expressed the *E. coli argE* gene, which encodes an *N*-acetylornithinase. The transgenic plants were selectively eliminated by the application of a pro-herbicide, *N*-acetyl-L-phosphinothricin (*N*-acetyl-PPT), since the *N*-acetylornithinase produced in transgenic plants converts *N*-acetyl-PPT into the herbicide phosphinothricin (PPT).

Conclusion

The important traits sought by breeders of *Lolium* and *Festuca* species are tolerance of abiotic and biotic stresses, high nutritional value (such as high digestibility and the accumulation of useful proteins and fructans), late flowering, the development of hypoallergenic pollen, and improved ease of management. Although most of these traits are commonly required in forage and turf grass species, it is difficult to improve them by means of conventional breeding methods, since many forage and turf grass species are outcrossing plants with high genetic diversity in their offspring. As another option, genetic engineering appears to be a promising technique and a powerful tool to produce cultivars that possess many of the abovementioned traits in outcrossing plants such as *Lolium* and *Festuca* species.

Considering the potential influence of escaped transgenic plants on natural ecosystems, careful judgment is necessary to permit the practical use of these plants. This need will be particularly important in wind-pollinated outcrossing *Lolium* and *Festuca* species because of the long-distance transport of their pollen, the large areas cultivated with these plants, and the fact that escaped volunteer transgenic plants of these species can become aggressive weeds that easily hybridize with natural vegetation, and transgenic seeds would then spread widely (Giddings 2000). In fact, Z.Y. Wang et al. (2004) reported that transgenic pollen of tall fescue pollinated recipient plants at distances of up to 150 m from a central transgenic plot. As a first step to resolve the issue of transgene flow through transgenic pollen, Takamizo et al. (2005) proposed the utilization of a maternal inheritance trait through cytoplasmic male sterility (CMS), in which viable pollen cannot be produced, and developed CMS lines of Italian ryegrass and tall fescue as the recipients of transgenes. If a complete CMS trait can be stably introduced into a transgenic line, there will no longer be a risk of hybridization by dispersal of transgenic pollen. Also, transgenes are inherited maternally when they are integrated into the plastid genome, thereby minimizing the possibility of transfer of transgenes into related weed species (Daniell et al. 2005). Thus, plastid transformation will be a key technology in the next generation of genetic engineering of *Lolium* and *Festuca* species.

References

Altpeter F, Xu J, Ahmed S (2000) Generation of large numbers of independently transformed fertile perennial ryegrass (*Lolium perenne* L.) plants of forage- and turf-type cultivars. *Mol. Breed.* 6:519-528

Bajaj S, Ran Y, Phillips J, Kularajathevan G, Pal S, Cohen D, Elborough K, Puthigae S (2006) A high throughput *Agrobacterium tumefaciens*-mediated transformation method for functional genomics of perennial ryegrass (*Lolium perenne* L.). *Plant Cell Rep.* 25:651-659

Bettany AJE, Dalton SJ, Timms E, Manderyck B, Dhanoa MS, Morris P (2003) *Agrobacterium tumefaciens*-mediated transformation of *Festuca arundinacea* (Schreb.) and *Lolium multiflorum* (Lam.). *Plant Cell Rep.* 21:437-444

Bhalla PL, Swoboda I, Singh MB (1999) Antisense-mediated silencing of a gene encoding a major ryegrass pollen allergen. *Proc. Natl. Acad. Sci. USA* 96:11676-11680

Boyer JS (1982) Plant productivity and environment. *Science* 218:443-448

Buanafina MM, Langdon T, Hauck B, Dalton SJ, Morris P (2006) Manipulating the phenolic acid content and digestibility of Italian ryegrass (*Lolium multiflorum*) by vacuolar-targeted expression of a fungal ferulic acid esterase. *Appl. Biochem. Biotechnol.* 129-132:416-426

Cao MX, Huang JQ, He YL, Liu SJ, Wang CL, Jiang WZ, Wei ZM (2006) Transformation of recalcitrant turfgrass cultivars through improvement of tissue culture and selection regime. *Plant Cell Tissue Organ Cult.* 85:307-316

Chalmers J, Lidgett A, Cummings N, Cao Y, Forster J, Spangenberg G (2005) Molecular genetics of fructan metabolism in perennial ryegrass. *Plant Biotechnol. J.* 3:459-474

Chen L, Auh CK, Dowling P, Bell J, Chen F, Hopkins A, Dixon RA, Wang ZY (2003) Improved forage digestibility of tall fescue (*Festuca arundinacea*) by transgenic down-regulation of cinnamyl alcohol dehydrogenase. *Plant Biotechnol. J.* 1:437-449

Chen L, Auh CK, Dowling P, Bell J, Lehmann D, Wang ZY (2004) Transgenic down-regulation of caffeic acid *O*-methyltransferase (COMT) led to improved digestibility in tall fescue (*Festuca arundinacea*). *Funct. Plant Biol.* 31:235-245

Chen X, Yang W, Sivamani E, Bruneau AH, Wang B, Qu R (2005) Selective elimination of perennial ryegrass by activation of a pro-herbicide through engineering *E. coli argE* gene. *Mol. Breed.* 15:339-347

Cho MJ, Ha CD, Lemaux PG (2000) Production of transgenic tall fescue and red fescue plants by particle bombardment of mature seed-derived highly regenerative tissues. *Plant Cell Rep.* 19:1084-1089

Dalton SJ, Bettany AJE, Timms E, Morris P (1995) The effect of selection pressure on transformation frequency and copy number in transgenic plants of tall fescue (*Festuca arundinacea* Schreb.). *Plant Sci.* 108:63-70

Dalton SJ, Bettany AJE, Timms E, Morris P (1998) Transgenic plants of *Lolium multiflorum, Lolium perenne, Festuca arundinacea* and *Agrostis stolonifera* by silicon carbide fibre-mediated transformation of cell suspension cultures. *Plant Sci.* 132:31-43

Dalton SJ, Bettany AJE, Timms E, Morris P (1999) Co-transformed, diploid *Lolium perenne* (perennial ryegrass), *Lolium multiflorum* (Italian ryegrass) and *Lolium temulentum* (darnel) plants produced by microprojectile bombardment. *Plant Cell Rep.* 18:721-726

Daniell H, Kumar S, Dufourmantel N (2005) Breakthrough in chloroplast genetic engineering of agronomically important crops. *Trends Biotechnol.* 23:238-245

Dong S, Shew H, Tredway L, Lu J, Sivamani E, Miller E, Qu R (2008) Expression of the bacteriophage T4 lysozyme gene in tall fescue confers resistance to gray leaf spot and brown patch diseases. *Transgenic Res.* 17:47-57

Dong S, Tredway LP, Shew HD, Wang G-L, Sivamani E, Qu R (2007) Resistance of transgenic tall fescue to two major fungal diseases. *Plant Sci.* 173:501-509

Gadegaard G, Didion T, Folling M, Storgaard M, Andersen CH, Nielsen KK (2008) Improved fructan accumulation in perennial ryegrass transformed with the onion fructosyltransferase genes 1-SST and 6G-FFT. *J. Plant Physiol.* doi:10.1016/j.jplph.2007.06.019

Ge Y, Cheng X, Hopkins A, Wang ZY (2007) Generation of transgenic *Lolium temulentum* plants by *Agrobacterium tumefaciens*-mediated transformation. *Plant Cell Rep.* 26:783-789

Giddings G (2000) Modelling the spread of pollen from *Lolium perenne*. The implications for the release of wind-pollinated transgenics. *Theor. Appl. Genet.* 100:971-974

Ha SB, Wu FS, Thorne TK (1992) Transgenic turf-type tall fescue (*Festuca arundinacea* Schreb.) plants regenerated from protoplasts. *Plant Cell Rep.* 11:601-604

Hisano H, Kanazawa A, Kawakami A, Yoshida M, Shimamoto Y, Yamada T (2004) Transgenic perennial ryegrass plants expressing wheat fructosyltransferase genes accumulate increased amounts of fructan and acquire increased tolerance on a cellular level to freezing. *Plant Sci.* 167:861-868

Hu Y, Jia W, Wang J, Zhang Y, Yang L, Lin Z (2005) Transgenic tall fescue containing the *Agrobacterium tumefaciens ipt* gene shows enhanced cold tolerance. *Plant Cell Rep.* 23:705-709

Jensen CS, Salchert K, Gao C, Andersen C, Didion T, Nielsen KK (2004) Floral inhibition in red fescue (*Festuca rubra* L.) through expression of a heterologous flowering repressor from *Lolium*. *Mol. Breed.* 13:1-11

Kuai B, Dalton SJ, Bettany AJE, Morris P (1999) Regeneration of fertile transgenic tall fescue plants with a stable highly expressed foreign gene. *Plant Cell Tissue Organ Cult.* 58:149-154

Lee SH, Ahsan N, Lee KW, Kim DH, Lee DG, Kwak SS, Kwon SY, Kim TH, Lee BH (2007) Simultaneous overexpression of both CuZn superoxide dismutase and ascorbate peroxidase in transgenic tall fescue plants confers increased tolerance to a wide range of abiotic stresses. *J. Plant Physiol.* 164:1626-1638

Li Q, Robson PR, Bettany AJ, Donnison IS, Thomas H, Scott IM (2004) Modification of senescence in ryegrass transformed with *IPT* under the control of a monocot senescence-enhanced promoter. *Plant Cell Rep.* 22:816-821

Miller LA, Moorby JM, Davies DR, Humphreys MO, Scollan ND, MacRae JC, Theodorou MK (2001) Increased concentration of water-soluble carbohydrate in perennial ryegrass (*Lolium perenne* L.): milk production from late-lactation dairy cows. Grass Forage Sci. 56:383-394

Petrovska N, Wu X, Donato R, Wang Z, Ong EK, Jones E, Forster J, Emmerling M, Sidoli A, O'Hehir R, Spangenberg G (2004) Transgenic ryegrasses (*Lolium* spp.) with down-regulation of main pollen allergens. *Mol. Breed.* 14:489-501

Qiao WH, Zhao XY, Li W, Luo Y, Zhang XS (2007) Overexpression of AeNHX1, a root-specific vacuolar Na^+/H^+ antiporter from *Agropyron elongatum*, confers salt tolerance to *Arabidopsis* and *Festuca* plants. *Plant Cell Rep.* 26:1663-1672

Sato H, Takamizo T (2006) *Agrobacterium tumefaciens*-mediated transformation of forage-type perennial ryegrass (*Lolium perenne* L.). *Grassland Sci.* 52:95-98

Spangenberg G, Wang ZY, Nagel J, Potrykus I (1994) Protoplast culture and generation of transgenic plants in red fescue (*Festuca rubra* L.). *Plant Sci.* 97:83-94

Spangenberg G, Wang ZY, Wu XL, Nagel J, Potrykus I (1995) Transgenic perennial ryegrass (*Lolium perenne*) plants from microprojectile bombardment of embryogenic suspension cells. *Plant Sci.* 108:209-217

Takahashi W, Fujimori M, Miura Y, Komatsu T, Nishizawa Y, Hibi T, Takamizo T (2005) Increased resistance to crown rust disease in transgenic Italian ryegrass (*Lolium multiflorum* Lam.) expressing the rice chitinase gene. *Plant Cell Rep.* 23:811-818

Takahashi W, Komatsu T, Fujimori M, Takamizo T (2004) Screening of regenerable genotypes of Italian ryegrass (*Lolium multiflorum* Lam.). *Plant Prod Sci.* 7:55-61

Takahashi W, Oishi H, Ebina M, Takamizo T, Komatsu T (2002) Production of transgenic Italian ryegrass (*Lolium multiflorum* Lam.) via microprojectile bombardment of embryogenic calli. *Plant Biotechnol.* 19:241-249

Takamizo T, Komatsu T, Fujimori M, Takahashi W, Arakawa A, Sugita S, Suginobu K, Sato H, Mano Y, Kiyoshi T (2005) Concept of amenable transgenic forage grasses possessing cytoplasmic male sterility. In: Zwierzykowski Z, Kosmala A (eds) Recent advances in genetics and breeding of the grasses. Polish Academy of Science, Poznan, pp 55-62

Van der Valk P, Proveniers MCG, Pertijs JH, Lamers JTWH, van Dun CMP, Smeekens JCM (2004) Late heading of perennial ryegrass caused by introducing an *Arabidopsis* homeobox gene. *Plant Breed.* 123:531-535

Wang GR, Binding H, Posselt UK (1997) Fertile transgenic plants from direct gene transfer to protoplasts of *Lolium perenne* L. and *Lolium multiflorum* Lam. *J. Plant Physiol.* 151:83-90

Wang ZY, Ge Y (2005) *Agrobacterium*-mediated high efficiency transformation of tall fescue (*Festuca arundinacea). J. Plant Physiol.* 162:103-113

Wang ZY, Lawrence R, Hopkins A, Bell J, Scott M (2004) Pollen-mediated transgene flow in the wind-pollinated grass species tall fescue (*Festuca arundinacea* Schreb.). *Mol. Breed.* 14:47-60

Wang ZY, Takamizo T, Iglesias VA, Osusky M, Nagel J, Potrykus I, Spangenberg G (1992) Transgenic plants of tall fescue (*Festuca arundinacea* Schreb.) obtained by direct gene transfer to protoplasts. *Nature Biotechnol.* 10:691-696

Wang ZY, Ye XD, Nagel J, Potrykus I (2001) Expression of a sulphur-rich sunflower albumin gene in transgenic tall fescue (*Festuca arundinacea* Schreb.) plants. *Plant Cell Rep.* 20:213-219

Wu G-T, Chen J-Q, Hu Z-H, Lang C-X, Chen X-Y, Wang F-L, Jin W, Xia Y-W (2006) Production of transgenic tall fescue plants with enhanced stress tolerances by *Agrobacterium tumefaciens*-mediated transformation. *Agric. Sci. China* 5:330-338

Wu Y-Y, Chen Q-J, Chen M, Chen J, Wang X-C (2005) Salt-tolerant transgenic perennial ryegrass (*Lolium perenne* L.) obtained by *Agrobacterium tumefaciens*-mediated transformation of the vacuolar Na^+/H^+ antiporter gene. *Plant Sci.* 169:65-73

Xu J, Schubert J, Altpeter F (2001) Dissection of RNA-mediated ryegrass mosaic virus resistance in fertile transgenic perennial ryegrass (*Lolium perenne* L.). *Plant J.* 26:265-274

Ye X, Wang ZY, Wu X, Potrykus I, Spangenberg G (1997) Transgenic Italian ryegrass (*Lolium multiflorum*) plants from microprojectile bombardment of embryogenic suspension cells. *Plant Cell Rep.* 16:379-384

Ye XD, Wu XL, Zhao H, Frehner M, Nösberger J, Potrykus I, Spangenberg G (2001) Altered fructan accumulation in transgenic *Lolium multiflorum* plants expressing a *Bacillus subtilis sacB* gene. *Plant Cell Rep.* 20:205-212

Zhao J, Ren W, Zhi D, Wang L, Xia G (2007a) Arabidopsis *DREB1A/CBF3* bestowed transgenic tall fescue increased tolerance to drought stress. *Plant Cell Rep*. 26:1521-1528

Zhao J, Zhi D, Xue Z, Liu H, Xia G (2007b) Enhanced salt tolerance of transgenic progeny of tall fescue (*Festuca arundinacea*) expressing a vacuolar Na^+/H^+ antiporter gene from *Arabidopsis*. *J. Plant Physiol.* 164:1377-1383

In: Genetically Modified Plants
Editors: Timm Wolf and Jonas Koch

ISBN 978-1-60456-696-3

Chapter 4

GENETICALLY ENGINEERED PLANTS FOR FEED AND FOOD

***Filippo Rossi*[1] *and Piero Morandini*[2]**

[1]Institute of Feed and Food Science and Nutrition, Faculty of Agriculture, Catholic University, Via Emilia Parmense 84, 29100 Piacenza, Italy
[2]Department of Biology "L. Gorini", Section of Physiology and Biochemistry of Plants, University of Milan, via Celoria 26, 20133 Milan, Italy

ABSTRACT

The main focus of first-generation GE (Genetically Engineered) plants was agronomic traits, such as herbicide tolerance or pest resistance, which benefited mainly farmers. The benefits for human or animal health, when present, were indirect. For instance, a reduced mycotoxin content, due to a lower plant colonization by mycotoxigenic fungi, or a reduction in commodity prices, due to higher yield or improved farming practices, translating into increased food availability.

At present GE plants with nutritionally interesting traits are becoming available, both for feed and food. Forages with improved digestibility, due to a reduction in lignin, have been obtained and also soybean and cereals with increased content of lysine, methionine and tryptophan have been developed. Other interesting GE plants in the pipeline are corn and soybean with increased resistance to mycotoxins-producing fungi or able to degrade mycotoxins. Another important research trend is the reduction of anti nutritional factors such as phytate, which reduces phosphorus availability and increases environmental impact of livestock, particularly of pigs. Oleaginous crops with altered oil composition are available since many years but, while in previous years the increase of oleic acid content was the main target of genetic engineering, recent research focused to improve the content of the cardiovascular protective omega 3 and 6 fatty acids in seed oil.

Other important topics of importance both for human and animal nutrition are the increased content of vitamins (Vit. E, B_9 and C), pro-vitamin (β-carotene) and flavonoids, an important component of a health promoting diet. The development of oilseed crops with increased sterol content could be an interesting way for reducing cholesterol uptake in humans.

The alteration of starch composition or the accumulation of sugars in different polysaccharides could be used in order to obtain processed foods with a more suitable glycemic index, an important characteristic, especially considering the increasing incidence of diabetes.

Introduction

In 1994 the commercialization of Flavr Savr® tomato marked the beginning of the 'public' era for genetically engineered (GE) plants. The main characteristic of this berry was a sharp extension of the shelf-life, a very interesting feature for the supermarket industry and customers, but deprived of nutritional significance.

In the same way, Bt corn and herbicide-resistant plants had no improvements in their nutritive value, with the exception of the lower fumonisins content in Bt corn, presumably an unforeseen positive side-effect. The only clear attempt to improve the nutritional quality of foods were performed in soybean and rapeseed varieties, modified in order to increase the content of palmitic, stearic and oleic acid content to the detriment of poly-unsaturated fatty acids (PUFA) (Neff e List, 1999) (Wiberg et al., 1997) (Hawkins e Kridl, 1998).

The consequence of this modification was a better oxidative stability due to the lower content of PUFA and a reduction of *trans*-fatty acid promoting cardio-vascular disease (Katan et al., 1995; Nelson, 1998) when the oil is hydrogenated for margarine production. However the reduction in PUFA content in itself should be considered negatively, because these fatty acids are important cardio-protective agents, therefore these GE crops did not provide any nutritional advantage to the consumer. Probably the first plant, specifically engineered for an improved nutritional profile was the Golden Rice, characterized by accumulation of β-carotene in rice endosperm, a tissue devoid of carotenoids (Ye et al., 2000) and whose content in this pro-vitamin has been significantly improved further by Paine et al., (2005).

In the last years the interest of plant genetists moved from agronomic to nutritional traits and now the results are at hand or close to be released. In this chapter we try to outline the future trends for the improvement of nutritional quality of plant using genetic engineering. We will deal mainly with new applications of genetic engineering which are already in the commercialization phase or at the approval stage, but we will also mention interesting research advances with great promise. Other reviews on this subject (Zhu et al, 2007; Hall et al., 2008; Morandini et al., 2005) are available.

Anti-Oxidant and Anti-Inflammatory Molecules

Resveratrol is an anti-fungal compound synthesized by grapevine in response to pathogen attack (Schroder and Schroder, 1990), but resveratrol is also considered to have protective effects on human health because it has anti-inflammatory, antiplatelet and anticarcinogenic properties (Calabrese, 1999; Manna et al., 2000; Wallerath et al., 2002; Morelli et al., 2006). Grapevine (*Vitis vinifera*) and red wine are the most important source of resveratrol in the human diet, however wine is also a carrier for ethanol, whose excessive consumption is dangerous. For this reason any increase in resveratrol content in grapevine can be regarded as positive because it increase the intake of this protective molecule. It is therefore noteworthy the work of Fan et al. (2007) who where able to generate a grapevine containing 5.5 times more resveratrol than conventional plant (0.469 μg/g *vs* 2.586 μg/g).

Another application for a grapefruit genetically engineered to increase the synthesis of flavonoids was presented to APHIS. (Animal and Plant Health Inspection Service) In this case the target of metabolic modification were three enzymes involved in the phenylpropanoid pathway. This metabolic route is the source both of anthocyanins and phytoestrogens, compounds exhibiting antioxidant, antiasthmatic, antimicrobial activity and have beneficial roles as component of human diet (Dixon and Steele, 1999; Davis et al., 1999; Clarkson, 2002).

The phenylpropanoid pathway starts with phenylalanine and has an important crossroad in *para*-Coumaryl-CoA. The catalytic activity of chalcone synthase alone allows the transformation of this compound in naringenin chalcone, while combining its catalytic activity with chalcone reductase the 4,2', 4'-trihyroxychalcone is produced. The naringenin chalcone can be further metabolized to naringenin by chalcone isomerase (CHI) type 1, naringenin is the precursor of many flavonoid compounds (Ralston et al. 2005). In legumes a CHI type 2 is present and catalyzes the synthesis of isoflavonoids, like daidzein in soybean.

The regulation of the genes producing the pathway enzymes is coordinated at the transcriptional level and a number of these transcription factors have been identified (Ni et al., 1996; Weisshaar and Jenkins, 1998; Yu et al., 2000). To engineer the production of economically important isoflavones in non legume plants the regulation of naringenin accumulation is critical, because heterologously expressed enzymes must compete with endogenous ones for naringenin (Yu and McGonigle, 2005)

Two enzymes were the target of the genetic modification of grapefruit: chalcone synthase and chalcone isomerase, catalyzing the two last steps in the synthesis of naringenin.

Another opportunity to modify the phenylpropanoid pathway and increase/modify the isoflavones biosynthesis, was proposed by Yu et al. (2003). The authors cloned into soybean the corn transcription factors C1 and R, leading to the synthesis of anthocyanins (Grotewold et al., 1998). This resulted in a decrease of genistein and an increase of daidzein levels, with a small overall increase in total isoflavones. Cosuppression of flavanone 3-hydroxylase to block the anthocyanins branch of the pathways, resulted in higher accumulation of isoflavones.

To increase the flavonoid content of potatoes, Lukaszewicz et al (2004) overexpressed chalcone synthase, chalcone isomerase, and dihydroflavonol reductase, bringing about an increase in the levels of phenolic acids and anthocyanins as well as an improved antioxidant capacity. When compared to control values, increases in both pelargonidin (4-fold) and petunidin (3-fold) were observed.

Antioxidant capacity showed a correlation with anthocyanins content and the best results were obtained when all three enzymes were expressed.

Considering the growing interest on the nutritional relevance of polyphenols, the metabolic engineering of phenylpropanoid pathway will receive much attention in the next years.

ANTHOCYANINS

Anthocyanins are red to violet pigments used by plants to attract insects and foraging animals. Hundreds of different anthocyanins exist in nature, all with slightly different chemical structures and composition. Anthocyanins also help to protect plants against environmental stresses and diseases and they have anti-cancer properties. For this reason there is growing interest for their inclusion or boost in the specific foods.

A promising research on the factors regulating anthocyanins' production is in progress at the John Innes Centre and the Institute of Food Research in Norwich. The key enzymes involved in the production of plant pigments have been identified and the respective genes have been modified and cloned from *Arabidopsis thaliana*. Anthocyanins conjugates are more stable than original pigments and could be used as natural food colourants to replace many artificial colours used in various foods.

Furthermore the improved understanding of the genetics of anthocyanins also provides a better platform modify their profile, which is important in the fight against cancer, cardiovascular disease and age-related degeneration. Therefore in the future we expect GE plants with significant increases in the content of different antocyanins.

Another interesting source of anthocyanins is an Alaskan variety of cranberry (*Vaccinium* spp.) that will be used by Agricultural Research Service (ARS) scientist for conventional improvement of American cranberry. The wild cranberry accumulates glucose-linked anthocyanins, contrary to marketed varieties that produce pigments bound to other sugars and for this reason less digestible. The final goal is to transfer the genes for glucose-linked anthocyanins from the experimental cranberry line into a variety acceptable to the market. Researcher from ARS will use conventional breeding, but the use of genetic engineering could speed up the process and allow the achievement of more precise and therefore more predictable genetic modification.

SWEETENERS

The metabolic engineering of anthocyanins pathway can be modified in order to obtain a compound with sweetener properties. The inclusion of a rhamnosyl transferase in a grapefruit can add a rhamnose residue to an anthocyanins leading to the synthesis of neohesperidin dihydrochalcone, a compounds about 2000 times sweeter than sucrose.

GLUCOSINOLATES

An interesting recent advance suggesting potential applications to food and feed, is the identification of the regulators of the glucosinolates (GLS) pathway of Brassicaceae. GLS are sulphur rich secondary metabolites derived from aminoacids, some of which have interesting anticancer properties while other are toxic to animals (for a review see Halkier and Gershenzon, 2006). Being abundant in crop plants such as rapeseed and broccoli there is interest in modulating the overall content and the composition of GLS in a precise manner. The isolation of mutants blocked in specific steps of GLS biosynthesis has allowed the generation of plants lacking some or all GLS (Halkier and Gershenzon, 2006). The identification of several regulators however allows now the upregulation of specific branches of the pathway. Some years ago, Myb34 was shown to be involved in the regulation of the aromatic GLS pathway. Now several Myb factors (Myb29, 28, 76, 122 and 51) belonging to the same Myb subfamily of Myb34 have been independently demonstrated to regulate either the aliphatic (processing methionine or elongated methionine) or aromatic metabolic branch (processing tryptophan or phenylalanine). Plants lacking either Myb28 or Myb 29 have a diminished content of aliphatic glucosinolates, with the double KO completely lacking this class, while plants mutated in Myb 34, 51 or 122 are altered in the aromatic branch (Hirai et al., 2007; Gigolashvili et al., 2007a; Gigolashvili et al., 2007b; Sønderby et al., 2007, Gigolashvili et al., 2008; Beekwilder at al., PLoS ONE, in press, A. Aharoni et al., pers. communication). Overexpression or inactivation of the different MYBs caused a different pattern of GLS, demonstrating that it is possible to modulate the flux directly by toggling one or more regulators. We expect many more examples of metabolic engineering performed using regulators (from the same or from other species) rather than targeting the enzymatic steps themselves; this approach should retain metabolite (and control) homeostasis (Morandini et al., 2005) and is therefore expected to have less side effects.

Vitamins and Carotenoids

Vitamin deficiencies are a major problem in many developing countries, but also in Western Countries dietary supplementation with vitamins is proposed as anti-aging treatment (using vitamin A, C and E) or for the prevention of neural tube defects (folic acid).

Vitamin A

In Asia poverty is associated with a predominant consumption of rice and because this cereal lacks β-carotene (the pro-vitamin A), poor people's diet often falls short of vitamin A. This nutritional deficiency can result in permanent blindness and increase the incidence and severity of infectious diseases.

In 2000 a GE rice, able to accumulate carotenoids in the endosperm thank to the insertion of a phytoene synthase (*psy*) gene from daffodil and a Pds (phytoene desaturase gene from bacteria), was developed (Ye et al., 2000) and because of its yellow colour it was named 'Golden Rice'. In this cereal a maximum level of 1.6 μg/g of total carotenoids was achieved. Recently Paine et al. (2005) substituting the *psy* gene of daffodil with an homologous gene from maize, were able to increase the total carotenoids up to 37 μg/g (84.% of which was β-carotene) equivalent to 5.14 IU/g of retinol.

This sharp increase in pro-vitamin A has a relevant nutritional importance. As a matter of fact Golden Rice consumption can fulfill a big part of daily intake for retinol recommended in Europe (Table 1).

Table 1. Vitamin A requirements (express as retinol equivalent, RE) according to the Italian Society of Human Nutrition (SINU, 1996) and intake of Golden Rice required for their fulfilment

Category	Age (years)	Requirements (μg RE)	Golden Rice intake (g/day)
Children	1-6	400	78.
	7-10	500	98
Male	11-14	600	117
	> 14	700	136
Female	> 11	600	117
	Pregnant[1]	700	136
	Milking	950	185

Other transgenic plants that have been developed with increased carotenoid/provitamin A content include:

- Canola (*Brassica rapa*) with increased carotenoids (50-fold) developed by introducing a gene coding for lycopene biosynthesis (Shewmaker et al., 1999);
- Tomato with increased β -carotene developed by inserting bacterial genes that code for enzymes in the carotenoid pathway (Bramley et al., 2000; Fraser et al., 2002).
- Potato (Diretto et al., 2007) by engineering a mini bacterial pathway using CrtB, CrtI and CrtY from Erwinia. Transgenic tubers contained 20-fold more carotenoids (114 μg/g dry weight), with β-carotene increasing 3600-fold (47 μg/g dry weight).

An increased carotenoid accumulation (6-fold) was obtained by Li and Van Eck (2007) transferring in potato a novel gene mutation (*Or*) discovered in cauliflower (Lu et al., 2006). The expression of the *Or* gene leads to the formation of chromoplasts in the cauliflower curd cells. In potato the *Or* gene determines the formation of orange bodies (intact chromoplasts, helical sheets and fragments of chromoplasts).

Such structures serve as deposition sink to sequester excess carotenoids and preventing the end product inhibition of synthesis. The result is an increase in carotenoids, not necessarily β-carotene, content.

Using association analysis, linkage mapping, expression analysis and mutagenesis at the lycopene epsilon cyclase (*lycE*) Harjes et al. (2008) obtained some (non transgenic) corn plants where the carotenids biosynthetic pathways was modified with reduction of α-carotene and lutein synthesis and improvement of β-carotene, β-cryptoxanthin and xanthin synthesis. At present no high-provitamin A corn is marketed, but these results are encouraging.

Carotenoids

Carotenoids constitute a vast group of pigments with several biological activities and there is increasing evidence that higher intakes of carotenoids are associated with reduced risk of cardio-vascular disease (CVD), age related macular degenerations and some cancers

(Cooper et al., 1999; Craig, 1997; Neuhouser et al., 2001; Christen et al., 2008). The precursor of vitamin A, β-carotene, and lutein are the predominant carotenoids in human plasma (22) and along with zeaxanthin are found selectively at the centre of the retina; higher intakes of these two carotenoids resulted associated with significant reduced risk of cataract (Christen et al., 2008). Jayaraman et al. (2007) engineered the carotenoid biosynthetic pathway in carrot tissue by introducing a β-carotene ketolase gene isolated from the alga *Haematococcus pluvialis*. Endogenous expression of carrot β-carotene hydroxylases was up-regulated in transgenic leaves and roots, and up to 70% of total carotenoids was converted to novel ketocarotenoids, with accumulation up to 2,400 μg/g root dry weight. Astaxanthin, adonirubin, and canthaxanthin were most prevalent, followed by echinenone, adonixanthin and β-cryptoxanthin. However the content of the most nutritional important carotenoids: β-carotene, lutein and zeaxanthin, decreased. Although these last results are obviously undesirable, this work shows that carrot can be suitable as bio-factory, and this put the basis for a future manipulation of carotenoids' synthesis aimed to the production of fortified foods.

Folic Acid

Deficiency in folate, also known as Vitamin B9, may cause various diseases in humans such birth defects, anemia, cardiovascular diseases, and some types of cancer. A diet with folate deficiency is a risk factor for the onset of neural tube defects. Because of the importance of folate especially to pregnant women, (during pregnancy folate requirement increases by 50%) food fortification programs are being implemented in many Countries. However food fortification can be difficult to implement in developing Countries due to high costs and distribution problems.

UNICEF estimates at least 200,000 birth defects in developing Countries due to folate deficiency (Adamson, 2004). Since plants are major folate sources for humans, to increase folate content by means of genetic engineering could be a suitable way to reduce these severe problems. Folate are synthesized from pteridine, p-aminobenzoic acid (PABA) and glutamate precursors, Diaz de la Garza et al. (2007) produced a double transgenic tomato with increased expression of GTP cyclohydrolase I, the first enzyme of pteridine synthesis and aminodeoxychorismate synthase, that catalyzes the first step of PABA synthesis. The transgenic tomato fruits accumulate 25-fold more folate (840 μg/100 g of raw product), so that a normal-size fruit contained enough folates to fulfill the whole dietary requirement for a pregnant woman (600 μg/day). Storozhenko et al. (2007) were even more successful in increasing folate content in rice (*Oryza sativa japonica*) achieving a 100-fold accumulation compared to conventional varieties.

These manipulations however caused large accumulation of intermediate metabolites, an effect easily foreseeable within the framework of Metabolic Control Analysis (Morandini et al, 2005, Morandini and Salamini 2003). How these large changes in metabolites (for instance the amount of pteridine was raised 140-fold in one of the transgenic lines) remains to be tested.

Vitamin E

Improving Vitamin E content has also witnessed major advances. Arabisopsis, soybean Brassica, lettuce and potato with increased α–tocopherol content were developed (Van Eenennaam et al., 2003; Valentin et al., 2006; Crowell et al., 2007; Della Penna and Pogson, 2006)

Vitamin C

Vitamin C (L-ascorbic acid) has important antioxidant and metabolic functions in both plants and animals, humans and several other mammalian species have lost the ability to synthesize it because of an inactivation of the last gene in the pathway, and so we are dependent on dietary vitamin C from vegetables and fruits.

Two biosynthetic pathways for vitamin C in plants are presently known. The D-mannose pathway appears to be predominant in leaf tissue, but a D-galacturonic acid pathway operates in developing fruits (Agius et al., 2003; Radzio et al., 2003).

An important progress in the elucidation of ascorbate biosynthesis through the D-mannose pathway was made by Laing et al. (2007) who cloned in tobacco a gene from *Actinidia chinensis*, coding for a L-galactose guanyltransferase able to convert GDP-galactose into a hexose-1P. Before of this research, the enzyme catalyzing this reaction was not known..

The expression of this gene in tobacco leaves resulted in a more than 3 fold increase in ascorbate content.

Other researches have been reported to increase ascorbate content in plant tissue by overexpressing an enzyme in a biosynthetic pathway presumably which uses myo-inositol (MI) as the initial substrate Lorence et al. (2004). They cloned an *Arabidopsis thaliana* myo-inositol oxygenase. mainly expressed in flowers and leaves. Its overexpression led to 2-3 fold higher Ascorbate level compared to control. Jain and Nessler (2000) created a new pathway by introducing a gene from animals and achieved a 2-fold increase in vitamin C leaf content compared to controls.

Jain and Nessler (2000) demonstrated that by cloning the terminal enzyme of the animal pathway, L-gulono-1,4-lactone oxidase (GLOase), in lettuce and tobacco, they were able to increase the vitamin C leaf content between 4- and 7-fold, suggesting that at least part of the animal pathway may be present in plants. When a wild-type of *A. thaliana* was transformed with a rat GLOase gene, ascorbic acid in leaves increases up to 3 fold (Radzio et al., 2003). These data suggest that an alternative pathway is present in plants, which can bypass the deficiency of GDP-mannose production, being this sugar used also in mannan biosynthesis (Piro et al., 1993) and as a precursor in cell wall synthesis (Seifert, 2004), and possibly circumvent other steps in the D-mannose pathway to synthesize vitamin C.

Ascorbate (Vitamin C) content was increased also by reducing consumption (Pignocchi et al., 2003; Nunes-Nesi et al., 2005). In human body vitamin C can be regenerated from its oxidated form in a reaction catalyzed by deydroascorbate reductase (DHAR), Chen et al. (2003) over-expressed DHAR gene in transgenic corn and tobacco obtaining an increase in ascorbic acid level of 2-4 fold. In addition the level of glutathione, the reductant used by DHAR, also increases (Chen et al., 2003). These results demonstrate that increasing expression of the enzyme responsible for recycling ascorbate is an effective tool to elevate vitamin C content in plant.

Nowadays the extra-nutritive effect of vitamins is becoming more and more important and plants genetically engineered to increase the production of vitamins are becoming available.

At present conventional or GE micro organisms are the major source of vitamins intended to be included in supplements, however a growing number of information on the regulation of metabolic pathways leading to vitamins synthesis in crops are becoming available (Raschke et al., 2007). In the next years the use of GE plants for the production of vitamins or for direct consumption will be probably widespread. We also anticipate that the modification of riboswitch structures sensing specific vitamin levels to increase vitamin content of plant tissues. For instance, a riboswitch able to sense thiamine has been demonstrated in Arabidopsis and other plants (Bocobza et al., 2007; Wachter et al., 2007).

As an interesting approach to increase the nutritive value could be to alter cell number or development. The hp-1 mutation in tomato which almost doubles carotenoid content in leaves and fruits, and increases 9-fold the anthocyanin content in seedlings, is due to defects in cell division and development (Cookson et al., 2003). The mutation dramatically increases the periclinal elongation of leaf palisade mesophyll cells, which results in increased leaf thickness; moreover, the total plan area of chloroplasts per cell was larger in the mutant.

IMPROVEMENT OF DIGESTIBILITY

Plant cell wall is the major constraint in forage digestibility by livestock, lignin in particular is the fibrous component with the highest detrimental effect on feed digestibility. Increasing forage digestibility will allow a reduction in the use of cereal and proteaginous feeds and consequently of feeding costs.

According to the Field Test Release Database for the US, there is an application for the release of an alfalfa with decreased lignin. In this case the fibre content was decreased by gene silencing, while a different approach was utilized by Buonafina et al (2007-8) in developing a tall fescue with high digestibility.

In this study, genetically modified *Festuca arundinacea* plants were produced expressing an *Aspergillus niger* ferulic acid esterase (FAEA) targeted to the vacuole. In the cell walls of forage grasses, ferulic acid is esterified to arabinoxylans and participates with lignin monomers in oxidative coupling pathways to generate ferulate–polysaccharide–lignin complexes that cross-link the cell wall. Such cross-links hinder cell wall degradation by ruminant microbes, reducing plant digestibility.

Following grass harvest and consequently cell death, vacuole-targeted FAEA resulted in the release of both monomeric and dimeric ferulic acids from the cell walls, and this was enhanced several fold by the addition of exogenous endo-1,4-β-xylanase. Most of the FAEA-expressing plants showed increased digestibility and reduced levels of cell wall esterified phenolics relative to non-transformed plants.

Another interesting case where genetic engineering can improve feeds and foods quality is represented by phosphate (P) digestibility; in plants about 2/3 of P is in form of phytic acid, a molecule characterized by a very low digestibility in mammals and poultry, thereby making its P unavailable. Undigested P in liquid waste is an environmental pollutant, particularly for lakes and sea; at present the feed industry tackles the problem supplementing concentrate feed with phytases, but this increases the costs. Moreover phytate is an antinutritional factor chelating essential metals like Ca, Zn and Fe.

In human the situation is less critical due to more diverse diet, however in vegan people P lack is a realistic possibility.

In 1993 the first GE plant expressing a phytase was produced (Pen et al., 1993). A variety of other crops have since then been transformed to express phytases from different sources (Hamada et al., 2005; Chen at al., 2007; Brich-Pedersen et al., 2000). Another group reported the development of maize and soybean with reduced phytate (Shi et al., 2007) due to impairment of a specific transporter.

Amino Acids Pattern of Feeds and Foods

Amino acids are an important issue both in human and animal nutrition. The WHO-FAO (1991) defined the requirements for amino acids in the human diet at different life stages but no amino acids were defined as most limiting, quite in contrast with animal production, where the most limiting amino acids are lysine and methionine (Schwab et al., 1992; Rulquin et al., 1993). In order to support the high performance of modern livestock production it is necessary to supplement animal diets with amino acids produced by fermentation (Leuchtenberger, 1996; Kircher and Pfefferle, 2001).

Different feeds and foods typically lack specific amino acids: cereals have a well-known lysine deficiency and partly of threonine, with corn lacking Trp as well. Methionine is the limiting aminoacid of legumes.

A mutant corn (Opaque-2), with higher level of lysine, was discovered in 1964 and its lysine content ranged from 0.34 to 0.37% compared to 0.26-0.30% for normal maize, however the yield of Opaque-2 is about 7-10% lower than conventional corn (Thomison, 2005) and this is a serious limitation to its widespread use. It is therefore clear that improving lysine levels is one of the main target of biotechnology research on corn. Requests for the field release of 9 GE maize varieties are filed in the APHIS database from 2004 to December 2007.

Although methionine is not the most limiting amino acid in corn, authorization for field test release were issued by APHIS in 2006 for corn with a high content of sulfur-containing amino acids.

In corn with Opaque-2 mutations, the increase in lysine content was obtained by means of 2-3 fold reduction in the activity of lysine-ketoglutarate reductase, a lysine catabolizing enzyme (Brochetto-Braga et al., 1992). A different approach was utilized for the high lysine corn LY038 developed by Monsanto, this maize express the *cordapA* gene from *Corynebacterium glutamicum*, encoding a lysine-insensitive dihydropicolinate synthase enzyme (cDHDPS). Its expression under the control of the corn Glb 1 promoter, predominantely active in the seed germ, results in accumulation of lysine in the grain (0.4% of lysine).

The same company modified the lysine content of corn combining the expression of a deregulated *cordapA* enzyme with the silencing of the lysine degradation enzyme lysine-ketoglutarate reductase/saccharophine dehydrogenase (LKR/SDH) (Frizzi et al., 2008) obtaining a lysine content of 0.4% on DM, a value 40 fold higher than control (0.01% on DM).

Assuming that grain represent 28% of corn plant at dough (Masoero et al., 2006) and assuming a corn silage intake of 20 kg by a dairy cow, high lysine corn is expected to provide 7.4 g of lysine more than control. An additional 4 kg/day of corn grain translates into a lysine intake of 15.6 higher than control. The total lysine supply coming from maize silage and corn grain would thus be 23 g, a value very close to the average amount supplemented to high producing dairy cows. Therefore high-lysine corn is a suitable tool for reducing or avoiding the addition of purified lysine.

An alternative solution for increasing corn lysine content through GE was developed by Houmard et al. (2007). They showed that by suppressing lysine catabolism transgenic corn kernels accumulated significant amounts of lysine. This was achieved by RNA interference (RNAi) by the endosperm-specific expression targeting of the maize bifunctional lysine degradation enzyme, lysine-ketoglutarate reductase/saccharopine dehydrogenase (ZLKR/SDH).

Research for the improvement of amino acids pattern in rice have been carried out at the beginning of ‘90s when grain methionine levels were increased by introduction of methionine-rich genes (Altenbach et al., 1989; De Clercq et al., 1990; Altenbach et al. 1992; Molvig et al., 1997).

Three further applications for field release of GE rice were presented. In all these crops the gene for a storage protein from pea was cloned into rice. This strategy is coherent with the goal of increasing the lysine content of the cereal, because pea, being a legume, has a good content of lysine. In 1999 Momma et al. expressed a glycinin from soybean in rice, obtaining an increase in protein (8% *vs* 6.5%), lysine (0.30% *vs* 0.25%) threonine (0.30% *vs* 0.23%) and triptophan (0.11% *vs* 0.08%). In more recent years Met content of corn was increased by manipulation of native mRNA stability (Lai and Messing, 2002), while the protein content was increased in pomato (Chakraborty et al., 2000) and maize (Yu et al., 2004) by expressing protein from other plants. Similarly Bicar et al. (2008) improved the amino acid pattern of corn developing transgenic lines that produce the milk protein α-lactalbumin in the endosperm. Total protein content in endosperm from transgenic kernels was not significantly different from total protein content in endosperm from control kernels in three out of four comparisons, whereas the lysine content of the lines examined was 29–47% greater in endosperm from transgenic kernels. The content of some other essential amino acids was increased to a lesser extent: Thr (+ 6.1 %), Ile (+ 9.8-17.9 %). The overall result means that the transgenic endosperms has an improved amino acid balance. Kernel appearance, weight, density and zein content did not exhibit substantial differences in kernels expressing the transgene when compared to non-expressing siblings. However since α-lactalbumin is a known human allergen, an assessment of the antigenicity is required in order to determine the impacts on animal and human health.

Soybean protein quality is limited by the low sulfur content of its protein, particularly of β-conglycinin (7S globulin). Glycinin, the other predominant storage protein, contains 3-4 times more sulfur. El-Shemy et al. (2007) successfully inserted a modified pro-glycinin gene, encoding for 4 methionine residues, into soybean. Although the insertion of this gene was demonstrated by Southern hybridization, no data on methionine content of transgenic soybean was provided, so it is not possible to quantify the nutritional improvement of this soybean.

In many developing Countries the daily intake of essential amino acids is often inadequate due to the lack of food of animal origin. Rice, cassava and potato are staple foods in many developing Countries, but have a low protein content. In order to improve the nutritive value of cassava, a gene coding for ASP1, an artificial storage protein rich in essential amino acids, was introduced into embryogenic suspensions of cassava via *Agrobacterium*-mediated gene transfer and successfully expressed (Zhang et al., 2003).

The results were disappointing, because, although the level of His and Arg increased by 18 and 27% respectively, the level of an essential amino acid like methionine decreased.

It is clear that the cloning of proteins rich in limiting amino acids will be one of the major goal of biotechnological research aimed at improving the protein value of feeds and foods. And usually, despite the claims of large increase in free aminoacid levels are formally true (20-50 fold for lysine), the total lysine content is often only marginally changed because most of the lysine present in seed is included in protein and free lysine is only a few % of the total lysine content. Moreover *Arabidopsis* seed with a very high lysine content showed problems at germination (Zhu et al., 2003).

Fatty Acids Composition

The studies on the modification of fatty acid pattern started at the beginning of the "biotech era", but the target of GE has changed during the years. At the end of '90s the increase of saturated fatty acid (SFA) or at least the increase in oleic acid content (a mono-unsaturated one) was the final goal for GE rapeseed or soybean. This higher content of SFA was obtained to detriment of poly-unsatureated fatty acids (PUFA) (Neff e List, 1999; Wiberg et al., 1997; Hawkins and Kridl, 1998) and would have allowed a reduction of *trans* fatty acid produced during the hydrogenation process of saturated fatty acid. Due to their promoting action toward cardio-vascular disease (Katan et al., 1995; Nelson, 1998), attempts to reduced *trans*-fatty acid intake are generally regarded as positive.

However the growing evidence of positive effect of PUFA, particularly the ω-3 and ω-6 series, has shifted the strategies of metabolic modification of oilseeds and now the increase in linoleic and linolenic acids are the target of genetic engineering in plants.

This change has had repercussions on the enzymes employed: 10 years ago the researchers exploited SFA thioesterase in order to increase FA accumulation (particularly laurate) in seeds or tried to obtain high-oleic soybean/rapessed oil, by cloning Δ-9 desaturase, the enzyme that transforms stearic acid in oleic acid or silencing the Δ-12 desaturase whose metabolic role is the addition of a second double bond to oleic acid transforming it in linoleic acid.

A high-oleic oil is an obvious alternative to olive oil, but the positive effect of olive oil is not due to its oleate content only, but also and primarily, to its content in polyphenols.

Now the focus is on the enzymes involved in the synthesis of essential PUFA (ω-6 and ω-3 desaturase) for modification of oilseed or corn (for oil production). In the future this trend will increase, due to growing interest towards essential PUFAs.

Molecules for the Reduction of Intestinal Absorption of Cholesterol

Phytosterols and phytostanols are cholesterol-like molecules found in all plant foods, nut, olive, oilseed, whole grains and legumes are good dietary sources of phytosterols. Several studies have demonstrated that these compounds can lower blood levels of cholesterol and the risk of cardiovascular disease. As phytostanols are more stable than phytosterols during food processing, genetic engineering has been applied for the development of rapeseed and soybean oils with modified ratios of phytosterols to phytostanols (Venkatramesh et al., 2003).

Plants were transformed with a gene from yeast, encoding the enzyme 3-hydroxysteroid oxidase, which converts phytosterols to phytostanols.

Alteration of Carbohydrate Composition

Starch content and quality has been modulated in different crop plants. The most interesting manipulations concern amylose/amylopectin ratio. High amylose potatoes were developed (Hofvander et al., 2004; Andersson et al., 2006) and sweet potatoes with both high and low amylose (Kitakara et al., 2007). The relevance of this is due to the possible effect on glycemic index. Food with low glycemic index can help in the prevention of diabetes both by reducing the hunger perception and insulin responses. Rats fed with a high amylose corn (translating into a low glycemic index) showed a lower feed intake compared to animals fed with high amylopectin corn diet (Sculati et al., 2007; Quilez et al., 2007; Behall et al., 2006). Using another approach Stoop et al., (2007) produced maize and potato plants accumulating inulin, a fructose polymer being used as a prebiotic for its capacity to stimulate lactic acid bacteria living in the gut.

Resistance to Fusarium Attack and Fumonisin Degradation

Mycotoxins are important toxic compounds produced by several type of moulds and several factors affect the severity of moulds' attack and mycotoxins contamination. One of the side effect of Bt technology is the reduction in corn susceptibility to *Fusarium* attack, however this characteristic is a secondly effect of resistance of stalk to ECB attack. The reduction of damages to stalk, reduces also the sites for the entrance of fungi into the plant.

Use of genetic engineering for the development of germoplasm resistant to pathogen fungi started about 15 years ago. In these years genes from various source have been tested for fungal resistance in transgenic plants: a bean chitinase gene conferred resistance to *Rhizoctonia solani* (Broglie et al., 2001), while an antimicrobial peptide, expressed in tobacco plants, determined resistance to fungal (*Aspergillus flavus*, *Fusarium moniliforme* and *Verticillium dahliae*) and bacterial (*Pseudomonas syringae*) pathogens (DeGray et al., 2001).

The discovery in corn of proteins that directly determine the susceptibility of plants to fungi attack (Chen et al., 2007; Motto et al., 2004) boosted the research of genetic traits related to fungi contamination of crop and mycotoxins contamination. At the end of 2007 six corn varieties described as *Fusarium* resistant were tested in open field, but no information on the genes involved is available into the APHIS database. Very interesting is also the ability to degrade fumonisin claimed for others four GE corn.

Aflatoxins are hepatic carcinogens produced mainly by *Aspergillus flavus* and *A. parasiticus* and at present no *Aspergillus*-resistant corn was reported, but the paper of Chen et al. (2007) refers to resistance to *Aspergillus flavus* and probably in the future transgenic corn with low susceptibility to aflatoxin-producing fungi will be available.

The enforcement of new strict Regulations on the maximum levels allowed for mycotoxins in foods and feeds, will increase the importance of a biotechnological research aimed to reduce plants' susceptibility to fungi.

CONCLUSIONS

As it can be gathered from this brief overview, the potential of genetic engineering for improving plants used for feed or food is large and rich. This will not be tapped in its depth or even lost altogether unless the regulatory burden is relieved. We advocate the use of a regulation based on product characteristics, not on the method used to achieve such characteristics, and a scrutiny proportional to risks. We propose that those kind of transgenics which have proven benefits and pose low or evanescent risks (e.g. Bt crops) to be regulated at a lower level than conventionally bred varieties.

REFERENCES

Adamson P. (2004) The micronutrient initiative. In "Vitamin and Mineral deficiency", UNICEF (PandLA Oxfordshire, UK), pp- 20-40.

Agius F., Gonzalez-Lamothe R., Caballero J.L., Munoz-Blanco J., Botella M.A., Valpuesta V. (2003) Engineering increased vitamin C levels in plants by overexpression of a D-galacturonic acid reductase. *Nature Biotechnology*, 21: 177-181.

Altenbach S.B., Kuo C.C., Staraci L.C., Pearson K.W., Wainwright C. Georgescu A., Towsend J. (1992) Accumulation of a Brazil nut albumin in seeds of transgenic canola results in enhanced levels of seed protein methionine. *Plant. Mol. Biol.*, 18: 235-245.

Altenbach S.B., Pearson K.W., Meeker G., Staraci L.C., Sun S.S.M. (1989) Enhancement of the methionine content of seed proteins by the expression of a chimeric gene encoding a methionine-rich protein in transgenic plants. *Plant. Mol. Biol.*, 13: 513-522.

Andersson M., Melander M., Pojmark P., Larsson H., Bülow L., Hofvander P. (2006) Targeted gene suppression by RNA interference: an efficient method for production of high-amylose potato lines. *J. Biotechnol.* 123:137-48.

Badejo A.A., Tanaka N., Esaka M. (2008) Analysis of GDP-D-Mannose Pyrophosphorylase gene promoter from Acerola (*Malpighia glabra*) and increase in ascorbate content of transgenic tobacco expressing the Acerola Gene. *Plant Cell Physiol.* 49:126-132.

Behall K.M., Scholfield D.J., Hallfrisch J.G., Liljeberg-Elmståhl H.G. (2006) Consumption of both resistant starch and beta-glucan improves postprandial plasma glucose and insulin in women. *Diabetes Care.* 29:976-981.

Bicar E.H., Woodman-Clikeman W., Sangtong V., Peterson J.M., Yang S.S., Lee M., Scott M.P. (2008) Transgenic maize endosperm containing a milk protein has improved amino acid balance. *Transgenic research, vol.* 17(1): 59-71.

Bocobza S., Adato A., Mandel T., Shapira M., Nudler E., Aharoni A. (2007) Riboswitch-dependent gene regulation and its evolution in the plant kingdom. *Genes Dev.* 21:2874-2879.

Brinch-Pedersen H., Olesen A., Rasmussen S.K., Holm P.B. (2000) Generation of transgenic wheat (*Triticum aestivum* L.) for constitutive accumulation of an *Aspergillus* phytase. *Mol. Breed.* 6, 195–206.

Brochetto-Braga M., Leite A., Arruda P. (1992) Partial purification and characterization of lesine-ketoglutarate reductase in normal and opaque-2 maize endosperms. *Plant Physiol.*, 98: 1139-1147.

Broglie K.E., Chet I., Holliday M., Cressman R., Biddle O., Knowlton S., Mauvais C.J., Broglie R. (1991) Transgenic plants with enhanced resistance to the fungal pathogen *Rhizoctonia solani*. *Science,* 254: 1194-1197.

Buonafina M.M.deO., Langdon T., Hauck B., Dalton S., Morris P. (2007) Expression of a fungal ferulic acid esterase increases cell wall digestibility of tall fescue (*Festuca arundinacea*) *Plant Biotechnology Journal*, published on line 18 december 2007. doi:10.1111/j.1467-7652.2007.00317.x

Calabrese G. (1999) Nonalcoholic compounds of wine: the phytoestrogen resveratrol and moderate red wine consumption during menopause. *Drugs Exp. Clin. Res.*, 25(2-3):111-114.

Chakraborty S., Chakraborty N., Datta A. (2000) Increase dnutritive value of transgenic potato by expressing a nonallergenic seed albumin gene from *Amaranthus hypochondriacus*. *Proc.Natl. Acad. Sci. USA*, 97: 3724:3729.

Chen Z., Young T.E., Ling J., Chang S.C., Gallie D.R. (2003) Increasing vitamin C content of plants through enhanced ascorbate recycling. *Proc. Natl. Acad. Sci. U. S. A.* 100:3525–3530

Chen Z.Y., Brown R.L., Damann K.E., Cleveland T.E. (2007) Identification of maize kernel endosperm proteins associated with resistance to aflatoxin contamination by *Aspergillus flavus*. *Phytopathology*, 97: 1094-1103.

Christen W.G., Liu S., Glynn R.J., Gaziano M.J., Buring J.E. (2008) Dietary carotenoids, vitamins C and E, and risk of cataract in women: a prospective study. *Arch. Ophthalmol.*, 126:102-109.

Clarkson T.B. (2002) Soy, soy phytoestrogens and cardiovascular disease. *J. Nutr.*, 132: 566S-569S.

Cookson P.J., Kiano J.W., Shipton C.A., Fraser P.D., Romer S., Schuch W., Bramley P.M., Pyke K.A. (2003) Increases in cell elongation, plastid compartment size and phytoene synthase activity underlie the phenotype of the high pigment-1 mutant of tomato. *Planta.* 217:896-903.

Cooper D.A., Eldrige A.L., Peters J.C. (1999) Dietary carotenoids and certain cancers, heart disease and age-related macular degeneration: a review of recent research. *Nutr. Rev.*, 57: 201-214.

Craig W.J. (1997) Phytochemicals: guardians of our health. *J. Am. Diet. Assoc.*, 97(Suppl. 2): S199-S204.

Crowell E.F., McGrath J.M., Douches D.S. (2007) Accumulation of vitamin E in potato (*Solanum tuberosum*) tubers. *Transgenic Res.* DOI: 10.1007/s11248-007-9091-1.

Davis S.R., Dalais F.S., Simpson E.R., Murkies A.L. (1999) Phytoestrogens in health and disease. Recent Prog. *Horm. Res.*, 54: 185-210.

De Clercq A., Vandewiele M., van Damme J., Guerche P., van Montagu M., Vandekerckove J., Krebbers E. (1990) Stable accumulation of modified 2S albumin seed storage protein with higher methionine contents in transgenic plants. *Plant Physiol.*, 94: 970-979.

DeGray G., Rajasekaran K., Smith F., Sanford J., Daniel H. (2001) Expression of an antimicrobial peptide via the chloplast genome to control phytopathogenic bacteria and fungi. *Plant Physiol.*, 127: 852-862.

DellaPenna D., Pogson B.J. (2006) Vitamin synthesis in plants: tocopherols and carotenoids. *Annu. Rev. Plant Biol.* 57:711-738.

Diaz de la Garza R.I., Gregory J.F. 3rd, Hanson A.D. (2007) Folate biofortification of tomato fruit. *Proc Natl Acad Sci U S A*, 104: 4218-4222.

Diretto G., Al-Babili S., Tavazza R., Papacchioli V., Beyer P., Giuliano G. (2007) Metabolic engineering of potato carotenoid content through tuber-specific overexpression of a bacterial mini-pathway. *PLoS ONE.* 2:e350.

Dixon R.A., Steele C.L. (1999) Flavonoids and isoflavonoids, a gold mine for metabolic engineering. *Trends Plant Sci.*, 4: 394-400.

Drakakaki G., Marcel S., Glahn R.P., Lund E.K., Pariagh S., Fischer R., Christou P., Stoger E. (2005) Endosperm-specific co-expression of recombinant soybean ferritin and *Asper*gillus phytase in maize results in significant increases in the levels of bioavailable iron. *Plant Mol. Biol.* 59:869-880.

Fan C., Pu N., Wang X., Wang Y., Fang L., Xu W., Zhang J. (2007) *Agrobacterium*-mediated genetic transformation of grapevine (*Vitis vinifera* L.) with a novel stilbene synthase gene from Chinese wild *Vitis pseudoreticulata. Plant Cell Tiss Organ Cult,* published on line.

FAO/WHO (1985) Energy and protein requirements. Report of a joint FAO/WHO expert consultation. WHO, Geneva, CH.

Frizzi A., Huang S., Gilbertson L.A., Armstrong T.A., Luethy M.H., Malvar T.M. (2008) Modifying lysine biosynthesis and catabolism in corn with a single bifunctional expression/silencing transgene cassette. *Plant Biotechnol. J.*, 6: 13-21.

Galpaz N., Wang Q., Menda N., Zamir D., Hirschberg J. (2008) Abscisic acid deficiency in the tomato mutant high-pigment 3 leading to increased plastid number and higher fruit lycopene content. *Plant J.* Jan 8;

Gigolashvili T., Yatusevich R., Berger B., Muller C., Flugge U.I. (2007a) The R2R3-MYB transcription factor HAG1/MYB28 is a regulator of methionine-derived glucosinolate biosynthesis in Arabidopsis thaliana. *Plant J.*, 51: 247-261.

Gigolashvili T., Berger B., Mock H.P., Muller C., Weisshaar B., Flugge U.I. (2007b) The transcription factor HIG1/MYB51 regulates indolic glucosinolate biosynthesis in Arabidopsis thaliana. *Plant J.* 50:886-901.

Gigolashvili T, Engqvist M, Yatusevich R, Müller C, Flügge UI. (2008) HAG2/MYB76 and HAG3/MYB29 exert a specific and coordinated control on the regulation of aliphatic glucosinolate biosynthesis in Arabidopsis thaliana. *New Phytol.* 177:627-42.

Giorio G., Stigliani A.L., D'Ambrosio C. (2007) Agronomic performance and transcriptional analysis of carotenoid biosynthesis in fruits of transgenic HighCaro and control tomato lines under field conditions. *Transgenic Res.* 16:15-28.

Grotewold E., Chamberlin M., Snook M., Siame B., Butler L., Swenson J., Maddoc S., Clair G.S., Bowen B. (1998) Engineering secondary metabolism in maize cells by ectopic expression of transcription factors. *Plant Cell*, 10: 721-740.

Halkier B.A., Gershenzon J. (2006) Biology and biochemistry of glucosinolates. *Annu. Rev. Plant Biol.* 57:303-33.

Hall R.D., Brouwer I.D., Fitzgerald M.A. (2008) Plant metabolomics and its potential application for human nutrition. *Physiol Plant.* 132:162-75.

Hamada A, Yamaguchi K, Ohnishi N, Harada M, Nikumaru S, Honda H. (2005) High-level production of yeast (Schwanniomyces occidentalis) phytase in transgenic rice plants by a combination of signal sequence and codon modification of the phytase gene. *Plant Biotechnol J.* 3:43-55.

Harjes C.E., Rocheford T.R., Bai L., Brutnell T.P., Bermudez Kandianis C., Sowinski S.G., Stapleton A.E., Vallabhaneni R., Williams M., Wurtzel E.T., Yan J., Buckler E.S. 2008) Natural genetic variation in *lycopene epsilon cyclase* tapped for maize biofortification. *Science,* 319:330-333.

Hawkins D.J., Kridl J.C. (1998) Characterization of acyl-ACP thioesterases of mangosteen (*Garcinia mangostana*) seed and high levels of stearate production in transgenic canola. *Plant J.*, 13: 743-752.

Hirai MY, Sugiyama K, Sawada Y, Tohge T, Obayashi T, Suzuki A, Araki R, Sakurai N,

Hofvander P., Andersson M., Larsson C.T., Larsson H. (2004) Field performance and starch characteristics of high-amylose potatoes obtained by antisense gene targeting of two branching enzymes. *Plant Biotechnol. J.*, 2:311-320.

Houmard N.M., *Mainville J.L., Bonin C.P., Huang S., Luety M.H., Malvar T.M.* (2007) High-lysine corn generated by endosperm-specific suppression of lysine catabolism using RNAi. *Plant Biotechnol. J.*, 5: 605-614.

Huang S. Frizzi A., Florida C.A., Kruger D.E., Luethy M.H. (2006) High lysine andhigh tryptophan transgenic maize resulting from the rduction of both 19- and 22-kD α-zein. *Plant Mol. Biol.*, 61: 525-535.

Jain, A.K. and Nessler, C.L. (2000) Metabolic engineering of an alternative pathway for ascorbic acid biosynthesis in plants. *Mol. Breed.* 6:73–78.

Jayaraj J., Devlin R., Punja Z. (2007) Metabolic engineering of novel ketocarotenoid production in carrot plants. *Transgenic Res.*, 16: 581-585.

Katan M.B., Zock P.L., Mensink R.P. (1995) *Trans* fatty acids and their effects on lipoproteins in humans. *Ann. Rev. Nutr.*, 15: 473-493.

Kircher M., Pfefferle W. (2001) the fermentative eproduction of L-lysine as an animal feed additive. *Chemosphere*, 43: 27-31.

Kitahara K., Hamasuna K., Nozuma K., Otani M., Hamada T., Shimada T., Fujita K., Suganuma T. (2007) Physicochemical properties of amylose-free and high-amylose starches from transgenic sweetpotatoes modified by RNA interference. *Carbohydrate Polymers*, 69:233-240.

Lai J.S., Messing J. (2002) Increasing maize seed methionine by mRNA stability. *Plant J.*, 30: 395-402.

Laing W.A., Wright M.A., Cooney J., Bulley S.M. (2007) The missing step of the L-galactose pathway of ascorbate biosynthesis in plants, an L-galactose guanyltransferase, increases leaf ascorbate content. *Proc Natl Acad Sci U S A.* 104:9534-9539.

Leuchtenberger W. (1996) Amino acids – technical production. In Rehm H.J., Reed G., eds., Biotechnology, vol. 6, VCH Weinheim, pp. 488-502.

Li L., Van Eck J. (2007) Metabolic engineering of carotenoid accumulation by creating a metabolic sink. *Transgenic Res.*, 16: 581-585.

Lorence A, Chevone B.I., Mendes P., Nessler C.L. (2004) myo-inositol oxygenase offers a possible entry point into plant ascorbate biosynthesis. *Plant Physiol.*, 134: 1200-1205.

Lu S., Eck Van J., Zhou X., Lopez A.B., O'Halloran D.M., Cosman K.M., Conlin B.J., Paolillo D.J., Garvin D.F., Vrebalov J., Kochian L.V., Kupper H., Earle E.D., Cao J., Li L. (2006) The cauliflower *Or* gene encodes a DNAJ cysteine-rich domain-containing protein that mediates high level of beta-carotene accumulation. *Plant Cell,* 18: 3594-3605.

Lukaszewicz M., Matysiak-Kata I., Skala J., Fecka I., Cisowski W., Szopa J. (2004) Antioxidant capacity manipulation in transgenic potato tuber by changes in phenolic compounds content. J. *Agric. Food Chem.*, 52: 1526-1533.

Manna S.K., Mukhopadhyay A., Aggarwal B.B. (2000) Resveratrol suppress TNF-induced activation of nuclear transcription factor NF-kB, activator protein-1 and apoptosis: potential role of reactive oxygen intermediates and lipid peroxidation. *J. Immunol.*, 164: 6509-6519.

Masoero F., Rossi F., Pulimeno A.M. (2006) Chemical composition and in vitro digestibility of stalks, leaves and cobs of four corn hybrids at different phenological stages. It. *J. Anim. Sci.*, 5: 19-27.

Molvig L., Tabe L.M., Eggum B.O., Moore A.E., Craig S., Spencer D., Higgins T.J.V. (1997) Enhanced methionine levels and increased nutritive value of seeds of transgenic lupins (*Lupinus angustifolius* L.) expressing a sunflower seed albumin gene. Proc. *Natl. Acad. Sci. USA*, 94: 8393-8398.

Momma K., Hashimoto W., Ozawa S., Kawai S., Katsube T., Takaiwa F., Kito M., Utsumi S., Murata K. (1999) Quality and safety evaluation of genetically engineered rice with soybean glycinin: analyses of the grain composition and digestibility of glycinin in transgenic rice. *Biosci. Biotechnol. Biochem.*, 63: 314-318.

Morandini P., Salamini F. (2003) Plant biotechnology and breeding: allied for years to come. *Trends Plant Sci.* 8:70-75.

Morandini, P., Salamini, F., Gantet P. (2005) Engineering of Plant Metabolism for Drug and Food. *Curr. Med. Chem. – Immun., Endoc. and Metab. Agents* 5:103-112.

Morelli R., Das S. Bertelli A., Bollini R., Scalzo R.L., Das D.K., Falchi M. (2006) The introduction of the stilbene synthase gene enhances the natural antiradical activity of *Lycopersicon esculentum* mill. *Mol. Cell. Biochem.*, 282: 65-73.

Motto M., Balconi C., Triulzi T., Lanzanova C., Gualdi L., Forlani F., Berardo N., Lupotto E. (2004) La proteina b32 di mais: ruolo nella protezione contro patogeni fungini. 1° Congr. Nazionale "Le micotossine nella filiera agro-alimentare", Istituto Superiore di Sanità, Roma, Italy, 29th-30th november, pag. 31-32.

Neff W.E., List G.R. (1999) Oxidative stability of natural and randomized high-palmitic and high-stearic-acid oils from genetically modified soybean varieties. *J. American Oil Chem. Soc.,* 76: 825-831.

Nelson G. (1998) Dietary fat, *trans* fatty acids and risk of coronary hearth disease. *Nutr. Rev.*, 56: 250-252.

Neuhouser M.L., Rock C.L., Eldridge A.L., Kristal A.R., Patterson R.E., Cooper D.A., Neumark-Sztainer D., Cheskin L.J., Thornquist M.D. (2001) Serum concentrations of retinol, alpha-tocopherol and the carotenoids are influenced by diet, race and obesity in a sample of healthy adolescents. *J. Nutr.*, 131: 2184-2191.

Ni W., Fahrendorf T., Balance G.M., Lamb C.J., Dixon R.A. (1996) Stress responses in alfalfa (*Medicago sativa* L.) trancriptional activation phenilpropanoid pathways genes in elicitor-induced cell suspension culture. *Plant. Mol. Biol.* 30: 427-438.

Nunes-Nesi A., Carrari F., Lytovchenko A., Smith A.M., Loureiro M.E., Ratcliffe R.G., Sweetlove L.J., Fernie A.R. (2005) Enhanced photosynthetic performance and growth as a consequence of decreasing mitochondrial malate dehydrogenase activity in transgenic tomato plants. *Plant Physiol.* 137:611–622.

Paine J.A., Shipton C.A., Chaggar S., Howells R.M., Kennedy M.J., Vernon G., Wright S.Y., Hinchliffe E., Adams J.L., Silverstone A.L., Drake R. (2005) Improving the nutritional value of Golden rice through increased pro-vitamin A content. *Nature Biotechnol.*, 23: 482-487.

Pen, J., Verwoerd, T.C., van Paridon, P.A., Beudeker, R.F., van den Elzen, P.J.M., Geerse, K., van der Klis, J.D., Versteegh, H.A.J., van Ooyen, A.J.J., Hoekema, A. (1993) Phytase-containing transgenic seeds as a novel feed additive for improved phosphorus utilization. *Biotechnology,* 11, 811– 814.

Pignocchi C., Fletcher J.M., Wilkinson J.E., Barnes J.D., Foyer C.H. (2003) The function of ascorbate oxidase in tobacco. *Plant Physiol.* 132 :1631–1641.

Piro G., Zuppa A., Dalessandro G., Northcote D.H. (1993) Glucomannan synthesis in pea epicotyls: the mannose and glucose transferases. *Planta,* 190(2): 206-220.

Quílez J., Bulló M., Salas-Salvadó J. (2007) Improved postprandial response and feeling of satiety after consumption of low-calorie muffins with maltitol and high-amylose corn starch. *J. Food Sci.* 72: S407-411.

Radzio J.A., Lorence A., Chevone B.I., Nessler C.L. (2003) L-Gulono-1,4-lactone oxidase expression rescues vitamin C-deficient Arabidopsis (vtc) mutants. *Plant Mol. Biol.*, 53: 837-844.

Ralston L., Subramanian S., Matsuno M., Yu O. (2005) Partial reconstruction of flavonoid and isoflavonoid biosynthesis in yeast using soybean type I and type II chalcone isomerases. *Plant Physiol.*, 137: 1375-1388.

Raschke M., Bürkle L., Müller N., Nunes-Nesi A., Fernie A.R., Arigoni D., Amrhein N., Fitzpatrick T.B. (2007) Vitamin B1 biosynthesis in plants requires the essential iron–sulfur cluster protein, THIC. *Proc. Natl. Acad Sci. USA*, published on line 28th November 2007.

Rulquin H., Pisulewski P.M., Vérité R., Guinard J. (1993) Milk production and composition as a function of postruminal lysine and methionine supply: a nutrient-response approach. *Livest. Prod. Sci.*, 37: 69-90.

Schroder J., Schroder G. (1990) Stylbene and chalcone synthase related enzymes with key functions in plants-specific pathways. Z. Naturforsch., 45: 1-8.

Schwab C.G., Bozak C.K., Whitehouse N.L., Olson V.M. (1992) Aminoacid limitation and flow to duodenum at four stages of lactation. 1. Sequence of lysine and methionine limitation. *J. Dairy Sci.*, 75: 3486-3502.

Sculati M., Rossi F., Cena H., Morlacchini M., Roggi C. (2007) Diets with low glycemic index limit weight loss in rats reared in a simulation of microgravity by hindlimb suspension. *Nutr. Res.*, 27: 802-807.

Seifert G.J. (2004) Nucleotide sugar interconversions and cell wall biosynthesis: how to bring the inside to the outside. *Curr. Opin. Plant Biol.*, 7: 277-284.

Shi J., Wang H., Schellin K., Li B., Faller M., Stoop J.M., Meeley R.B., Ertl D.S., Ranch J.P., Glassman K. (2007) Embryo-specific silencing of a transporter reduces phytic acid content of maize and soybean seeds. *Nat. Biotechnol.* 25:930-937.

Siritunga D, Sayre R. (2007) Transgenic approaches for cyanogen reduction in cassava. *J AOAC Int.*, 90:1450-1455.

Sønderby IE, Hansen BG, Bjarnholt N, Ticconi C, Halkier BA, Kliebenstein DJ. (2007) A Systems Biology Approach Identifies a R2R3 MYB Gene Subfamily with Distinct and Overlapping Functions in Regulation of Aliphatic Glucosinolates. *PLoS ONE.* 2:e1322.

Storozhenko S., De Brouwer V., Volckaert M., Navarrete O., Blancquaert D., Zhang G., Lambert W., Van Der Straeten D. (2007) Folate fortification of rice by metabolic engineering. *Nat. Biotechnol.*, 25: 1277-1279.

Suzuki H, Aoki K, Goda H, Nishizawa OI, Shibata D, Saito K. (2007) Omics-based identification of Arabidopsis Myb transcription factors regulating aliphatic glucosinolate biosynthesis. *Proc Natl Acad Sci U S A.* 104:6478-83.

Thomison P. (accessed 2007) Specialty corns: waxy, high-amylose, high-oil and high lysine corn. Ohio State University Extension Fact Sheet. Dept. of Horticolture and Crop *Science*. http://ohioline.osu.edu/agf-fact/0112.html

Valentin H.E., Lincoln K., Moshiri F., Jensen P.K., Qi Q., Venkatesh T.V., Karunanandaa B., Baszis S.R., Norris S.R., Savidge B., Gruys K.J., Last R.L. (2006) The Arabidopsis vitamin E pathway gene5-1 mutant reveals a critical role for phytol kinase in seed tocopherol biosynthesis. *Plant Cell.* 18:212-224.

Van Eenennaam A.L., Lincoln K., Durrett T.P., Valentin H.E., Shewmaker C.K., Thorne G.M, Jiang J., Baszis S.R., Levering C.K., Aasen E.D., Hao M., Stein J.C., Norris S.R., Last R.L. (2003) Engineering vitamin E content: from Arabidopsis mutant to soy oil. *Plant Cell.* 15:3007-3019.

Venkatramesh M., Karunanandaa B., Sun B., Gunther C.A., Boddupalli S., Kishore G.M. (2003) Expression of a Streptomyces 3-hydroxy teroid oxidase gene in oilseeds for converting phytosteols to phytostanols. *Phytochemistry,* 62: 39-46.

Wachter A., Tunc-Ozdemir M., Grove B.C., Green P.J., Shintani D.K., Breaker R.R. (2007) iboswitch control of gene expression in plants by splicing and alternative 3' end processing of mRNAs. *Plant Cell.* 19:3437-3450.

Wallerath T., Deckert G., Terres T., Anderson H., Li H., Witte K., Förstermann U. (2002) Resveratrol, a polyphenolic phytoalexin present in red wine, enhances expression and activity of endothelial nitric oxide synthase. *Circulation*, 106:1652-1658.

Wiberg E., Banas A., Stymne S. (1997) Fatty acid distribution and lipid metabolism in developing seeds of laurate-producing rape (*Brassica napus* L.). *Planta,* 203: 341-348.

Ye X., Al-Babili S., Kloti A., Zhang J., Lucca P., Beyer P., Potrykus I. (2000) Engineering the provitamin A (β-carotene) biosynthetic pathway into (carotenoid-free) rice endosperm. *Science,* 287: 303-305.

Yu O., Yung W., Shi J., Croes R.A., Fader G.M., McGonigle B., Odell J.T. (2000) Production of the isoflavones genistein and daidzein in non legume monocot and dicot tissues. *Plant Physiol.*, 124: 781-794.

Yu O., McGonigle B. (2003) Metabolic engineering of isoflavone biosynthesis. *Adv. Agron.*, 86: 147-190.

Yu J., Peng P., Zhang X., Zhao Q., Zhy D., Sun X., Liu J., Ao G. (2004) Seed-expression of a lysine rich protein sb401 gene significantly increases both lysine and total protein content in maize seeds. *Mol. Breed.*, 5: 345-356.

Zhang P., Jaynes J.M., Potrykus I., Gruissem W., Puonti-Kaerlas J. (2003) Transfer and expression of an artificial storage protein (ASP1) gene in cassava (Manihot esculenta Crantz). *Transgenic Res.*, 12: 243-250.

Zhang W., Gruszewski H.A., Chevone B.I., Nessler C.L. (2008) An Arabidopsis purple Acid phosphatase with phytase activity increases foliar ascorbate. *Plant Physiol.* 146:431-440.

Zhu X., Galili G. (2003) Increased lysine synthesis coupled with a knockout of its catabolism synergistically boosts lysine content and also transregulates the metabolism of other amino acids in Arabidopsis seeds. *Plant Cell.* 15:845-853.

Zhu C., Naqvi S., Gomez-Galera S., Pelacho A.M., Capell T., Christou P. (2007) Transgenic strategies for the nutritional enhancement of plants. *Trends Plant Sci.* 12:548-555.

In: Genetically Modified Plants
Editors: Timm Wolf and Jonas Koch
ISBN 978-1-60456-696-3

Chapter 5

AGROBACTERIUM RHIZOGENES-MEDIATED GENETIC TRANSFORMATIONS: A POWERFUL TOOL FOR THE PRODUCTION OF METABOLITES

***M. Georgiev* [*a], *V. Georgiev* [a], *J. Weber* [b], *Th. Bley* [b], *M. Ilieva* [a], and *A. Pavlov* [a]**

[a] Department of Microbial Biosynthesis and Biotechnologies – Laboratory in Plovdiv, Institute of Microbiology, Bulgarian Academy of Sciences, 26 Maritza Blvd., 4002 Plovdiv, Bulgaria

[b] Institute of Food Technology and Bioprocess Engineering, Technische Universität Dresden, 120 Bergstrasse, 01062 Dresden, Germany

ABSTRACT

"Hairy root" – an infectious disease caused by the soil bacteria *Agrobacterium rhizogenes* – is a natural phenomenon, which has existed for centuries. The development of molecular biology and genetic engineering tools at the end of the 1970s and in the early 1980s revealed the principles and mechanisms of infection and allowed scientists to "copy" this phenomenon. Today, genetic transformations using the "natural genetic engineer" *Agrobacterium rhizogenes* are widely used for induction of the so-called "hairy root" plant *in vitro* systems. Such systems are valuable for the production of economically important biologically-active substances. Furthermore the opportunity to express and integrate foreign genes into plant cells using *Agrobacterium rhizogenes* plasmids has allowed the mass production of desirable phytochemicals, medicinally important emzymes and foreign proteins either in *in vitro* conditions (bioreactors) or through plant regeneration (in greenhouses or in the field).

This chapter summarizes recent progress in the techniques and methods relating to the genetic transformation of plant cells and the further maintenance and cultivation of hairy root cultures, as well as recent results from the authors' laboratories on using flow cytometry for investigating the polysomaty of transformed root *in vitro* systems, and the

[*] Phone: +359 32 603 831; E-mail: lbpmbas@yahoo.com.

bioreactor systems (submerged and temporary immersion) used for the cultivation of hairy roots.

Introduction

Agrobacterium rhizogenes-mediated genetic transformations for the induction of *in vitro* transformed root cultures (also frequently called "hairy root", in analogy to the natural disease of the same name) are known for more than two decades [40]. To date, hairy root cultures have been induced from more than 450 different species of plants, including a diverse range of dicotyledonous (the natural hosts of *Agrobacterium*) and, more recently, monocotyledonous plant families and some gymnosperms [117].

During this period of research it was realized that hairy root cultures are not just a good model system for the investigation of hairy root-disease, but are valuable producers of a wide range of biologically-active substances. The transformed root cultures possess high genetic and biochemical stability (theoretically initiated from a single cell), fast growth in media free of plant regulators, and exhibit high productivity [40]. Understanding the mechanisms of DNA transfer from *Agrobacterium* to the host plant cells has allowed the integration of foreign gene(s) into hairy root cultures and, thus, the *de novo* production of high-value compounds and foreign proteins has become possible [47]. The application of several strategies to exploit the biochemical machinery of hairy root cells has led to significant improvements in the yields of the desired metabolites, a development which would have been very unlikely in field grown plants. Several bioreactor systems and designs at different scales were developed for the cultivation of transformed root cultures, and these eventually allowed the production of the required metabolites and proteins under aseptic conditions and fully independent of any seasonal and geographic influences [40].

From Plant Cell to Hairy Root Culture

The interest in the genus *Agrobacterium* dates from the beginning of the last century and was the result of significant financial losses from vineyards, almond, plum, apple and peach orchards [76]. It was found that the causative agents that led to a profusion of root hairs (so-called "hairy root" disease) and unorganized tumors (so-called "crown gall" disease) are two soil bacteria: *Agrobacterium rhizogenes* (formerly *Phytomonas rhizogenes*) and *Agrobacterium tumefaciens*, respectively.

The ability of *Agrobacterium* strains to cause plant diseases is due to their relatively large plasmids (200-800 kbp). Their names, given by the discoverers in Ghent (Belgium), are derived from the disease caused, namely the Ti-(tumor inducing) plasmid of *A. tumefaciens* [10] and the Ri-(root inducing) plasmid of *A. rhizogenes* [34]. A fragment of T-DNA is transferred to the host plant cell when bacteria come into contact with damaged plant tissue [113]. It has been found that injured plant tissue produces and secretes phenolics and sugars into the soil as a defense reaction. These compounds, particularly acetosyringone, activate the virulent gene's expression in the bacteria, and are responsible for the gene transfer. The T-DNA transferred to the host plant cell is steadily integrated into the plant genome [112; 123] and goes on to cause uncontrolled cell proliferation in the form of neoplastic tumor growth (in the case of *A. tumefaciens*) or finely branched root tissue (in the case of *A. rhizogenes*) [30; 76]. In this way, *Agrobacterium* accomplishes "genetic colonization" of susceptible plant species and forces them into the production/secretion of compounds, called opines, which are not specific to the host [95]. Opines are metabolized as a source of carbon and nitrogen exclusively by *Agrobacterium,* conferring on them an advantage over other soil microbes [35].

AGROBACTERIUM TAXONOMY

Bacteria of the genus *Agrobacterium* are gram negative, rod-shaped, non-spore-forming soil bacteria. The taxonomy of the *Agrobacterium* genus is confusing and has still not been conclusively determined. In addition to *Agrobacterium rhizogenes* and *A. tumefaciens*, *A. vitis* (the causal agent of galls on grape plants), *A. rubi* (the causal agent of cane gall disease) and *A. radiobacter* (which is avirulent) have been added to the genus [35]. The taxonomic difficulties arise from the fact that the replacement of the Ti-plasmid of *A. tumefaciens* (causing crown gall) with the Ri-plasmid "converts" the bacteria into *A. rhizogenes* (causing hairy root), while the removal of the plasmid converts these bacteria into non-pathogenic species [35]. More recently *Agrobacterium* strains have been divided into three biotype/biovars based on their physiological and biochemical differences [117]. Biotype I includes *A. tumefaciens* and *A. rubi*, Biotype II includes *A. rhizogenes* and Biotype III contains *A. vitis* [35]. The genome sequencing of different *Agrobacterium* strains is now complete (*A. tumefaciens* C58, *A. radiobacter* K84, *A. vitis* S4 and currently *A. rhizogenes* A4) and has proved that the biotype classification is the best means of organizing the taxonomy of *Agrobacterium* [117]. Another classification of the virulent *Agrobacterium* strains, which has practical applications, is based on the plasmid present and the respective opines produced after infection. This classification makes use of the fact that different *Agrobacterium* strains possessing Ti- and Ri-plasmids biosynthesize different opines [20]. Opines are low molecular weight compounds, formed by condensation of an amino acid, either with a keto acid or a sugar; they are not normally synthesized in plant cells [20]. Each Ti- or Ri-plasmid encodes the information for the biosynthesis of a specific opine [11; 65; 95]. Based on the types of opines produced, *A. rhizogenes* can be divided into four lines: agropine, mannopine, cucumopine and mikimopine [90]. This classification, however, is only applicable to laboratory situations, since *Agrobacterium* could transfer the genetic element, responsible for the utilization of a specific opine to other bacteria through conjugation [11]. Furthermore *Agrobacterium* could transfer the whole plasmid to other non-pathogenic *Agrobacterium* strains or to closely related species of other α-*Proteobacteria*. Root mat disease of cucumbers and tomatoes, for example, is caused by *A. radiobacter*, *Rhizobium* sp., *Ochrobactrum* sp. and *Sinorhizobium* sp., harboring the Ri-plasmid of *A. rhizogenes* [121].

Mechanism of Genetic Transformation

The host range of *Agrobacterium* is quite broad and the species can transfer DNA to numerous dicot and monocot angiosperm species [35; 64] and gymnosperms [64; 69; 128]. In addition, *Agrobacterium* can transform several types of eukaryotic cells, including phytopathogenic fungi [93], yeasts, ascomycetes, basidiomycetes [64; 113], as well as animal cells [85]. Recently, *Agrobacterium* has been reported to transfer DNA to human cells as well [59].

The molecular mechanism of hairy root induction is not completely understood [99; 117]. However, the process of hairy root formation can be divided into five steps:

(1) Bacteria-to-plant cell attachment and activation by chemoattractants,
(2) Processing in the bacterial cell,
(3) Bacteria-to-plant cell transport of the T complex,
(4) DNA integration and expression,
(5) Subsequent hairy root induction.

The transferred DNA (T-DNA) is a segment of the *Agrobacterium* Ri-plasmid, which is flanked by 25-bp imperfect direct border repeats – both left and right borders. The T-DNA is transferred as a single-stranded intermediate (T-strand) from the agrobacteria into the host plant cells, where it is integrated into the nuclear genome [5]. The mechanism of T-DNA transfer is similar for *A. tumefaciens* and *A. rhizogenes* [2; 35; 76; 113]. It has also been found that the genes on the T-DNA are not responsible for the transfer and integration and thus any fragment DNA, placed between the left and right borders, could be transferred to the host cell [115]. This phenomenon makes it possible to clone the gene of interest between the border sequences and for it to be transferred to the host cells [115]. Furthermore, *Agrobacterium*-mediated transformations are considered preferable to genetic transformation relying on artificial approaches because of the straightforward, low cost procedure and the relatively low complexity of intact transgenes integrated into the plant genome [36].

The transformation begins with the bacteria–plant attachment [114]. However, little is known about the assembly of the transfer apparatus or the molecular details of its operation. It is known that the apparatus has a pilus and may form a transmembrane channel for cell-to-cell trafficking of the T complex by a type IV secretion system [65; 113]. For successful transformation the following must be present: the major loci (*virA*, *virB*, *virC*, *virD*, *virE*, *virG*, and *virH*) of the *vir* region of the Ti- or Ri-plasmids, T-DNA with left and right borders and some virulent *chv* genes, localized in the bacterial chromosome [112]. The products of the *virD2* and *virE2* genes, i.e. VirD2 and VirE2 proteins, play a key role in the processes of transfer, nuclear localization and integration into the host plant chromosome [5; 112; 119]. Experiments have shown that the integration of T-DNA into the plant genome is performed in unspecified regions, but tends to occur in regions of plant DNA that are A-T-rich [8].

Most of the *Agrobacterium* strains transfer a single fragment, however there are some (the agropine type of *A. rhizogenes* for example), which transfer two independent T-DNAs, denoted T_L-DNA and T_R-DNA [76]; these are separated by a non-integrated region [2]. T_R-DNA is highly homologous to the T-DNA of the Ti-plasmid of *A. tumefaciens*, while T_L-DNA is completely different and possesses similarities to the T-DNA of the Ri-plasmid of mannopine *A. rhizogenes* strains [76]. Both fragments (T_L-DNA and T_R-DNA) are known to be independently transferred and integrated to the host plant genome; however the transfer of T_L-DNA is both necessary and sufficient to induce hairy root syndrome [2; 76; 99]. Homologues of both the auxin biosynthesis genes (*iaaM and iaaH*) and the genes encoding synthases for the opines mannopine (*mas1'* and *mas2'*) and agropine (*ags*) have been found in the T_R-DNA genes [2; 76]. The auxin gene homologues in *A. rhizogenes* are known as *aux1* and *aux2*; they are considered to play an accessory role in hairy root induction [117]. The sequence analysis of T_L-DNA identified 18 open read frames (ORF), of which four are essential for hairy root induction. These loci were denoted root locus A, B, C and D (*rolA*, *rolB*, *rolC* and *rolD*) and assigned to ORF10, ORF11, ORF12 and ORF15, respectively [76].

The molecular mechanism of root formation after infection with *A. rhizogenes* has still not been fully elucidated; in the case of *A. tumefaciens*, the formation of tumor growth is due to changes in the phytohormonal balance in the plant cells. The neoplastic root formation in plants infected with *A. rhizogenes* is not, however, due to changes in phytohormonal balance in plant cells and is probably because the cells become more sensitive to auxins [2; 76]. The products of the four *rol* (*A*, *B*, *C* and *D*) genes play key roles in the process of hairy root formation. The *rolA* gene contains an ORF of about 300 bp, encoding a 100-amino acid protein [76] and is believed to be involved in the generation of a functional imbalance in phytohormone levels [19]. The *rolB* gene contains an ORF of about 777 bp, encoding a 259-amino acid protein, and appears to be the most important in hairy root induction, since mutation in this locus renders the plasmid avirulent [2; 76]. In fact, when *rolB* is introduced into the host plant genome as a single gene, it is capable of hairy root induction [2]. It has been proposed that *rolA* and *rolB* effects are antagonistic to each other [117]. The *rolC* gene contains an ORF of 540 bp, encoding a cytokinin-β-glucosidase [76; 117] and plays an important role in shoot formation [45]. The *rolD* gene contains an ORF of 1032 bp, encoding a protein of 344 amino acids [76]. RolD protein bears sequence homology with ornithine cyclodeaminase and acts as a stress-related osmoprotectant and it is also involved in the production of proline at the flowering stage [2].

Induction of Transformed Root Cultures

The induction of transformed root culture does not require any specific equipment or facilities and could be easily performed in an ordinary microbiological laboratory [18; 49]. The procedure for hairy root induction (Figure 1) includes aseptic cultivation of a wounded sterile explant (part of the mother plant: hypocotyls, leaf, stem, stalk, petiole, shoot tip, protoplast, storage root and tuber) [117] with *A. rhizogenes* suspension [40]. The infection process is generally performed by direct inoculation with a bacterial suspension then incubation on a solid medium, or by co-cultivation in a liquid medium (Figure 1) [40]. Using the method of co-cultivation in the authors' laboratories, transformed root cultures from several plant species were induced (Figure 2). In both methods, the *Agrobacterium* is subsequently destroyed by the use of antibiotics. The application of suitable antibiotics is of great significance, since they should eliminate the bacteria effectively without influencing the plant explant. Cefotaxime [84] and penicillin derivates [87] are the antibiotics most frequently used.

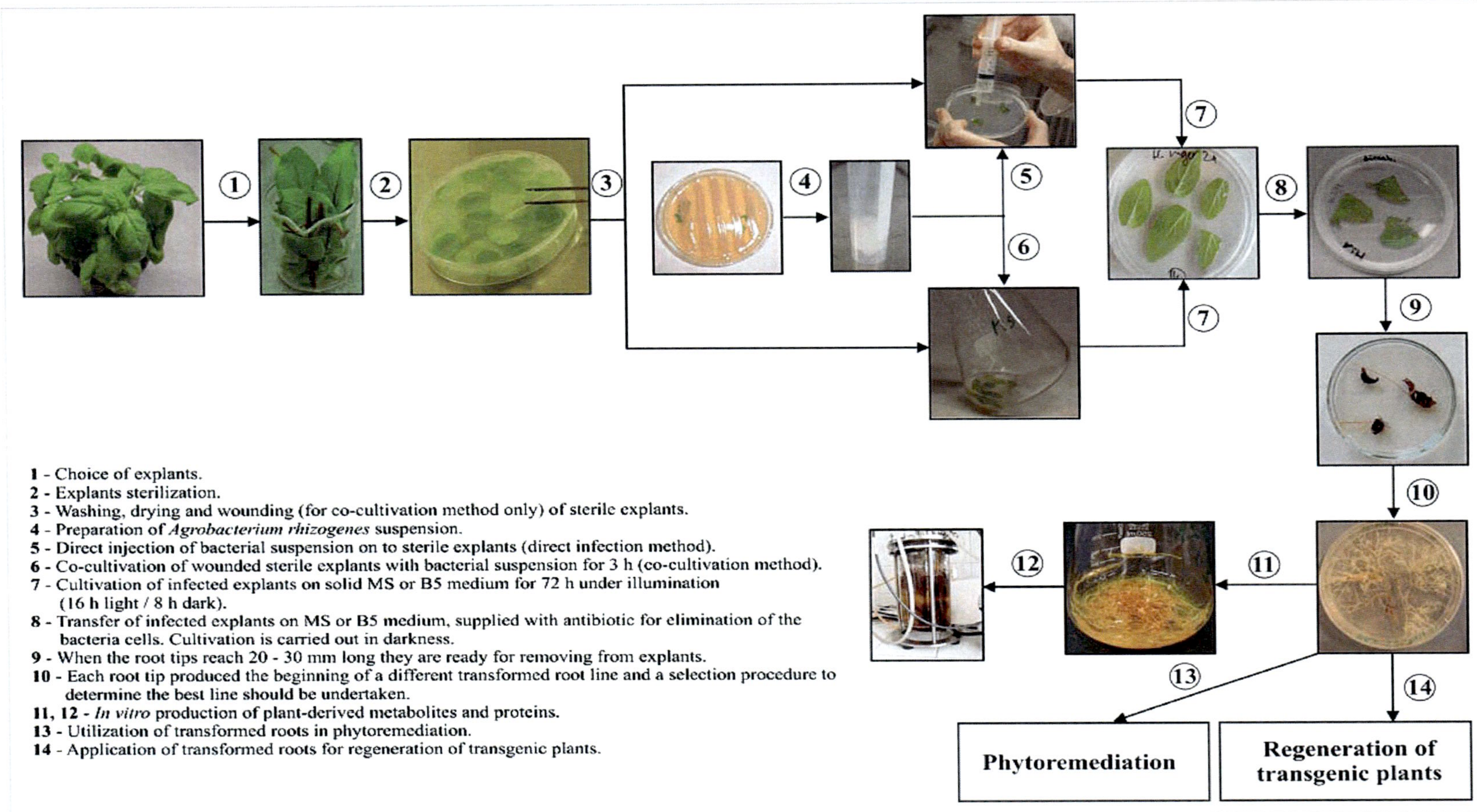

Figure 1. Overview of the induction of transformed root cultures and some applications.

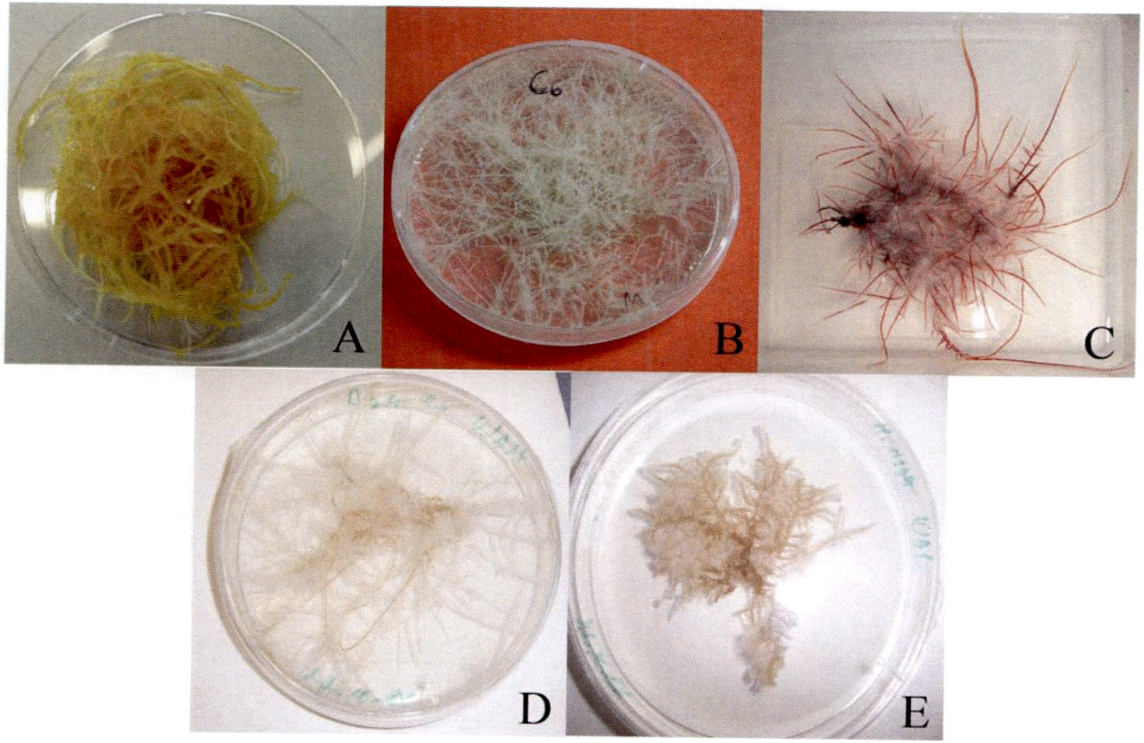

Figure 2. Examples from the authors' laboratories: transformed root cultures of *Harpagophytum procumbens* (A), *Ocimum basilicum* (B), *Beta vulgaris* (C), *Datura stramonium* (D) and *Hyoscyamus niger* (E).

The resultant neoplastic hairy roots grow in a profusely branched manner, with many lateral branches on Murashige and Skoog [38; 82; 84] or Gamborg's B5 [99] hormone-free media. These hairy root lines could be further used for the production of biologically-active substances [40; 99], phytoremediation [43; 107; 108] or the regeneration of transgenic plants [12; 14; 96].

There are several ways to prove the success of the genetic transformation. In the past, the production of opines was used as being indicative of a successful transformation process. However, in some cases opine production disappears after several cell division cycles [31; 99], thus it is not reliable. The other possibility is to determine T-DNA directly [40], either by polymerase chain reaction or southern blot hybridization [61; 127]. The engineered strains of *Agrobacterium*, harboring the β-glucuronidase (GUS) reporter gene, can easily confirm transformation using PCR or northern blot [17]. A green fluorescent protein (GFP) marker gene, cloned into a binary vector (under the control of a constitutive promoter) and inserted into *Agrobacterium* can also be used as an indicator of cell transformation [73]. The benefits are the easy screening or selection of the transformed (fluorescent) cells and lines.

Successful transformation and further hairy root induction depend upon the species, age and type of plant explant, the strain of *Agrobacterium* used, the density of the bacterial suspension and the addition of some chemicals promoting the infection process, such as acetosyringone [54; 80; 99]. Recently, a new technique – sonication assisted *Agrobacterium* mediated transformation, SAAT – for induction of hairy root cultures from difficult-to-transform plant species was introduced [111]. Using the SAAT procedure, hairy root cultures of *Atropa belladonna* [7], *Papaver somniferum* [61], and *Leucojum aestivum* [21] have been obtained.

UTILIZATION OF TRANSFORMED ROOT CULTURES

During the last two decades, transformed root cultures have been shown to have great potential for the production of phytochemicals, medicinally important enzymes and foreign proteins. Interest in the utilization of transformed root cultures for mass production of useful compounds is the result of a number of advantages, including genetic and biochemical stability (compared to dedifferentiated plant *in vitro* systems), fast growth (compared to adventitious root cultures), growth in hormone-free media, and the ability to produce metabolites in the same or even greater quantities than intact plants. Lorence et al. [66], for example, reported that the production of camptothecin (an anticancer and antiviral alkaloid) from hairy root cultures of *Camptotheca acuminata* was equal to and sometimes greater than that by intact plants. Table 1 and Figure 3 show some secondary metabolites of significant value that are produced by transformed root cultures. Some medicinally important emzymes have also been produced by transformed root cultures. Superoxide dismutase (SOD; EC 1.15.11.11), for example, was produced by transformed root cultures of carrot [56-57] in batch and fed-batch modes. Peroxidase (EC 1.11.1.7), which in tandem with SOD scavenges harmful oxygen radicals, was produced intracellularly by hairy root cultures of *Beta vulgaris*. Furthermore the purified red beet peroxidase showed better thermal stability than the commercial source from horseradish, hinting at its wider potential applications [94].

Table 1. Examples of important secondary metabolites produced by hairy root cultures

Biological properties	Metabolites	Plant specie/hairy root culture	Reference
Anticancer	Paclitaxel	*Taxus* x *media* var. *Hicksii*	[33]
	6-Methoxy-podophyllotoxin	*Linum album; Linum persicum*	[124]
	Camptothecin	*Ophiorrhiza pumila*	[102]
	Justicidin B	*Linum leonii*	[116]
	Resveratrol	*Arachis hypogaea*	[71]
Antimalarial	Artemisinin	*Artemisia annua*	[120]
Anti-inflammatory	Asiaticoside	*Centella asiatica*	[58]
	Rosmarinic acid	*Coleus forskohlii*	[63]
	Iridoid glycosides	*Harpagophytum procumbens*	[38; 67]
Therapeutic agents for treating Parkinson's disease	L-DOPA	*Stizolobium hassjoo*	[106]
Antimicrobial	Thiarubrine A	*Ambrosia artemisiifolia*	[6]
	Acetylshikonin	*Lithospermum canescens*	[86]
	Verbascoside	*Catalpa ovata*	[126]

Metabolic engineering tools could be used to improve the production of desired metabolites, as well as for the creation of *de novo* metabolic pathways [47]. The expression of foreign genes, encoding key enzymes in biosynthetic pathways offers a good opportunity for yield improvement and/or *de novo* synthesis [40]. One of the most remarkable examples is the enzyme hyoscyamine 6β-hydroxylase (H6H; EC 1.14.11.11), which catalyzes the convertion of hyoscyamine to the medicinally important scopolamine, in two consecutive steps (hydroxylation followed by epoxidation) [129]. An engineered *A. rhizogenes* strain, harboring the 35S-H6H gene construct, was introduced into *Hyoscyamus muticus*; this finally resulted in significant improvements in scopolamine yields (the best clone produced over 100-times more scopolamine than the control one) [55]. Using the same approach, Palazon et al. [78] transformed *Duboisia* hybrids (a plant with an initially high scopolamine content) using *A. rhizogenes* carrying the 35S-H6H gene and further regenerated plants with an even higher scopolamine content. The authors also observed some morphological changes, known as a T-phenotype changes, in the regenerated plants (short, wide, wrinkled and dark green leaves, and elevated rooting capacity associated with root plagiotropism) [78]. In more recent work, Moyano et al. [74] reported on the establishment of an *N. tabacum* cell suspension culture, containing the 35S-H6H gene from *Hyoscyamus niger*. The resultant transgenic *N. tabacum* cell suspension culture showed a considerable capacity for the bioconversion of hyoscyamine into scopolamine. This is a good example of how plant cells that do not produce secondary compounds of interest are able to do so by the conversion of a suitable precursor, if they overexpress the gene coding the key enzyme involved in the biosynthesis of such compounds [74].

However, because only a few metabolic pathways in plants are well known, the progress in the field of metabolic engineering remains slow and is frequently hampered [77; 118]. Thus, functional genomic approaches (including genomics, transcriptomics, proteomics and metabolomics) are used to speed up comprehensive investigations of biological systems [77]. Using a genomics approach, all genes in a plant can be identified; by applying transcriptomics and proteomics, the active genes in particular pathways could be identified. Finally the incorporation of this knowledge to metabolomics data allows better exploitation of plants' biochemical machinery [44].

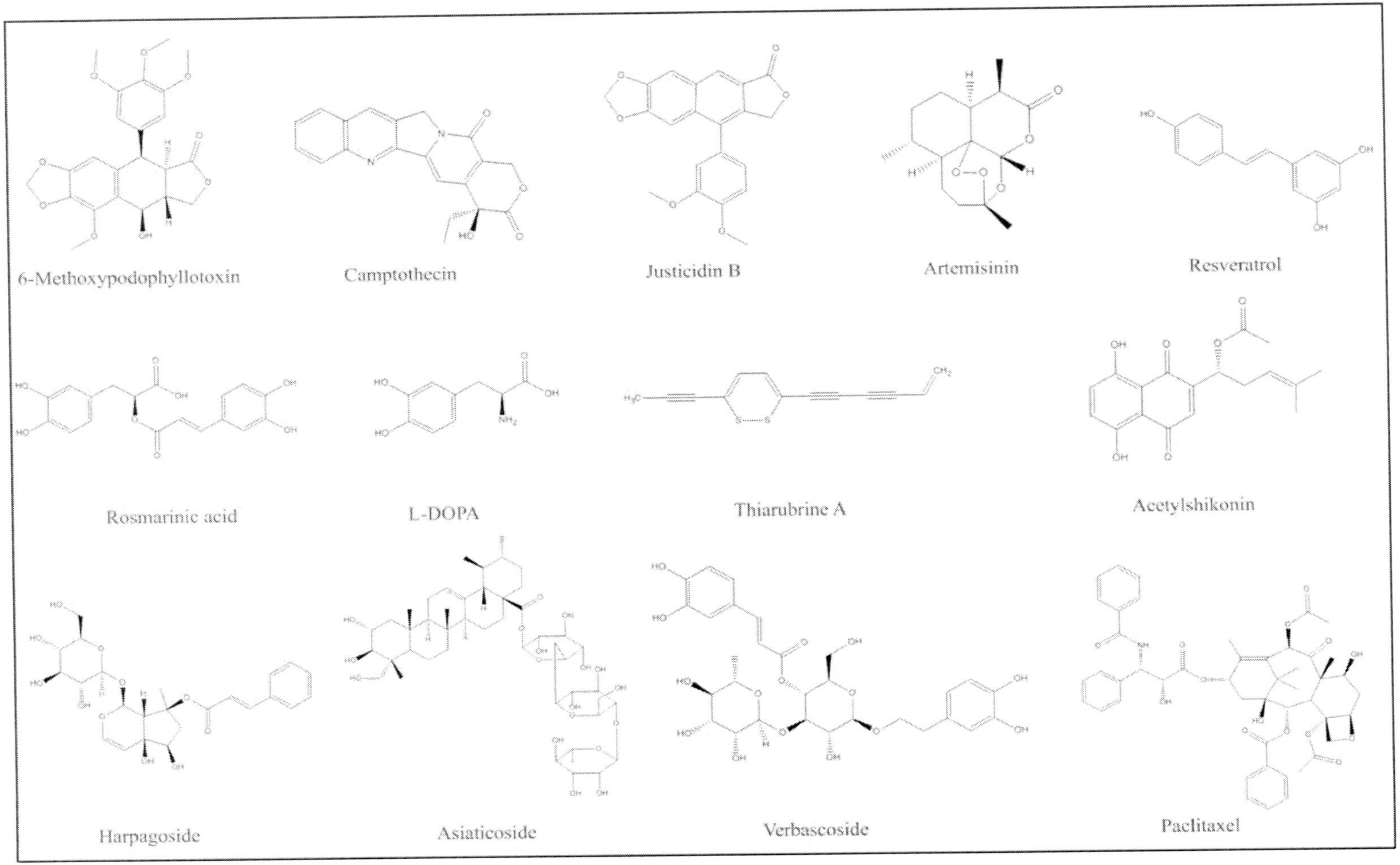

Figure 3. Structures of high-value secondary metabolites produced by hairy root cultures.

Recently, this type of combined genome-wide transcript profiling (by cDNA-amplified fragment-length polymorphism) and metabolic profiling has been applied to *Catharanthus roseus* cell cultures and has yielded a collection of both known and previously undescribed transcript tags and metabolites associated with terpenoid indole alkaloids [92]. Such gene-to-metabolite networks seem to be quite promising for the enhanced exploitation of plant cells' biosynthetic potential; they could lead to improvements in the production of useful compounds and to the design of new compounds with novel and/or superior biological activities [44].

De novo synthesis of poly(3-hydroxybutyrate) has been reported for hairy root cultures of *Beta vulgaris*, transformed with *A. tumefaciens*, harboring the root inducing plasmid (pRi15834) together with the pBI ABC plasmid, containing the three *Ralstonia eutropha* (today reclassified into *Cupriavidus necator*) genes *phaA*, *phaB* and *phaC* [72]. Foreign protein production by transgenic hairy root cultures also is also developing rapidly. Sharp and Doran [100] investigated the production of a monoclonal antibody (Guy's 13 mouse IgG1) by transgenic *Nicotiana tabacum* hairy root cultures in bioreactors and outlined the possibilities for the enhancement of antibody accumulation. Ribosome-inactivating protein, which possesses interesting pharmacological properties, has been produced by hairy roots of *Trichosanthes kirilowii* [109]. During the search for ways to improve foreign protein stability recently, chloroplast expression was recognized [26]. Because each plant cell contains many chloroplasts, thousands of transgene copies could be expressed simultaneously. Positional and gene silencing effects could be also avoided, and co-transcription of multiple genes facilitates the expression of multiple proteins [26 and the literature cited therein].

A wide range of strategies and techniques have been developed to boost the yields of useful metabolites. These include: selection of high-producing hairy root clones; nutrient medium optimization; elicitation and bioreactor culture environment optimization [40]. Such techniques have allowed the mass production of secondary metabolites and foreign proteins in significantly higher quantities than is possible using intact mother plants, in sterile conditions; the process is also completely independent of seasonal and geographic conditions [40].

In the middle of the 1980s, it was realized that transformed root cultures have the potential to be regenerated into whole plants. Plants can be directly regenerated from hairy root cultures either spontaneously or by transferring roots to a medium containing hormones [41]. Plants regenerated from hairy roots do not produce chimeras or somaclonal variants because they originate from single cells, a condition not necessarily expected for plants transformed by infection with *A. tumefaciens* [12; 96]. *A. rhizogenes*-mediated transformations allow the production of "composite plants", a term derived from the fact that transformed roots are induced on a non-transformed plant [14]. Saxena et al. [96] produced transgenic plants of rose-scented geranium (*Pelargonium graveolens*) with increased concentrations of geraniol and geranyl esters; this significantly improved oil quality with respect to the citronellol/geraniol ratio. Christey [13] summarized that transgenic plants from 89 different taxa have been produced so far using *A. rhizogenes*-mediated transformations; more than half of these have been transformed with foreign genes in addition to the Ri-plasmid, and these include agronomically useful traits. Plants regenerated from hairy roots frequently exhibited altered plant morphology, such as dwarfing, increased root branching, altered flowering, and wrinkled leaves due to *rol* genes expression. These altered phenotypic features could have potential applications for plant improvement, especially in the horticultural industry where such morphological variations may be desirable [13]. *A.*

rhizogenes-mediated transformation of trees and the subsequent regeneration of transgenic plants could be a useful alternative to the classical breeding programs used for trees; it would provide a rapid and direct way of introducing and enhancing the expression of specific traits. Furthermore, this would significantly reduce the time required for tree improvement and would facilitate new gene combinations not possible using traditional methods [41].

Another interesting potential use for transformed root cultures, with respect to ecological applications, is phytoremediation. Phytoremediation makes use of the ability of plants/plant cells to absorb (directly or indirectly), sequester, and/or degrade inorganic and organic contaminants in soils, sediments, surface waters, and groundwater [15]. Plants are also able to hyper-accumulate various heavy metals through the action of phytochelatins and metallothioneins, which form complexes with heavy metals that are then stored in the vacuoles [107]. Gonzales et al. [43] reported on the successful phytoremediation of phenol from wastewater using hairy root cultures of tomato (*Lycopersicon esculentum*) and determined the association between tomato peroxidases and the removal of phenol. In another study, Suresh et al. [108] described the degradation of DDT (a chemical used for pest control) using hairy root cultures of *Cichorium intybus* and *Brassica juncea*. The rapid disappearance of tetracycline and oxytetracycline antibiotics from aqueous media has been achieved using transformed root cultures of sunflower (*Helianthus annuus*). Increasing knowledge about the factors affecting the phytoremediation process will provide a basis for genetic modification of plants/plant *in vitro* systems for improved performance and more efficient environmental clean up.

Cultivation Systems for Hairy Roots

Cultivation of transformed root cultures in bioreactors represents a critical step in the development of scale-up technologies for the production of target metabolites. Although hairy root cultures have several advantages (as mentioned above) they are extremely sensitive to a number of stress factors that reduce their growth and/or biosynthetic capacity [81]. Thus, the commercial exploitation of the hairy root culture biochemical machinery requires the development of suitable bioreactor systems and operational regimes in which key physical and chemical parameters need to be optimized [81]. Many types of bioreactor design have been successfully used for the cultivation of transformed root cultures: stirred tank reactors with or without a separate impeller [39; 60]; bubble column reactors [81]; trickle-bed reactors [88]; mist [110] and biowave reactors [79] (Figure 4). The choice of the most suitable bioreactor design should be consistent with the physiological peculiarities of the particular hairy root clone. Disposable biowave reactor systems have developed rapidly in recent years and a number of examples of their successful application to hairy root culture can be found. The absence of mechanical agitation (the working principle is based on wave-induced agitation) ensures a low-stress environment and the use of plastic disposable chambers facilitates easy inoculation and adherence to GMP (Good Manufacturing Practice) requirements [27; 28]. The successful cultivation of *Harpagophytum procumbens*, *Hyoscyamus muticus* and *Panax ginseng* hairy root cultures has been achieved in wave bioreactors of different volumes [29].

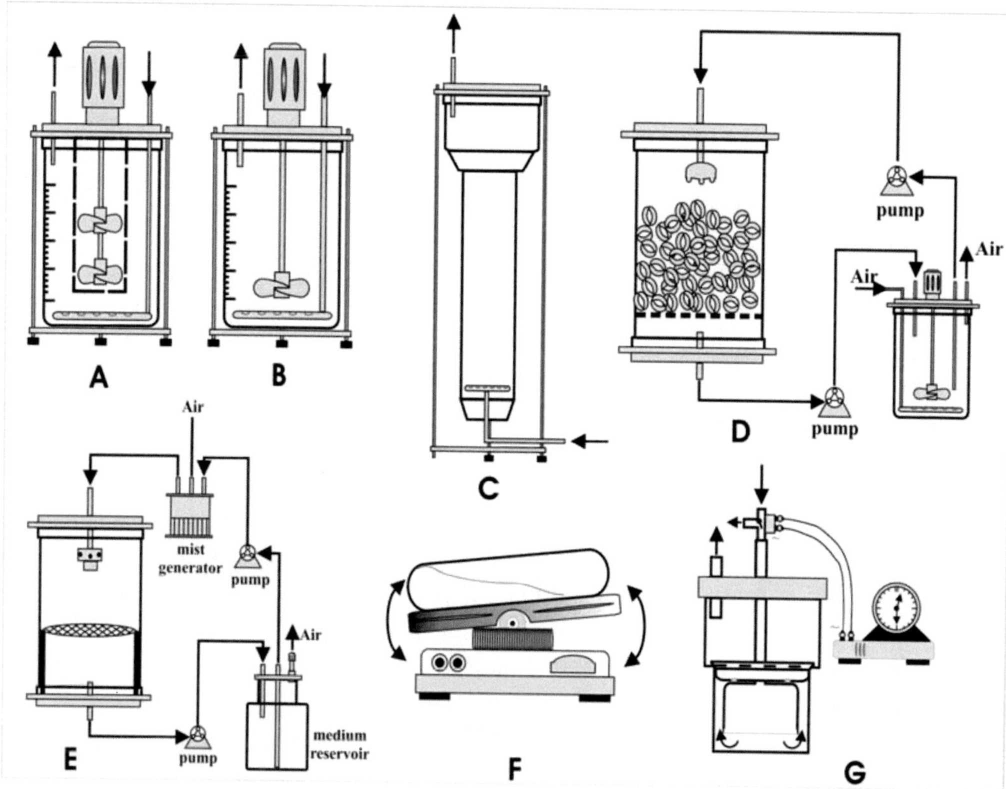

Figure 4. Design of bioreactors used for hairy root cultivation [stirred tank reactor with separate impeller (A) and without separate impeller (B); bubble column reactor (C); trickle-bed reactor (D); mist bioreactor (E); biowave reactor (F) and RITA® temporary immersion system (G)].

Moreover, the commercially available wave systems with high volumes (600 L geometric volume; Wave Biotech AG®, Switzerland) will allow the up-scaling of hairy root based processes.

The only successful large-scale cultivation of hairy root culture was reported a decade ago by Wilson [122], who used a 500 L bioreactor for the cultivation of a hairy root culture of *Datura stramonium* immobilized on barbs. The author also proposed a solution for the problems associated with the inoculation of large-scale reactors, namely using a seed vessel (10 L capacity) and a transfer tube incorporating a conical stainless steel helical screw with a blade attached to the leading edge [122]. Ramakrishnan et al. [89] proposed another possible inoculation method, based on homogenizing the bulk root culture (at 3515 rpm for 3 s) followed by an aseptic transfer of the resultant slurry to the reactor. Their data showed that the accumulated hairy root mass was not significantly different from the control variant. Chopping the roots and the homogenizing treatment, however, causes damage in different ways, which may lead to a range of effects on the biosynthesis of secondary metabolites [60]. One alternative for large-scale cultivation of transformed root culture, which could solve the difficulties associated with the inoculation procedure, is to change the cultivation mode from batch to fed-batch [81]. A single feed application (1 L Murashige and Skoog media 13 days from the beginning of the cultivation) resulted in 13.3 g accumulated dry biomass/L, dividing the feed into five separate applications (200 mL Murashige and Skoog media 13 to 17 days after the beginning of the cultivation) lead to 11 % more betalains (red-violet pigments with antioxidant and cancer-preventing properties) in a *Beta vulgaris* hairy root culture [81]. Although commercial applications of hairy root culture have not yet been achieved, the recent commercial production of ginsenoside (10000 L bioreactor scale), in South Korea, from *in vitro* adventitious root culture (normal non-transformed roots) of *Panax ginseng* [48] has shown that it might be possible to cultivate plant organ cultures (e.g. hairy roots) at a commercial scale. Furthermore, the growing interest in hairy root systems by commercial organizations (e.g. ROOTec bioactives GmbH, Switzerland), will facilitate knowledge transfer from laboratories to the pharmaceutical, food and cosmetic industries [40].

Cultivation of Devil's claw hairy root cultures in submerged bioreactor and temporary immersion systems.

Harpagophytum procumbens (Devil's claw; Pedaliaceae family) is an important medicinal plant species, which grows in southern Africa and is abundant in the Kalahari dessert region [70; 105]. Extracts from *H. procumbens* tubers, exhibiting analgesic and antidiabetic properties are used as a drug for the treatment of different inflammatory disorders, such as arthritis, lumbago and muscular pain [62]; the remarkable biological activities of the extracts are due to the presence of iridoid glycosides (harpagoside, 8-*p*-coumaroyl-harpagide, harpagide, prucumbide etc) [68; 70]. Transformed root cultures of *H. procumbens* were induced using the co-cultivation method [38] and were maintained for several weeks on solid MS media with the aim of producing roots with stable growth and morphological characteristics. After this period the hairy root cultures were adapted to submerged cultivation conditions through alternating cultivation in liquid and on solid media (this period took about 3 months) and the adapted cultures were grown in the submerged conditions in shaken-flasks [38], a 3-L bubble column bioreactor [37] and RITA® temporary immersion systems with an immersion frequency of 15 min flooding/45 min standby [this work] (Figure 5).

The results obtained from the cultivation of *H. procumbens* in the different cultivation systems are summarized in Table 2.

Figure 5. Transformed root cultures of Devil's claw (*Harpagophytum procumbens*), cultivated in shaken-flasks (A), 3-L bubble column bioreactor (B) and RITA® temporary immersion system (C).

Table 2. Accumulated dry biomass, total iridoid content and production rates from *in vitro* hairy root culture of *Harpagophytum procumbens* and micropropagated Devil's claw plants

Cultivation mode	Cultivation time [days]	ADB [g/L][a]	Total iridoids [mg HS equivalents/g dry weight][b]	Production rate [µg HS equivalents/g dry weight/day][b]	Reference
Shaken-flasks	21	11.6	0.66	31.43	[38]
Bubble column bioreactor (air flow rate 10 L/h)	21	6.6	0.67	31.90	[37]
Temporary immersion systems (air flow rate 60 L/h)	30	6.3	0.95	31.67	This work
Propagated *H. procumbens* plants:	~ 540				[62]
- leaves of plants		-	1.00	1.85	
- tubers of plants		-	10.00	18.52	

[a] ADB – accumulated dry biomass

[b] HS - harpagoside

The best conditions for root growth and biomass accumulation were achieved in shake-flasks and the accumulated biomass (~0.58 g/flask and a growth index of 81.3) [38] was significantly higher than the other cultivation systems studied. However the accumulated biomasses in the bubble column reactor (6.6 g ADB/L) and temporary immersion systems (6.3 g ADB/L) are comparable to those reported for other transformed root cultures from *Atropa belladonna* and *Panax ginseng*, grown in bioreactors [16]. In a recent investigation, Ludwig-Mueller et al. [67] found that the levels of stress-related hormones (abscisic acid, salicylic acid, and aminocyclopropane carboxylic acid) were significantly higher in shaken-flasks than in a bubble column bioreactor, indirectly supporting the belief that bubble columns provide low-stress environments and are, therefore, suitable for the cultivation of transformed root cultures. All of this imposes the optimization of the bioreactor culture environment as a powerful tool for enhancing the yields of target metabolites [83]. In addition, stress related hormones could be used to evaluate stress levels in transformed root cells in different cultivation systems [67].

Total iridoid production was also influenced by different cultivation systems. The iridoid content achieved during cultivation in shake-flasks and a bubble column was the same (~0.66-0.67 mg harpagoside equivalents/g dry weight), while the accumulated iridoids in temporary immersion systems were found to be about 40 % higher (0.95 mg harpagoside equivalents/g dry weight). HPLC analysis showed greater harpagoside levels in a bioreactor than in shake-flasks and significantly greater harpagide levels [67]. Since the amounts of total iridoids in both cultivation regimes did not differ, this enhancement is probably on other iridoids account (from the total iridoid pool). These amounts are lower than those detected in the tubers of micropropagated *H. procumbens* plants and about 24-34 % lower than the amounts detected in the leaves of the same species (Table 2). However, the production of total iridoids from micropropagated plants requires a much longer cultivation time (~18 months), which leads to lower production rates (about 17-fold and 1.7-fold lower, respectively, compared to the production rates calculated for the hairy root culture; see Table

2). Furthermore, these initial yields indicate the potential value of further investigations, addressing better exploitation of the biochemical machinery of the transformed *Harpagophytum* roots. Recent GC-MS analysis has revealed the production of other high value metabolites, namely β-sitosterol (possessing anti-tumor, chemopreventive properties and anti-inflammatory, anti-microbial and cholesterol-lowering properties) and astaxanthin (possessing immunostimulating properties) [67].

Flow Cytometry Investigations of Hairy Root Culture of *Beta Vulgaris* Cv. Egypt

Flow cytometry is a technique in which cells or particles flowing individually in front of a light source are illuminated and the resultant signals are detected and correlated [91]. As flow cytometry is a method for examining individual cells, these have to be suspended in a liquid medium. First, the cytometrist has to obtain a suspension. Historically, blood cells were analyzed as they are perfectly adapted to this application: they occur in an individual form and do not have to be processed before analysis [42]. Recently, cultured cells, yeast and bacteria have been the focus of flow cytometric analysis [101]. To support this, methods for the separation of tissue cells by mechanical or enzymatic treatments have been developed [97].

As plant cells do not occur in the form of single cells in suspension, the tissue has to be processed before analysis can be performed. For this purpose, a variety of protocols have been developed [24]. All these procedures include the extraction of intact nuclei from the plant cell, staining with a DNA-specific dye and the analysis of the fluorescence intensity by flow cytometry.

Usually, suspensions of nuclei are obtained by chopping up the tissue [22]. In this case, the cells of the plant are cut to open them and allow access to the nuclei. Then, specific buffers extract the nuclei, preserve their integrity and create optimal conditions for staining. Many different buffers are referred to in the literature [46]; they vary in their efficiency of extraction and staining with respect to different species and tissues. Unfortunately, there is no *a priori* rule for the selection of the buffer and the staining conditions for a given tissue and species, so that a trial and error approach is necessary to achieve the best results.

For the analysis of DNA-content, several fluorochromes can be applied. The most widely used are 4',6-diamidino-2-phenylindole (DAPI) and propidium iodide (PI). While DAPI specifically stains DNA, PI intercalates between base pairs of all double stranded nucleic acids. For this reason, the enzymatic digestion of double stranded RNA has to be performed prior to staining. While staining with DAPI results in very narrow and symmetrical peaks, due to its high specificity, it does exhibit a strong preference for A-T base pairs in the DNA sequence. This characteristic makes DAPI a good tool for estimating the base composition of DNA, but not very suitable for the analysis of genome sizes, as the intensity of fluorescence not only depends on the amount of DNA in a cell but on its base pair composition. Therefore, and because of its favorable spectral properties, PI was the fluorochrome of choice for our analysis.

Flow cytometry affords a great opportunity for the development of algorithms to select high-secondary-metabolite-producing plant *in vitro* culture lines, based on the assessment of their nuclear DNA (ploidy levels, genetic stability, effects of endoreduplication etc.). Compared to conventional methods (chromosome counting and densitometry), the analysis of plant nuclear DNA is significantly faster, more accurate and more convenient. Nowadays, flow cytometry investigations of the intact plant genome are well established [51; 53]. However, there are few data from such investigations of *in vitro* plant systems [50].

Red beet (*B. vulgaris* cv. Egypt) hairy roots have been obtained after transformation of young leaves from intact plants with *A. rhizogenes*. The best four hairy root lines, with respect to their growth rate and morphology, were selected for further work. *B. vulgaris* hairy roots cultures have been widely investigated for their function as producers of betalains pigments [75; 82] and as a model system for basic biochemical studies [98; 104]. The hairy root cultures are reported to be genetically stable, in contrast to undifferentiated calli and suspension, and their karyotype, as well as the number of chromosomes, has been found to be identical to their parent plants [1; 4]. However, our preliminary studies of other *Beta vulgaris* cultivars and their *in vitro* systems (Pavlov et al., unpublished results) showed significant differences in ploidy levels of hairy roots and the parent leaf tissue from which they were derived. On this basis, we investigated the ploidy levels and genome sizes of *Beta vulgaris* cv. Egypt and hairy roots obtained from it by means of flow cytometry.

The results of the flow cytometric investigations of leaves of *Beta vulgaris* cv. Egypt (Figure 6A) showed a small 2C (33 %) peak corresponding to diploid cells in phases G0 and G1. The main peak was 4C (56 %) resulting from a mixture consisting of diploid cells in G2 phase and tetraploid cells, and a small fraction (3 %) of 8C.

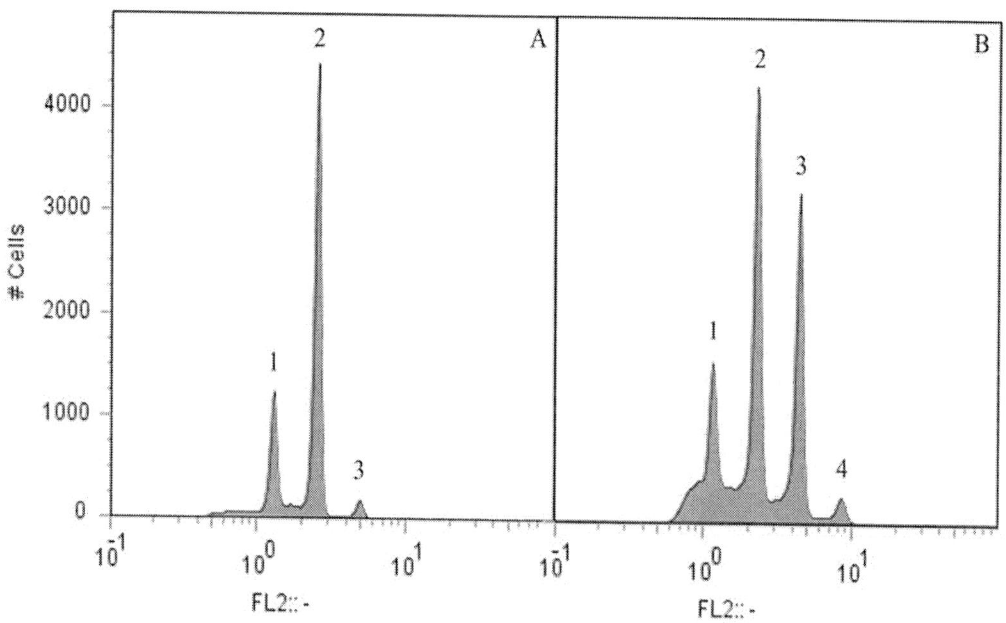

Figure 6. Histograms of genome of plant (A) and hairy root culture (B) of *B. vulgaris* cv. Egypt: 1- 2C peak; 2- 4C peak; 3- 8C peak; 4- 16C peak.

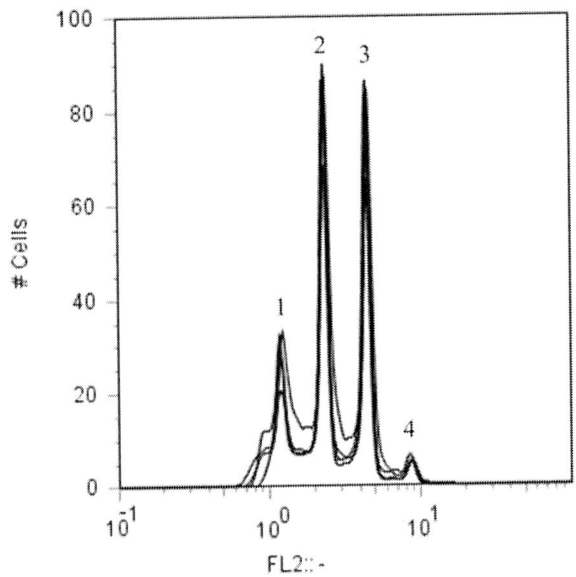

Figure 7. Overlays of four hairy root lines from *B. vulgaris* cv. Egypt: 1- 2C peak; 2- 4C peak; 3- 8C peak; 4- 16C peak.

This distribution of cells in different phases indicates that cells from leaf tissue undergo two cycles of endoreduplication. In contrast, the data from flow cytometric analyses of hairy roots showed that they undergo three cycles of endoreduplication (Figure 6B); the cells exhibited four different levels of ploidy. The majority of cells were 4C (36 %) and 8C (32 %). The fractions of 2C and 16C cells were 17 % and 4 %, respectively. Endoreduplication is a process of replication of the nuclear DNA content without cell division [9]. It is widespread in plants and may occur in any cell type except the gametes and the meristematic and guard cells [9]. Since this correlates with the stage of differentiation, it could serve as an indicator of differentiation [32; 52].

All four investigated hairy root lines exhibited similar patterns of ploidy levels, so that no significant changes in the DNA profiles were observed (Figure 7).

It is important to note that the differences in ploidy levels found in intact plants and hairy root culture (Figure 6) contradict the widespread hypothesis that hairy roots have the same genomic organization as the plants from which they were obtained [1; 4]. Moreover, these results support preliminary results obtained in our laboratories that transformed root lines from diploid *Datura stramonium* and *Hyoscyamus niger* plants were mixoploid, in contrast to the primary plant material from which the lines were obtained. It must be emphasised that we are at an early stage of investigating this phenomenon.

Flow cytometric investigations showed that the first DNA- peaks for both the leaf tissue and the hairy roots, corresponding to the diploid level, exhibited identical fluorescence intensities (Figure 6); this is in agreement with the observation that the size of the genome is not influenced by transformation to *in vitro* cultures [25]. For the determination of the genome size of *Beta vulgaris* cv. Egypt, intact plant- and hairy root- internal standardization was used [23]. A sample of known standard (in this particular case *Pisum sativum*, kindly provided by Dr. Dolezel, Czech Republic) is processed together with samples of unknown genome sizes, then the amounts of DNA in the unknown samples are determined by interpolation from the fluorescence signals generated from the known genome (Figure 8) using the following equation:

$$2C_{sample} = \frac{G1_{sample} * 2C_{standard}}{G1_{standard}},$$

where, $2C_{sample}$ is the amount of 2C DNA in the investigated sample (pg), $G1_{sample}$ is the mean DNA content of the sample's G1 peak, $C_{standard}$ is the amount of 2C DNA in the standard (pg) and $G1_{standard}$ is the standard G1 peak.

The genome sizes (2C-values) for the plant and the hairy roots all were determined to have values of 1.6 pg DNA. The values for the leaf tissue and the hairy roots confirm the results we have obtained previously for *Beta vulgaris* cv. Detroit Dark Red (Pavlov et al., unpublished results) as well as in published investigations of *Beta vulgaris* genome size [3; 103]; there were only minor differences, which are due to the different cultivars investigated. These values do not correspond to the C-values in the database of the Royal Botanic Garden, Kew [125]. In their database, the genome size of *Beta vulgaris* is recorded as 2.50 pg for the 2C value, which is almost twice the amount we determined. As consequence of numerous investigations mentioned in different laboratories, into the genome size of *Beta vulgaris* intact plants, as well as its *in vitro* cultures, we suggest that the data deposited in the database of the Royal Botanic Garden, Kew be brought up-to-date.

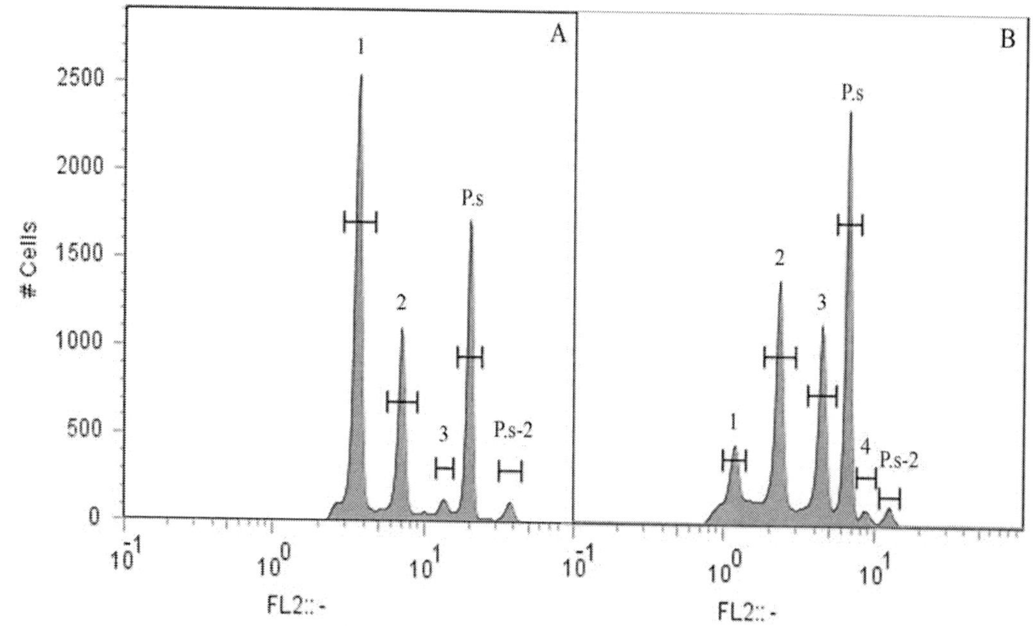

Figure 8. Flow cytometric analyses of the genome size of plant (A) and hairy roots (B) of *B. vulgaris* cv. Egypt: P.s- 2C peak of standard *Pisum sativum* cv. Ctirad (9.09 pg); P.s-2- 4C peak of *P. sativum* cv. Ctirad; 1- 2C peak; 2- 4C peak; 3- 8C peak; 4- 16C peak.

Conclusion

Hairy root culture is already in its third decade. Over this relatively short period, interest in transformed root cultures has steadily increased and has become focused on their enormous biosynthetic capacity. In addition, hairy root cultures have been used as a model system for the investigation of different strategies for the induction of metabolite yields, metabolic engineering and functional genomics. The high yields of economically important plant derived metabolites and/or proteins that have been obtained have already attracted private companies, suggesting that production at the laboratory scale will be exceeded in the near future. The ever-expanding knowledge of the theoretical mechanism of hairy root induction has allowed the transformation of an increasing number of plant species, making a significant contribution to the preservation of global biodiversity and the alleviation of ecological problems, connected with over-harvesting plant species for the supply of high-value plant-derived metabolites. Furthermore, the tremendous potential of transformed root cultures for phytoremediation will provide new tools for the elimination of contaminants from polluted soils and for environmental clean-up.

Acknowledgement

This work has been supported by a Marie Curie Fellowship from the European Community program "Development Host Fellowships" (HPMD-CT-2001-00092), Marie Curie European Reintegration Grant (MERG-CT-2007-046427), DAAD (D/05/01303) and the Bulgarian National Science Fund (D-01-84/2006).

References

[1] Aird, ELH; Hamill, JD; Rhodes, MJC. Cytogenetic analysis of hairy root cultures from a number of plant species transformed by *Agrobacterium rhizogenes*, *Plant Cell, Tissue and Organ Culture*, 1988 15, 47-57.

[2] Altamura, MM. *Agrobacterium rhizogenes rolB* and *rolD* genes: regulation and involvement in plant development. *Plant Cell, Tissue and Organ Culture*, 2004, 77, 89-101.

[3] Arumuganathan, K; Earle, ED. Nuclear DNA content of some important plant species. *Plant Molecular Biology Reporter*, 1991 9, 208-218.

[4] Baiza, AM; Quiroz-Moreno, A; Ruiz, JA; Loyola-Vargas, VM. Genetic stability of hairy root cultures of *Datura stramonium*. *Plant Cell, Tissue and Organ Culture*, 1999 59, 9-17.

[5] Bako, L; Umeda, M; Tiburcio, AF; Schell, J; Koncz, C. The VirD2 pilot protein of *Agrobacterium*-transferred DNA interacts with the TATA box-binding protein and a nuclear protein kinase in plants. *Proceedings of the National Academy of Sciences of the USA*, 2003 100, 10108-10113.

[6] Bhagwath, SG; Hjortsø, MA. Statistical analysis of elicitation strategies for thiarubrine A production in hairy root cultures of *Ambrosia artemisiifolia*. *Journal of Biotechnology*, 2000 80, 159-167.

[7] Bonhomme, V; Laurain-Mattar, D; Lacoux, J; Fliniaux, MA; Jacquin, A. Tropane alkaloid production by hairy roots of *Atropa belladonna* obtained after transformation with *Agrobacterium rhizogenes* 15834 and *Agrobacterium tumefaciens* containing rol A,B,C genes only. *Journal of Biotechnology*, 2000 81, 151-158.

[8] Brunaud, V; Balzergue, S; Dubreucq, B; Aubourg, S; Samson, F; Chauvin, S; Bechtold, N; Cruaud, C; DeRose, R; Pelletier, G; Lepiniec, L; Caboche, M; Lecharny, A. T-DNA integration into the *Arabidopsis* genome depends on sequences of pre-insertion sites. *EMBO Reports*, 2002 12, 1152-1157.

[9] Cebolla, A; Vinardell, JM; Kiss, E; Olah, B; Roudier, F; Kondorosi, A; Kondorosi, E. The mitotic inhibitor ccs52 is required for endoreduplication and ploidy dependent cell enlargement in plants. *EMBO Journal,* 1999 18, 4476-4487.

[10] Chilton, M-D. *Agrobacterium*. A memoir. *Plant Physiology*, 2001 125, 9-14.

[11] Cho, K; Fuqua, C; Winans, SC. Transcriptional regulation and locations of *Agrobacterium tumefaciens* genes required for complete catabolism of octopine. *Journal of Bacteriology*, 1997 179, 1-8.

[12] Choi, PS; Kim, YD; Chung, HJ; Choi, DW; Liu, JR. Plant regeneration from hairy-root cultures transformed by infection with *Agrobacterium rhizogenes* in *Catharanthus roseus*. *Plant Cell Reports*, 2004 22, 828-831.

[13] Christey, MC. Use of Ri-mediated transformation for production of transgenic plants. *In Vitro Cellular and Developmental Biology - Plant*, 2001 37, 687-700.

[14] Crane, C; Wright, E; Dixon, RA; Wang, Z-Y. Transgenic *Medicago truncatula* plants obtained from *Agrobacterium tumefaciens*-transformed root and *Agrobacterium rhizogenes*-transformed hairy roots. *Planta,* 2006 223, 1344-1354.

[15] Cunningham, SD; Ow, DW. Promises and prospects of phytoremediation. *Plant Physiology*, 1996 110, 715-719.

[16] Curtis, WR. Hairy roots, bioreactor growth. In: Spier, RE, editor. Encyclopedia of Cell Technology. New York: John Wiley and Sons; 2000; 827-841.

[17] Cseke, LJ; Cseke, SB; Podila, GK. High efficiency poplar transformation. *Plant Cell Reports*, 2007 26, 1529-1538.

[18] Dandekar, AM; Fisk, HJ. Plant transformation: *Agrobacterium*-mediated gene transfer. *Methods in Molecular Biology*, 2004 286, 35-46.

[19] Dehio, C; Grossmann, K; Schell, J; Schmulling, T. Phenotype and hormonal status of transgenic tobacco plants overexpressing the *rolA* gene of *Agrobacterium rhizogenes* T-DNA. *Plant Molecular Biology*, 1993 23, 1199-1210.

[20] Dessaux, Y; Petit, A; Tempe, J. Chemistry and biochemistry of opines, chemical mediators of parasitism. *Phytochemistry*, 1993 34, 31-38.

[21] Diop, MF; Hehn, A; Ptak, A; Chretien, F; Doerper, S; Gontier, E; Bourgaud, F; Henry, M; Chapleur, Y; Laurain-Mattar, D. Hairy root and tissue cultures of *Leucojum aestivum* L. - relationships to galanthamine content. *Phytochemisrty Reviews,* 2007 6, 137-141.

[22] Dolezel, J; Bartos, J. Plant DNA flow cytometry and estimation of nuclear genome size. *Annals of Botany*, 2005 95, 99-110.

[23] Dolezel, J; Greilhuber, J; Lucretti, S; Meister, A; Lysak, MA; Nardi, L; Obermayer, R. Plant genome size estimation by flow cytometry: Inter-laboratory comparison. *Annals of Botany*, 1998 82, 17-26.

[24] Dolezel, J; Greilhuber, J; Suda, J. Flow cytometry with plants: an Overview. In: Dolezel, J; Greilhuber, J; Suda, J, editors. Flow Cytometry with Plant Cells. Weinheim, Germany: Wiley-VCH Verlag GmbH and Co KGaA; 2007; 41-66.

[25] Dolezelova, M; Dolezel, J; Van den Houwe, I; Roux, N; Swennen, R. Focus on the Musa collection: Ploidy levels revealed. *InfoMusa,* 2005 14, 34-36.

[26] Doran, PM. Foreign protein degradation and instability in plants and plant tissue cultures. *Trends in Biotechnology*, 2006 24, 426-432.

[27] Eibl, R; Eibl, D. Design and use of the wave bioreactor for plant cell culture. In: Gupta SD, Ibaraki, Y, editors. Plant Tissue Culture Engineering. Focus on Biotechnology. Dordrecht, The Netherlands: Springer; 2006; 203-227.

[28] Eibl, R; Eibl, D. Disposable bioreactors for pharmaceutical research and manufacturing. In: Sorvari, S; Toldi, O, editors. Proceedings of Second International Congress on Bioreactor Technology in Cell, Tissue Cultures and Biomedical Applications. Turku, Finland: Karhukopio OY; 2006; 12-21.

[29] Eibl, R; Eibl, D. Design of bioreactors suitable for plant cell and tissue cultures. *Phytochemistry Reviews*, 2008, in press.

[30] Escobar, MA; Dandekar, AM. *Agrobacterium tumefaciens* as an agent of disease. *Trends in Plant Science,* 2003 8, 380-386.

[31] Flores, HE; Hoy, MW; Pickard, JJ. Secondary metabolites from root cultures. *Trends in Biotechnology,* 1987 5, 64-69.

[32] Francis, D. The plant cell cycle - 15 Years on. *New Phytologist*, 2007 174, 261-278.

[33] Furmanowa, M; Syklowska-Baranek, K. Hairy root cultures of *Taxus* x *media* var. *Hicksii* Rehd. as a new source of paclitaxel and 10-deacetylbaccatin III. *Biotechnology Letters*, 2000 22, 683-686.

[34] Gartland, JS. *Agrobacterium* virulence. In: Gartland, KMA; Davey, MR, editors. Methods in molecular biology, vol. 44: *Agrobacterium* protocols. Totowa, New Jersey: Humana Press Inc; 1995; 15-28.

[35] Gelvin, SB. *Agrobacterium*-mediated plant transformation: The biology behind the "Gene-Jockeying" tool. *Microbiology and Molecular Biology Reviews*, 2003 67, 16-37.

[36] Gelvin, SB. Improving plant genetic engineering by manipulating the host. *Trends in Biotechnology*, 2003 21, 95-98.

[37] Georgiev, M; Heinrich, M; Kerns, G; Bley, Th. Transformed root cultures from Devil's claw in a bubble column bioreactor: growth and secondary metabolites production, *Scientific Works UFT*, 2006 53, 242-247.

[38] Georgiev, M; Heinrich, M; Kerns, G; Pavlov, A; Bley, Th. Total iridoids and phenolics by transformed *Harpagophytum procumbens* root cultures. *Engineering in Life Sciences*, 2006 6, 593-596.

[39] Georgiev, M; Pavlov, A; Bley, Th. Betalains by transformed *Beta vulgaris* roots in stirred tank bioreactor: batch and fed-batch processes. In: Sorvari, S; Toldi, O, editors. Proceedings of Second International Congress on Bioreactor Technology in Cell, Tissue Cultures and Biomedical Applications. Turku, Finland: Karhukopio OY; 2006; 22-28.

[40] Georgiev, M; Pavlov, A; Bley, Th. Hairy root type plant in vitro systems as sources of bioactive substances. *Applied Microbiology and Biotechnology*, 2007 74, 1174-1185.

[41] Giri, A; Narasu, ML. Transgenic hairy roots: Recent trends and applications. *Biotechnology Advances*, 2000 18, 1-22.

[42] Givan, AL. Flow Cytometry an Introduction. In: Hawley, TS; Hawley, RG, editors. Methods in Molecular Biology, vol. 263: Flow Cytometry Protocols, 2nd Edition. Totowa, New Jersey: Humana Press; 2004; 2-31.

[43] Gonzales, PS; Capozucca, CE; Tigier, HA; Milrad, SR; Agostini, E. Phytoremediation of phenol from wastewater, by peroxidases of tomato hairy root cultures. *Enzyme and Microbial Technology*, 2006 39, 647-653.

[44] Goossens, A; Rischer, H. Implementation of functional genomics for gene discovery in alkaloid producing plants. *Phytochemistry Reviews*, 2007 6, 35-49.

[45] Gorpenchenko, TY; Kiselev, KV; Bulgakov, VP; Tchernoded, GK; Bragina, EA; Khodakovskaya, MV; Koren, OG; Batygina, TB; Zhuravlev, YN. The *Agrobacterium rhizogenes rolC*-gene-induced somatic embryogenesis and shoot organogenesis in *Panax ginseng* transformed calluses. *Planta*, 2006 22, 2323-2328.

[46] Greilhuber, J; Temsch, EM; Loureiro, JCM. Nuclear DNA content measurement. In: Dolezel, J; Greilhuber, J; Suda, J, editors. Flow Cytometry with Plant Cells. Weinheim, Germany: Wiley-VCH Verlag GmbH and Co KGaA; 2007; 67-99.

[47] Guillon, S; Tremouillaux-Guiller, J; Pati, PK; Rideau, M; Gantet, P. Hairy root research: recent scenario and exciting prospects. *Current Opinion in Plant Biology*, 2006 9, 341-346.

[48] Hahn, EJ; Yu, KW; Jeong, CS; Paek, KY. Commercial scale production of bioactive compounds through large-scale bioreactor culture of *Panax ginseng* C.A. Meyer. In: Sorvari, S, editor. Proceedings of the First International Congress on Bioreactor Technology in Cell, Tissue Culture and Biomedical Applications. Turku, Finland: Karhukopio OY; 2003; 30-40.

[49] Hamill, JD; Lidgett, AJ. Hairy root cultures - opportunities and key protocols for studies in metabolic engineering. In: Doran, PM, editor. Hairy roots culture and applications. Amsterdam, the Netherlands: Harwood Academic Publishers; 1997; 1-31.

[50] Hendrix, B; Stewart, JM. Estimation of the nuclear DNA content of *Gossypium* species. *Annals of Botany*, 2005 95, 789-797.

[51] Hopping, ME. Preparation and preservation of nuclei from plant-tissues for quantitative DNA analysis by flow-cytometry. *New Zealand Journal of Botany*, 1993 31, 391-401.

[52] Inze, D; De Veylder, L. Cell cycle regulation in plant development. *Annual Review of Genetics*, 2006 40, 77-105.

[53] Johnston, JS; Bennett, MD; Rayburn, AL; Galbraith, DW; Price, HJ. Reference standards for determination of DNA content of plant nuclei. *American Journal of Botany*, 1999 86, 609-613.

[54] Joubert, P; Beaupere, D; Lelievre, P; Wadouachi, A; Sangwan, RS; Sangwan-Norreel, BS. Effect of phenolic compounds on *Agrobacterium vir* genes and gene transfer induction - a plausible molecular mechanism of phenol binding protein activation. *Plant Science*, 2002 162, 733-743.

[55] Jouhikainen, K; Lindgren, L; Jokelainen, T; Hiltunen, R; Teeri, TH; Oksman-Caldentey, K-M. Enhancement of scopolamine production in *Hyoscyamus muticus* L. hairy root cultures by genetic engineering. *Planta*, 1999 208, 545-551.

[56] Kim, JH; Lee, YW; Yoo, YJ. Biosynthesis of SOD by carrot hairy roots. *Biotechnology Letters*, 1994 16, 581-586.

[57] Kim, JH; Yoo, YJE. Optimization of SOD biosynthesis by controlling sucrose concentration in the culture of carrot hairy root. *Journal of Microbiology and Biotechnology*, 2002 12, 617-621.

[58] Kim, O-T; Bang, K-H; Shin, Y-S; Lee, M-J; Jung, S-J; Hyun, D-Y; Kim, Y-C; Seong, N-S; Cha, S-W; Hwang, B. Enhanced production of asiaticoside from hairy root cultures of *Centella asiatica* (L.) Urban elicited by methyl jasmonate. *Plant Cell Reports*, 2007 26, 1941-1949.

[59] Kunik, T; Tzfira, T; Kapulnik, Y; Gafni, Y; Dingwall, C; Citovsky, V. Genetic transformation of *HeLa* cells by *Agrobacterium*. *Proceedings of the National Academy of Sciences of the USA*, 2001 98, 1871-1876.

[60] Lee, K-T; Suzuki, T; Yamakawa, T; Kodama, T; Igarashi, Y; Shimomura, K. Production of tropane alkaloids by transformed root cultures of *Atropa belladonna* in stirred bioreactors with a stainless steel net. *Plant Cell Reports*, 1999 18, 567-571.

[61] Le Flem-Bonhomme, V; Laurain-Mattar, D; Fliniaux, MA. Hairy root induction of *Papaver somniferum* var. *album*, a difficult-to-transform plant by *A. rhizogenes* LBA 9402. *Planta,* 2004 218, 890-893.

[62] Levieille, G; Wilson, G. *In vitro* propagation and iridoid analysis of the medicinal species *Harpagophytum procumbens* and *H. zeyheri*. *Plant Cell Reports*, 2002 21, 220-225.

[63] Li, W; Koike, K; Asada, Y; Yoshikawa, T; Nikaido, T. Rosmarinic acid production by *Coleus forskohlii* hairy root cultures. *Plant Cell, Tissue and Organ Cultures*, 2005 80, 151-155.

[64] Li, M; Leung, DWM. Root induction in radiata pine using *Agrobacterium rhizogenes*. *Electronic Journal of Biotechnology*, 2003 6, 251-258.

[65] Llosa, M; Zupan, J; Baron, C; Zambryski, P. The N- and C-terminal portions of the *Agrobacterium* VirB1 protein independently enhance tumorogenesis. *Journal of Bacteriology*, 2000 182, 3437-3445.

[66] Lorence, A; Medina-Bolivar, F; Nessler, CL. Camptothecin and 10-hidroxycamptothecin from *Camptotheca acuminate* hairy roots. *Plant Cell Reports*, 2004 22, 437-441.

[67] Ludwig-Mueller, J; Georgiev, M; Bley, Th. Metabolite and hormonal status of hairy root cultures of Devil's claw (*Harpagophytum procumbens*) in flasks and in a bubble column bioreactor. *Process Biochemistry*, 2008 43, 15-23.

[68] Mahomed, IM; Ojewole, JAO. Analgesic, antiinflammatory and antidiabetic properties of *Harpagophytum procumbens* DC (Pedaliaceae) secondary root aqueous extract. *Phytotherapy Research*, 2004 18, 982-989.

[69] McAfee, BJ; White, EE; Pelcher, LE; Lapp, MS. Root induction in pine (*Pinus*) and larch (*Larix*) spp. using *Agrobacterium rhizogenes*. *Plant Cell, Tissue and Organ Culture*, 1993 34, 53-62.

[70] McGregor, G; Fiebich, B; Wartenberg, A; Brien, S; Lewith, G; Wegener, T. Devil's claw (*Harpagophytum procumbens*): An anti-inflammatory herb with therapeutic potential. *Phytochemistry Reviews*, 2005 4, 47-53.

[71] Medina-Bolivar, F; Condori, J; Rimando, AM; Hubstenberger, J; Shelton, K; O'Keefe, SF; Bennett, S; Dolan, MC. Production and secretion of resveratrol in hairy root cultures of peanut. *Phytochemistry*, 2007 68, 1992-2003.

[72] Menzel, G; Harloff, H-J; Jung, C. Expression of bacterial poly(3-hydroxybutyrate) synthesis genes in hairy roots of sugar beet (*Beta vulgaris* L.). *Applied Microbiology and Biotechnology*, 2003 60, 571-576.

[73] Merritt, CD; Raina, S; Fedoroff, N; Curtis, WR. Direct *Agrobacterium tumefaciens*-mediated transformation of *Hyoscyamus muticus* hairy roots using green fluorescent protein. *Biotechnology Progress*, 1999 15, 278-282.

[74] Moyano, E; Palazon, H; Bonfill, M; Osuna, L; Cusido, RM; Oksman-Caldentey, K-M; Pinol, MT. Biotransformation of hyoscyamine into scopolamine in transgenic tobacco cell cultures. *Journal of Plant Physiology*, 2007 164, 521-524.

[75] Mukundan, UB; Bhide, V; Dawda, H. Production of betalains by hairy root cultures of *Beta vulgaris* L. In: Fu, T-J; Singh, G; Curtis, WR, editors. Plant Cell and Tissue Culture for the Production of Food Ingredients. New York, Boston, Dordrecht, London, Moscow: Kluwer Academic/Plenum Publishers; 1999; 121-127.

[76] Nilsson, O; Olsson, O. Getting to the root: The role of the *Agrobacterium rhizogenes rol*-genes in the formation of hairy roots. *Physiologia Plantarum*, 1997 100, 463-473.

[77] Oksman-Caldentey, K-M; Inze, D. Plant cell factories in the post-genomic era: new ways to produce designer secondary metabolites. *Trends in Plant Science*, 2004 9, 433-440.

[78] Palazon, J; Moyano, E; Cusido, RM; Bonfill, M; Oksman-Caldentey, K-M; Pinol, MT. Alkaloid production in *Duboisia* hybrid hairy roots and plants overexpressing the *h6h* gene. *Plant Science*, 2003 165, 1289-1295.

[79] Palazon, J; Mallol, A; Eibl, R; Lettenbauer, C; Cusido, RM; Pinol, MT. Growth and ginsenoside production in hairy root cultures of *Panax ginseng* using a novel bioreactor. *Planta Medica*, 2003 69, 344-349.

[80] Park, S-U; Facchini, PJ. *Agrobacterium rhizogenes*-mediated transformation of opium poppy, *Papaver somniferum* L., and California poppy, *Eschscholzia californica* Cham., root cultures. *Journal of Experimental Botany*, 2000 347, 1005-1016.

[81] Pavlov, A; Georgiev, M; Bley, Th. Batch and fed-batch production of betalains by Red beet (*Beta vulgaris*) hairy roots in a bubble column reactor. *Zeitschrift für Naturforschung*, 2007 62, 439-446.

[82] Pavlov, A; Georgiev, V; Ilieva, M. Betalain biosynthesis by Red beet (*Beta vulgaris* L.) hairy root culture. *Process Biochemistry*, 2005 40, 1531-1533.

[83] Pavlov, A; Georgiev, M; Panchev, I; Ilieva, M. Optimisation of rosmarinic acid production by *Lavandula vera* MM plant cell suspension in a laboratory bioreactor. *Biotechnology Progress*, 2005 21, 394-396.

[84] Pavlov, A; Kovatcheva, P; Georgiev, V; Koleva, I; Ilieva, M. Biosynthesis and radical scavenging activity of betalains during the cultivation of Red beet (*Beta vulgaris*) hairy root cultures. *Zeitschrift für Naturforschung*, 2002 57, 640-644.

[85] Pelczar, P; Kalck, V; Gomez, D; Hohn, B. *Agrobacterium* proteins VirD2 and VirE2 mediate precise integration of synthetic T-DNA complexes in mammalian cells. *EMBO Reports*, 2004 5, 632-637.

[86] Pietrosiuk, A; Sykłowska-Baranek, K; Wiedenfeld, H; Wolinowska, R; Furmanowa, M; Jaroszyk, E. The shikonin derivatives and pyrrolizidine alkaloids in hairy root cultures of *Lithospermum canescens* (Michx.) Lehm. *Plant Cell Reports*, 2006 25, 1052-1058.

[87] Rahman, L; Ikenaga, T; Kitamura, Y. Penicillin derivatives induce chemical structure-dependent root development, and application for plant transformation. *Plant Cell Reports*, 2004 22, 668-677.

[88] Ramakrishnan, D; Curtis, WR. Trickle-bed root culture bioreactor design and scale-up: growth, fluid-dynamics and oxygen mass transfer. *Biotechnology and Bioengineering*, 2004 88, 248-260.

[89] Ramakrishnan, D; Salim, J; Curtis, WR. Inoculation and tissue distribution in pilot-scale plant root culture bioreactors. *Biotechnology Techniques*, 1994 8, 639-644.

[90] Rhodes, MJC; Robins, RJ; Hamill, JD; Parr, AJ; Hilton, MG; Walton, NJ. Properties of transformed root cultures. In: Charlwood, BV; Rhodes, MJC, editors. Secondary products from plant tissue culture. New York: Oxford University Press; 1990; 201-226.

[91] Rieseberg, M; Kasper, C; Reardon, KF; Scheper, Th. Flow cytometry in biotechnology. *Applied Microbiology and Biotechnology*, 2001 56, 350-360.

[92] Rischer, H; Oresic, M; Seppanen-Laakso, T; Katajamaa, M; Lammertyn, F; Ardiles-Diaz, W; Van Montagu, MCE, Inze, D; Oksman-Caldentey, K-M; Goossens, A. Gene-to-metabolite networks for terpenoid indole alkaloid biosynthesis in *Catharanthus roseus* cells. *Proceedings of the National Academy of Sciences of the USA*, 2006 103, 5614-5619.

[93] Rolland, S; Jobic, C; Fevre, M; Bruel, C. *Agrobacterium*-mediated transformation of *Botrytis cinerea*, simple purification of monokaryotic transformants and rapid conidia-based identification of the transfer-DNA host genomic DNA flanking sequences, *Current Genetics*, 2003 44, 164-171.

[94] Rudrappa, T; Lakshmanan, V; Kaunain, R; Singara, NM; Neelwarne, B. Purification and characterization of an intracellular peroxidase from genetically transformed roots of red beet (*Beta vulgaris* L.). *Food Chemistry*, 2007 105, 1312-1320.

[95] Saito, K; Yamazaki, M; Murakoshi, I. Transgenic medicinal plants: *Agrobacterium*-mediated foreign gene transfer and production of secondary metabolites, *Journal of Natural Products*, 1992 55, 149-162.

[96] Saxena, G; Banerjee, S; Laiq-ur-Rahman; Verma, PC; Mallavarapu, GR; Kumar, S. Rose-scented geranium (*Pelargonium* sp.) generated by *Agrobacterium rhizogenes* mediated Ri-insertion for improved essential oil quality. *Plant Cell, Tissue and Organ Culture*, 2007 90, 215–223.

[97] Schapiro, HM. Practical Flow Cytometry. 4th Edition. New York: John Wiley and Sons; 2003.

[98] Schliemann, W; Kobayashi, N; Strack, D. The decisive step in betaxanthin biosynthesis is a spontaneous reaction. *Plant Physiology*, 1999 119, 1217-1232.

[99] Sevón, N; Oksman-Caldentey, K-M. *Agrobacterium rhizogenes*-mediated transformation: Root cultures as a source of alkaloids. *Planta Medica*, 2002 68, 859-868.

[100] Sharp, JM; Doran, PM. Strategies for enhancing monoclonal antibody accumulation in plant cell and organ cultures. *Biotechnology Progress*, 2001 17, 979-992.

[101] Sincock, SA; Robinson, JP. Flow cytometric analysis of microorganisms. *Methods in Cell Biology*, 2001 64, 511-537.

[102] Sirikantaramas, S; Sudo, H; Asano, T; Yamazaki, M; Saito, K. Transport of camptothecin in hairy roots of *Ophiorrhiza pumila*. *Phytochemistry*, 2007 68, 2881-2886.

[103] Sliwinska, E; Zielinska, E; Jedrzejczyk, I. Are seeds suitable for flow cytometric estimation of plant genome size?. *Cytometry* A, 2005 64, 72-79.
[104] Steiner, U; Schliemann, W; Böhm, H; Strack, D. Tyrosinase involved in betalain biosynthesis of higher plants. *Planta*, 1999 208, 114-124.
[105] Stewart, KM; Cole, D. The commercial harvest of devil's claw (*Harpagophytum* spp.) in southern Africa: The devil's in the details. *Journal of Ethnopharmacology*, 2005 100, 225-236.
[106] Sung, L-S; Huang, S-Y. Lateral root bridging as a strategy to enhance L-DOPA production in *Stizolobium hassjoo* hairy root cultures by using a mesh hindrance mist trickling bioreactor. *Biotechnology and Bioengineering*, 2006 94, 441-447.
[107] Suresh, B; Ravishankar, GA. Phytoremediation - a novel and promising approach for environmental clean-up. *Critical Reviews in Biotechnology*, 2004 24, 97-124.
[108] Suresh, B; Sherkhane, PD; Kale, S; Eapen, S; Ravishankar, GA. Uptake and degradation of DDT by hairy root cultures of *Cichorium intybus* and *Brassica juncea*. *Chemosphere*, 2005 61, 1288-1292.
[109] Thorup, JE; McDonald, KA; Jackman, AP; Bhatia, N; Dandekar, AM. Ribosome-inactivating protein production from *Trichosanthes kirilowii* plant cell cultures. *Biotechnology Progress*, 1994 10, 345-352.
[110] Towler, MJ; Wyslouzil, BE; Weathers, PJ. Using an aerosol deposition model to increase hairy root growth in a mist reactor. *Biotechnology and Bioengineering*, 2007 96, 881-891.
[111] Trick, HN; Finer, JJ. SAAT: sonication-assisted *Agrobacterium*-mediated transformation. *Transgenic Research*, 1997 6, 329-336.
[112] Tzfira, T; Citovsky, V. From host recognition to T-DNA integration: the function of bacterial and plant genes in the *Agrobacterium*–plant cell interaction. *Molecular Plant Pathology*, 2000 1, 201-212.
[113] Tzfira, T; Citovsky, V. The *Agrobacterium*-plant cell interaction. Taking biology lessons from a bug. *Plant Physiology*, 2003 133, 943-947.
[114] Tzfira, T; Citovsky, V. *Agrobacterium*-mediated genetic transformation of plants: Biology and biotechnology. *Current Opinion in Biotechnology*, 2006 17, 147-154.
[115] Valentine, L. *Agrobacterium tumefaciens* and the plant: the David and Goliath of modern genetics. *Plant Physiology*, 2003 133, 948-955.
[116] Vasilev, N; Elfahmi, BR; Kayser, O; Momekov, G; Konstantinov, S; Ionkova, I. Production of justicidin B, a cytotoxic arylnaphthalene lignan from genetically transformed root cultures of *Linum leoni*. *Journal of Natural Products*, 2006 69, 1014-1017.
[117] Veena, V; Taylor, CG. Agrobacterium rhizogenes: recent developments and promising applications. *In Vitro Cellular and Developmental Biology - Plant*, 2007 43, 383-403.
[118] Verpoorte, R; Contin, A; Memelink, J. Biotechnology for the production of plant secondary metabolites. *Phytochemistry Reviews*, 2002 1, 13-25.
[119] Ward, DV; Zambryski, PC. The six functions of *Agrobacterium* VirE2. *Proceedings of the National Academy of Sciences of the USA*, 2001 98, 385-386.
[120] Weathers, P; Bunk, G; McCoy, MC. The effect of phytohormones on growth and artemisinin production in *Artemisia annua* hairy roots. *In Vitro Cellular and Developmental Biology - Plant*, 2005 41, 47-53.

[121] Weller, SA; Stead, DE; Young, JPW. Acquisition of an *Agrobacterium* Ri plasmid and pathogenicity by other a-proteobacteria in cucumber and tomato crops affected by root mat. *Applied and Environmental Microbiology,* 2004 70, 2779-2785.
[122] Wilson, PDG. The pilot-scale cultivation of transformed roots. In: Doran, PM, editor. Hairy Roots and Culture Applications. Amsterdam, the Netherlands: Harwood Academic Publishers; 1997; 179-190.
[123] Winans, SC. Two-way chemical signaling in *Agrobacterium*-plant interactions. *Microbiological Reviews*, 1992 56, 12-31.
[124] Wink, M; Alfermann, AW; Franke, R; Wetterauer, B; Distl M; Windhoevel, J; Krohn, O; Fuss, E; Garden, H; Mohagheghzadeh, A; Wildi, E; Ripplinger, P. Sustainable bioproduction of phytochemicals by plant *in vitro* cultures: Anticancer agents. *Plant Genetic Resources,* 2005 3, 90-100.
[125] www.rbgkew.org.uk/cval/homepage.html.
[126] Wysokinska, H; Lisowska, K; Floryanowicz-Czekalska, K. Transformation of *Catalpa ovata* by *Agrobacterium rhizogenes* and phenylethanoid glycosides production in transformed root cultures. *Zeitschrift für Naturforschung*, 2001 56, 375-381.
[127] Xie, D; Zou, Z; Ye, H; Li, H; Guo, Z. Selection of hairy root clones of *Artemisia annua* L. for artemisinin production. *Israel Journal of Plant Science,* 2001 49, 129-134.
[128] Yibrah, HS; Gronroos, R; Lindroth, A; Franzen, H; Clapham, D; von Arnold, S. *Agrobacterium rhizogenes*-mediated induction of adventitious rooting from *Pinus contorta* hypocotyls and the effect of 5-azacytidine on transgene activity. *Transgenic Research*, 1996 5, 75-85.
[129] Yun, D-J; Hashimoto, T; Yamada, Y. Metabolic engineering of medicinal plants: transgenic *Atropa belladonna* with an improved alkaloid composition. *Proceedings of the National Academy of Sciences of the USA*, 1992 89, 11799-11803.

Reviewed by: Dr. S. Berkov, Departament de Productes Naturals, Biologia Vegetal i Edafologia, Facultat de Farmàcia, Universitat de Barcelona, Av. Joan XXIII s/n, 08028 Barcelona, Spain.

In: Genetically Modified Plants
Editors: Timm Wolf and Jonas Koch

ISBN 978-1-60456-696-3

Chapter 6

GENETICALLY MODIFIED PLANTS CONTAINING PLANT-DERIVED GENES FOR BROAD SPECTRUM INSECT CONTROL TO REDUCE MYCOTOXINS: BIOACTIVE PROTEINS

Patrick F. Dowd[1]*, Eric T. Johnson[1], T. Scott Pinkerton[1], and Stephen R. Hughes[2]

1.Crop Bioprotection Research Unit, National Center for Agricultural Utilization Research, USDA, Agricultural Research Service, 1815 N. University St., Peoria, IL, USA

2.Bioproducts and Biocatalysis Research Unit, National Center for Agricultural Utilization Research, USDA, Agricultural Research Service, 1815 N. University St., Peoria, IL, USA

ABSTRACT

Mycotoxins are acutely toxic or carcinogenic compounds produced primarily by *Aspergillus* and *Fusarium* molds that infect seeds of high oil content in the field, such as maize, cottonseed, peanuts, wheat and tree nuts. Damage by insects facilitates entry of the molds, and maize hybrids that express the *Bacillus thuringiensis* (Bt) crystal protein at sufficiently high levels to provide near immunity to the European corn borer can have greatly reduced levels of mycotoxins such as fumonisin in the presence of these pests. However, the number of different species of ear-damaging insect pests requires many different Bts, and the recent reports of field collected European corn borers with high levels of resistance to Bt CryIAb suggest nonBt strategies for broad spectrum insect control may be valuable in the future.

Our approach toward solving the mycotoxin problem involves the investigation of plant-derived genes, alone and in combination with each other, that have the potential to provide high levels of resistance to insects, including those genes that can be used as selectable markers. We have investigated directly active proteins, such as ribosome-inactivating proteins and *N*-acetyl hexosaminidases, and demonstrated significant effects against both beetles and caterpillars when these genes were expressed transgenically. Broad spectrum resistance to species in several different insect orders has been noted for plants producing enhanced levels of peroxidases, including in the field. A model anti-cell-death protein enhanced insect resistance when expressed in transgenic plants.

* Corresponding author, phone: 309-681-6242, FAX 309-681-6593, E-mail: Patrick.Dowd@ars.usda.gov.

Robotics has been used to develop antiinsectan peptides with enhanced oral activity over original forms. We have recently begun to introduce combinations of genes for directly active proteins in different functional classes, and have identified plants with greatly enhanced resistance over that seen with introductions of individual genes.

INTRODUCTION - THE MYCOTOXIN PROBLEM

Mycotoxins are toxins produced by different species of fungi, primarily molds, that are acutely toxic and/or carcinogenic [CAST 2003]. Several are regulated by many countries, both internally and as imports. The most highly regulated mycotoxin is aflatoxin in its various forms, as it is one of the most potent carcinogens known. Acceptable contamination levels are often in the range of single digit parts per billion or less. Aflatoxin has been attributed as the cause of incidences of human liver cancer in certain areas of the world through epidemiological studies, and its carcinogenicity has been demonstrated in feeding studies on several animals [Wang and Tang 2005]. Aflatoxin can be a problem in peanuts, tree nuts, figs, cottonseed, and maize; and is a more frequent problem in tropical and semitropical areas. Another widespread toxin is fumonisin, which is produced by several species of *Fusarium* fungi, including *Fusarium verticillioides* and *F. proliferatum*. Although less active than aflatoxins, fumonisins are also carcinogens. Additionally, they can interfere with sphingolipid metabolism in some animals, such that brain liquefaction can occur in horses, and neural tube defects can occur in human embryos [Wu et al. 2005]. Fumonisin can be found in maize essentially worldwide.

Although many environmental factors can contribute and interact to promote problems with mycotoxins, insects are one indirect factor that has been recognized, at least with the mold association, for around 100 years [Dowd 1998]. A large diversity of types of insects have been reported to be associated with enhanced mycotoxin levels, including insects in the orders Coleoptera (beetles), Lepidoptera (caterpillars), Hemiptera (true bugs), and Thysanoptera (thrips) that attack maize kernels. Additionally, insect feeding that stresses the plant can also potentially aggravate the problem, including that by root-feeding insects such as rootworms or wireworms. Although the relative importance of insect damage was somewhat controversial for many years, the commercial development and release of maize (and to a lesser extent cotton) plants expressing different forms of the Bt crystal protein provided material that could be tested to better determine the relative importance of insect damage in promoting the mycotoxin problems in maize. Under natural conditions, and when the protein is expressed throughout the plant at levels promoting near immunity to target pests (such as the European corn borer, *Ostrinia nubilalis*), maize expressing various Bt constructs has had several fold less fumonisin compared with nonBt isogenic lines under conditions of high infestation by the target insect [Dowd et. 2005]. Since then, a variety of other Bt genes, targeting several different insect pests, have been placed into maize, but they have not been as effective against the target pest, and as would be expected, have not provided the same degree of reduction in mycotoxins under isogenic conditions with high target pest levels [Dowd et al. 2005]. An additional problem with Bt corn is the potential to develop resistance by the target insect not only to the introduced Bt gene protein form, but also to other forms [e.g. Shelton et al. 2002]. As Bt genes targeting different insect species continue to be combined, or "stacked" to a greater extent in plants, the potential for widespread resistance increases [e.g. Ferre' and

Van Rie 2002], although protein modification may help with some forms of resistance [Sobero'n et al. 2007]. The first Bt maize was released and planted in the US in 1996, and resistant field- collected individuals were first reported about 10 years later [Siegfried et al. 2007]. Additional concerns are voiced by a significant consumer population, in Europe and elsewhere, about the "foreignness" of a bacterial gene being introduced into a food plant, as well as the selectable markers conferring resistance to antibiotics [NRC 2000].

Potential Candidate Insect Resistance Proteins

The essential physiological processes involved in basic metabolism, neurotransmission, digestion, growth, and reproduction are all potential targets in insects for plant defensive proteins. There are known plant proteins that can potentially affect many of these processes. For example, digestion and absorption can be affected by proteins that inhibit digestive enzymes (protease inhibitors), and block absorption (lectins). These roles may also be played by enzyme products (such as those of peroxidases, discussed below), or secondary metabolites (discussed by Johnson et al., 2008) which bind to nutrients and inhibit degradation and/or absorption. Other potentially active proteins or peptides which have not yet been described from plants but that are known to occur from microbial sources include cholesterol oxidase inhibitors and neuroactive cyclic peptides [Dowd 2002a].

Physical properties of plant tissues may deter or prevent insects from feeding on them (e.g. Sadasivam and Thayumanavan 2003). These properties may include smooth surfaces that prevent egg adhesion or larval grips and penetration, rough or hairy surfaces that prevent sucking insects from anchoring and then puncturing tissues, or material that is too tough or gummy for insects to chew through it. Proteins may be involved in indirectly producing gummy environments (such as the polyphenol oxidases in trichomes that polymerize contents into sticky aphids traps [e.g. Medeiros et al. 2005]), or tougher tissues (such as lignifying peroxidases). Extensins are proteins that toughen tissue. An advantage of these types of proteins involved in forming physical barriers is that the U.S. EPA may register them more quickly as less review is required (provided they naturally occur in food products)[Hagmann 1999].

Once the insect egg hatches (or live young are brought forth), the phytophagous insect must begin feeding on the plant tissue, and risk being eaten or parasitized if it moves around. Various chemical cues are used by the insect to initiate feeding, and protein components are one of the cues. The insect must then digest the material using various proteases, amylases, lipases and related enzymes. The digested material must then be absorbed. Plant proteins can act as both digestion and absorption inhibitors, as well as degrade essential amino acids. Enzymes from plants or insect pathogens that affect the insect gut lining (peritrophic membrane) or gut cells themselves include those that degrade particular components, such as chitin (chitinases, hexosaminidases), protein components (proteases), and mucin (viral enhancin). Removal of these barriers can cause naturally toxic components in the insect's diet to more readily penetrate to target sites (such as gut tissue) more effectively.

Insects also have general intracellular basic metabolic processes, as well as those unique to insects, that can be targeted by plant defensive proteins provided they can penetrate to the target site. Traditionally, insecticidal compounds that penetrate to intracellular target sites (such as the nervous system) are soluble in acetone. Acetone-soluble cyclic peptides (typically microbially derived) that are highly active against insects include roseotoxin B, valinomycin and destruxins [Dowd 2002a]. However, there are studies that demonstrate relatively large non-lipophilic proteins can penetrate insect cells [Fishman and Zlotkin 1984, Casartelli et al. 2005] and retain function [Casartelli et al. 2005]. Examples of plant defensive proteins that appear to be capable of penetrating to intracellular target sites based on mode of action are the ribosome-inactivating proteins, which occur widely in plants. Because of the nature of insects, they have particular physiological processes that may be useful targets, such as the production of uric acid, the storage of sugar as trehalose, and enzymes and hormones involved in the molting process.

Insects have specific target sites for plant resistance proteins that they share in common with fungal or other plant pathogens, such as chitin. When resistance genes have been mapped, it has not been uncommon for insect and fungal resistance genes to map to the same sites [McMullen and Simcox 1995]. From a evolutionary standpoint, it would be cost effective for plants to make individual defensive proteins that would be effective against as many potential pests as possible. Thus, introducing "insect" resistance proteins may also confer resistance to plant pathogens, and vice versa. Utilization of these types of multitarget pest organism proteins becomes more attractive when attempting to solve the mycotoxin problem through use of resistant transgenic plants, considering that it is fungal pathogens that make mycotoxins.

Gene mapping studies have also typically indicated that resistance to plant pests is due to multiple genes that presumably code for proteins directly or indirectly affecting different target sites [e.g. McMullen and Simcox 1995]. Studies with insecticides have indicated that when multiple products affecting different target sites are used, resistance development can be delayed [e.g. Gould 1998, Shelton et al. 2002]. This concept has led (perhaps obviously) to the concept of combining, or "pyramiding" resistance genes in plants for effective, stable resistance. Questions remain as to how regulatory agencies will interpret introduction of multiple genes affecting different target sites in the same target organism. Perhaps reintroduction of genes from the target plant itself in a constitutive or inducible manner will help overcome potential regulatory concerns. The critical determination will be which genes to employ so that the desired level of resistance will occur, and to use as few as possible, and to make sure that they will not interfere with each other. For example, a combination of genes coding for a protease and another active protein may result in the degradation of the second gene product by the protease.

If effective resistance alleles are not available, identification of active protein derivatives that might otherwise develop over long time frames through natural evolutionary processes can potentially be sped up through man-made molecular means. For example, a Bt gene was recently modified so that calmodulin binding was not necessary, resulting in efficacy against insect strains resistant to the original protein form [Soberón et al. 2000]. More advanced molecular evolution may result in the discovery of protein forms that are much improved and more insect-specific compared to the original forms.

The following discussion will elaborate on the different functional classes of proteins just mentioned. We admittedly will not be comprehensive and tend to emphasize our own work in these areas due to greater familiarity. We will also include other examples that appear relevant, due to their prior efficacy or presence in plants such as maize (which is our ultimate target plant), and ones that potentially can contribute to mycotoxin control through acting on both insects and plant pathogens. We will also provide some thoughts on new opportunities for discovery of plant genes coding for insect resistance proteins.

PROTEINS ACTING AS PHYSICAL BARRIERS

The number of proteins known to act as physical barriers to insects or pathogens is relatively limited, although the toughening of plant tissues through lignification catalyzed by peroxidases has been recognized for many years (Dowd and Lagrimini 1997b). One example of a type of protein that does appear to be a direct participant as a physical barrier are the extensins. These cell wall proteins typically contain high levels of less common amino acids, such as hydroxyproline, and are highly glycosylated [Wilson and Fry 1986] Multiple forms can occur in different plant species, including maize [Kieliszewski et al. 1990, 1992, Murphy and Hood 1993]. They can be cross-linked through leucine or tyrosine residues [Heid et al. 2004]. This cross-linking appears to be performed by peroxidases that may be specific and induced by wounding [Brownleader et al. 1995, Schnabelrauch et al. 1996]. Although no insect studies appear to have been performed with plants expressing high levels of extensin, *Arabidopsis thaliana* plants engineered to overexpress the EXT1 extensin gene had smaller lesions that did not spread compared to wild type plants when challenged with a virulent strain of *Pseudomonas syringe* [Wei and Shirsat 2006]. Because more insect resistant tomato fruit, which appears to be due to tougher skin, was produced by expression of tobacco anionic peroxidase in tomato [Dowd et al. 1998b - see below], plants that overexpress extensins may also be more resistant to insects.

PROTEINS INTERFERING WITH INSECT NUTRITION

A variety of proteins have been identified as being involved in antinutritive defenses in plants [e.g. Felton and Gatehouse 1996]. Protease inhibitors have been studied to a great extent in transgenic plant systems, and were one of the earlier gene types introduced into plants for insect resistance [Reeck et al. 1997]. However, it was found that Lepidoptera larvae would soon express proteases resistant to the protease inhibitors, thereby circumventing this resistance strategy. Multiple forms of the protease/amylase inhibitors are known for plants such as maize [e.g. Malehorn et al. 1994], so it may be possible that forms that cannot be circumvented by the insect's digestive process may be discovered and effectively utilized. One such example of a protease inhibitor that did not stimulate protease production in *Helicoverpa armigera* was recently reported from *Cicer arietinum* [Srinivasan et al. 2005]. Transgenic expression of protease inhibitors in plants can also increase resistance to fungi such as *Trichoderma reesei* [Malehorn et al. 1994] and *Fusarium verticillioides* [Quilis et al. 2007]. Plants that expressed arginase were more resistant to the tobacco hornworm (*Manduca sexta*) [Chen et al. 2005].

Lectins are carbohydrate-binding proteins that are widely distributed in the plant kingdom, including in food plants such as beans and grains [Pusztai 1991, Sadasivam and Thayumanavan 2003]. Lectins presumably interfere with absorption of nutrients by binding to sites where absorption occurs, although some are capable of passing through the human small intestine [Pusztai 1991]. Some are toxic to humans or animals (especially those from non-food plants), while others are allergens [Pusztai 1991].

A series of studies with snowdrop (*Galanthus nivalis*) lectin has demonstrated activity against a wide variety of insects including sucking insects, and to a more limited extent, caterpillars [e.g. Loc et al. 2002]. Grain lectins such as those from barley and wheat have been studied against insects as well [Czapla 1997]. Although not as active as snowdrop lectin, significant reductions in feeding, and in some cases, mortality, have been reported in assays involving insects and grain lectins [Balasubramaniam et al. 1991, Hopkins and Harper 2001]. Some enhanced resistance to European corn borers, corn rootworms [Czapla 1997] and aphids [Kanrar et al. 2002] has been reported for plants transformed to express different grain lectins. Using wheat germ lectin (as a model protein to investigate functional activity of this class of grain lectins), we have also found activity against caterpillar and beetle pests [Dowd 2002b, Dowd et al. 2006b]. Our initial studies with transformed tobacco expressing the WGA gene driven by a CaMV 35S promoter have indicated some plants that appear to express the protein based on antibody assays will significantly reduce feeding by corn earworms within a 2-3 day period [Johnson et al. unpublished].

Proteins that Degrade Insect Barriers

The insect cuticle (skin) is a complicated layered tissue composed of many different structural components, often crosslinked, that includes protein, chitin, lipid, and other factors [Hepburn 1985]. The cuticle lines the gut of the insect, and forms the peritrophic membrane in the midgut, where much digestion and absorption occurs. This membrane serves to protect the underlying gut cells from unwanted physical, chemical, and biochemical disruption. Because insect microbial pathogens typically need to penetrate the cuticle in order to gain entry into insect tissues, many studies have investigated the enzymes that these pathogens use [Dowd 2002a]. Some effective resistance has been engineered into insect pathogens [Narayanan 2002, Hayakawa et al. 2000]] or plants [Hayakawa et al. 2004] utilizing the genes coding for these enzymes [Hayakawa et al. 2000, 2004]. Chitinases (see below), are one class of enzymes produced by insect pathogens that has been investigated in transgenic plant systems, and insect-derived chitinases (used during the molting process) have also been engineered into plants and demonstrated some effective increases in resistance [Ding et al. 1998, Fitches et al. 2004b]. There are a wide diversity of chitinases produced by individual plant species [Collinge et al. 1993], including those reported from maize [e.g. Nucere et al. 1991, Zhe-Fu et al. 1992, Wu et al. 1994, Moore et al. 2004].

Comparable enzymes to those produced by insect pathogens or the insects themselves occur in plants, but few have been found to be active against insects so far. A cysteine protease from maize enhanced resistance to insects when expressed in plant tissue [Pechan et al. 2000]. Several studies have demonstrated that plants that overexpress different plant or fungal degradative enzymes (such as chitinases or glucanases) have enhanced resistance to various plant pathogens, although it appears combining different ones can result in greatly enhanced (or synergized) resistance to fungi such as *Rhizoctonia solani* compared to introduction of the individual genes [Jach et al. 1995]. Although chitinases are prevalent in plants, it was not until recently that an endochitinase active against insects (Colorado potato beetle, *Leptinotarsa decemlineata*) was reported from poplar [Lawrence and Novak 2006]. The lack of activity of plant chitinases against insects has been attributed to their having a pH optimum that is too low for them to be active in the alkaline gut of most caterpillars [Gongora and Broadway 2002]. The poplar chitinase was found to be active against a leaf-feeding beetle, which typically have a neutral gut pH [Purcelle et al. 1992], so it may be that the lack of reports on plant chitinase activity against insects is because they have only been tested on Lepidoptera (which are much easier to use in bioassays than beetle larvae). Another potential disadvantage in using chitinases for insect (or plant pathogen) control is that high expression of microbial chitinases can alter plant growth parameters [Gongora and Broadway 2002, Bolar et al. 2000].

Our experiences with a plant "exochitinase" began with examination of activity of some commercially available N-acetyl-hexosaminidases (NAHAs), although it is generally assumed this form is less likely to be active on its own compared to endochitinases. A NAHA from jack bean (*Canavalia ensiformis)* was particularly active, causing significant mortality of neonate fall armyworm (*Spodoptera frugiperda*) larvae at 75 ppm in a few days [Dowd et al. 2007a]. A NAHA from *Aspergillus niger* was much less active [Dowd et al. 2007a]. Once the *Arabidopsis* genome was sequenced, several putative NAHAs became available. We found that one apparent full length clone that was available from the *Arabidopsis* stock center did have the three regions that are typically conserved by NAHAs in other organisms [Peterbauer et al. 1996]. This clone was adapted and inserted into plasmid pAHC25 [Christensen and Quail, 1996] after removal of the GUS gene, and multiple series of callus types were transformed biolistically, some lines of which were regenerated into plantlets [Dowd et al. 2007a]. Even at the highest expression levels of over 100 ppm jack bean NAHA equivalents in antibody assays, no abnormal plant growth associated with expression levels was noted. Mortality of cigarette beetle (*Lasioderma serricorne*) larvae fed black Mexican sweet corn (BMS) callus expressing the *Arabidopsis* NAHA was significantly higher than that of larvae fed GUS or an inactive NAHA-transformant callus. Significantly smaller fall armyworm larvae were also noted when fed on the active NAHA callus vs. the inactive form. Both Hi-II and Oh43 NAHA transformed callus were regenerated to plants and evaluated for insect resistance. Due to the small size of the plants and plans to cross with B73, the tassel of the Hi-II plants was used in bioassays. In two separate batches of plant assays, plant tassels from NAHA transformants causing high mortality compared to limited mortality of GUS plants toward fall armyworms were noted. Overall mortality was significantly negatively correlated with NAHA expression levels detected enzymatically. For Oh43 regenerated plants, again some transformant plants expressing the gene caused high mortality compared to limited mortality of null transformants, and leaf disk feeding levels were significantly negatively correlated with expression levels detected with antibody (higher phenolic content of Oh43 compared to Hi-II plants apparently interfered with the enzyme assay used with Hi-II plants). A fungal NAHA expressed in plants enhanced resistance to the fungal pathogen *Venturia inaequalis*, and resistance was greatly enhanced when both a NAHA and chitinase were used together [Bolar et 2001]. We continue to explore the potential utility of plant NAHAs for plant resistance to insects and plant pathogens.

PROTEINS THAT AFFECT INTRACELLULAR TARGETS

A large variety of proteins can affect different intracellular targets. Potentially any enzyme, receptor, regulatory protein, or structural protein could be targeted. Potential targets are described in a number of reviews, [e.g. Christou et al. 2006, Ferry et al. 2006]. Despite this potential, these defensive proteins must still need to be able to penetrate the insect gut to reach their ultimate target site. We will discuss two types of proteins that have some unique properties and/or appear to be active against multiple insect groups.

Ribosome-inactivating proteins (RIPs) are site-specific *N*-glycosidases that inactivate ribosomes by removing a single adenine from large rRNA [Stirpe et al. 1992, Barbieri et al. 1993, Nielsen and Boston 2001]. They are produced by a variety of plants, including those that are used for human food or related uses [Stirpe et al. 1992, Barbieri et al. 1993, Nielsen and Boston 2001]. Their toxicity to vertebrates is highly variable. Ricin, from castor beans (*Ricinus communis*) is very toxic (as it contains a cell binding chain as well as a toxic lectin chain), but the forms from edible grains such as wheat, barley, and maize are not reported to be toxic [Stirpe et al. 1992]. Initial tests of some from dicots such as castor bean, soapwort (*Saponaria officinalis*) and bitter gourd (*Momordica cochinchinenis*) showed toxicity to beetles examined [Gatehouse et al. 1990], but not caterpillars [Gatehouse et al. 1990, Brandhorst et al. 1996].

The maize endosperm RIP is rather unusual among RIPs in that it requires removal of both terminal and internal peptides to produce two large peptides that associate noncovalently to the active form [Walsh et al. 1991, Bass et al 1992]. It can be activated *in vivo* by thiol protease treatment such as with papain [Walsh et al. 1991, Bass et al. 1992]. Interestingly, beetles typically contain thiol proteases in their guts [Murdock et al. 1987]. When activated RIP was added to insect diets at naturally occurring levels of 1000 ppm, significant reductions in feeding, and in some cases significant mortality, was noted for a non-maize feeder (cabbage looper, *Trichoplusia ni*) and maize-feeding caterpillars such as the corn earworm (*Helicoverpa zea*), European corn borer, and fall armyworm after 2 days of feeding by neonate larvae. Although no effect was noted on larvae of the Freeman sap beetle *Carpophilus freemani*), both pro- and active RIP significantly deterred feeding by adult Freeman sap beetles, dusky sap beetles (*Carpophilus lugubris*), maize weevils (*Sitophilus zeamais*) and strawberry sap beetles (*Stelidota geminata*). This same antifeedant activity was noted for a fungal RIP (restrictocin) which was also toxic to Freeman sap beetle larvae [Brandhorst et al. 1996].

The gene for an activated form of maize RIP (MOD1) [Krawetz and Boston 2000] driven by a 35S CaMV promoter was introduced into tobacco using *Agrobacterium* Ti plasmid [Dowd et al. 2003]. Two transformation events appeared to have good expression of the protein, and progeny (R2) plants were evaluated in several insect assays. Greenhouse studies indicated one of the lines was significantly more resistant to greenhouse whiteflies (*Trialeurodes vaporariorum*) as mortality was significantly greater for leaf caged individuals compared to the second line and wild type plants [Dowd et al. 2003]. Resistance of the transformed lines to the tobacco hornworm (*Manduca sexta*) and cigarette beetle (*Lasioderma serricorne*) larvae was variable, but some plants within an individual line were significantly more resistant to both insect species compared to others. Seed was obtained from these more resistant R2 plants, and progeny (R3) were again evaluated for insect resistance. Multiple sets of assays on two different ages of plants using corn earworms indicated significantly higher mortality in some cases, and significantly reduced feeding in most cases, of both transgenic maize RIP lines compared to wild type plants. Feeding damage by *H. zea* was significantly negatively correlated with RIP levels for both lines. For cigarette beetles, although no significant mortality differences were noted, feeding rates were often significantly lower for both lines, and feeding damage was significantly negatively correlated with RIP levels for one line, and the same trend was noted for the other line. Although expression levels, (which ranged up to 100 ppm) were lower than would naturally be present in maize kernels, apparently other plant factors were sufficient to interact and cause significant resistance with maize RIP present compared to when the RIP was introduced into more optimized insect diets during earlier assays [Dowd et al. 1998c]. However, the maize RIP was active against both a representative beetle (cigarette beetle), caterpillar (corn earworm), and sucking/rasping insect (whiteflies); indicating a potential broad-spectrum use. It is possible that other plant-derived RIPs may have activity against multiple groups of insects, as the fungal RIP restrictocin is also active against some caterpillar and beetle larvae [Brandhorst et al. 1996], and has shown activity against insects when expressed in *Aspergillus nidulans* as well [Brandhorst et al. 2000]. Maize RIP is active against *Aspergillus flavus* [Nielsen et al. 2001] and other plant RIPS are active in transgenic plant systems against pathogens such as tobacco mosaic virus [Desmyter et al. 2003] and *Fusarium culmorum* [Balconi et al. 2007].

The regulation of cell numbers in multicellular organisms is important to normal growth and development. Programmed cell death (apoptosis) is a progression of complicated events. Both insects [Narayanan 2002, Clem 2005] and plants [e.g. Heath 2000] will trigger cells to die prematurely when they are invaded by pathogens. Species of insect pathogenic viruses that have a relatively wide host range (such as *Autographa californica* MNPV) produce proteins such as p35 that inhibit capsases, a family of proteases involved in the signal cascade that initiates programmed cell death [Stennicke et al. 2002]. *In vitro* feeding assays with p35 indicated significant negative effects on caterpillars when tested at 100-300 ppm, both in the presence and absence of virus that is considered both pathogenic (to corn earworms) and nonpathogenic (to fall armyworms)[Pinkerton et al. 2008].

The gene for p35 was inserted into the maize ubiquitin driven pAHC25 vector after the GUS gene was removed, and the plasmid was introduced into maize line Oh43 by biolistic transformation [Pinkerton et al. 2008]. Mortality above 20% was noted for some T_0 plants for both corn earworms and fall armyworms [Pinkerton et al. 2008]. Significant feeding reductions by caterpillars were noted for several transformant lines, and reductions in feeding were significantly negatively correlated with the relative strength of the mRNA signal (the protein was not detectible by antibody, as has been reported for other plant expression systems [Lincoln et al. 2002]). Significant feeding reductions were noted in studies with T_1 plants that contained vs. didn't contain the selectable marker (as an indicator of transgene presence). When both caterpillar species were fed on leaves coated with *Ag*MNPV (which is considered an effective pathogen of corn earworms but not fall armyworms [Gröner 1986]), mortality was significantly enhanced for individuals of both species that were fed on positive selectable marker vs. negative selectable marker leaves.

We presume that the enhanced toxicity of the transformed plants to the two caterpillar species tested was due to activation of unknown species of latent virus that we found was present in some moribund individuals of both insect colonies. Activation of latent virus infections occurs after exposure to various environmental stress factors [Burand et al. 1986].

However, because programmed apoptosis appears important to proper insect growth and development [Hepburn 1985, Lockshin 1985] we cannot rule out at this point that the p35 is also interfering with the natural apoptotic process in the insects and it is by this means (also) that toxicity results.

Insect viruses also attack beetles and other insect groups, but different classes of viruses are more common in some orders of insects than others [Adams and Bonami 1991]. MNPVs primarily attack Lepidoptera (caterpillars), and to a lesser extent Diptera (flies) and Hymenoptera (bees and wasps) [Adams and Bonami 1991], so it would be interesting to see what effects p35 has on other groups of insects in the presence and absence of different viruses. Expression of p35 protein and other antiapoptotic proteins in plants has resulted in enhanced resistance to a number of plant pathogens, including *Alternaria alternata* [Lincoln et al. 2002] tobacco mosaic virus [del Pozo and Lam 2003] and *Xanthomonas axonopodis* [Scandiucci de Freitas et al. 2007].

Proteins that Produce Multiple Resistance Factors

There are a number of proteins groups that might be considered to produce multiple resistance factors. Among these groups of proteins are the regulatory proteins capable of turning on pathways that produce multiple resistance factors (such as protease inhibitors or pathways leading to bioactive secondary metabolites). Examples of overexpression of regulatory proteins are described in the chapter by Johnson et al. [2008]. Overexpression of proteins leading to the production of regulatory secondary metabolites such as salicylic acid or jasmonic acid, in combination with the disruption or downregulation of pathways that catabolize those compounds, is another potential example.

Another group of proteins that can potentially produce multiple resistance factors are oxidative enzymes such as polyphenoloxidases, tyrosinases, and peroxidases. All of these enzymes can potentially produce reactive quinones from the appropriate substrates, with the peroxidases being the most versatile [Dowd and Lagrimini 1997b]. Additionally, reactive oxygen species may be produced, and quinones may further complex or cause reactive cascades to form quinones from other molecules [Dowd and Lagrimini 1997b]. There is an extremely wide range of potential substrates for peroxidases [Gaspar et al. 1982]. Quinones may then bind with proteins, causing enzyme inhibition (if the bound protein is an enzyme) or resistance to nutrient digestion or absorption (if the bound material is a nutrient)[Felton et al. 1992, Dowd and Lagrimini 1997b]. Cross linking and polymerization reactions can also occur to form tougher plant tissue, as has been discussed earlier.

Initial indirect studies involving the combination of peroxidases with different materials *in vitro* were somewhat contradictory, as in some cases peroxidases appeared to increase insect resistance, and in other cases, reduce resistance [Dowd and Lagrimini 1997b]. This is perhaps not surprising, as activity would be dependent on which peroxidases are involved, and which substrates are available for them to react with. Maize peroxidases appear to have a wide range of capabilities, but also some substrate "preferences" [Dowd 1994]. Maize secondary metabolites acted on by peroxidases can have varying toxicity to both insects [Dowd and Vega 1996] and the plant pathogen *Fusarium graminearum* [Dowd et al. 1997].

Studies with insects as affected by transgenic plants expressing high levels of tobacco anionic peroxidase (TOBAP) involved examining several different plant varieties and species, allowing for some assessment in different plant secondary metabolite contexts. Laboratory studies were followed by some field tests with one of the high expressing tobacco lines, which allowed for some evaluations under more natural conditions and led to the realization that this peroxidase could lead to relatively broad spectrum increases in insect resistance that did not adversely affect beneficial insects present. Ultimately this gene was introduced into maize and led to the development of lines that were highly resistant to some insect pests, including in field studies.

Initial evaluations involved tobacco lines that both under-expressed and over-expressed the TOBAP (up to 4-fold higher levels of tobacco anionic peroxidase activity were noted in leaves). Leaf disks from underexpressing lines were generally more damaged, and overexpressing lines, less damaged by corn earworms [Dowd and Lagrimini 1997a]. A similar trend was noted for intact plants. Sections of stem browned rapidly when cut (an indicator of high peroxidase activity due to formation of colored quinones and coproducts), and were also more resistant to feeding by corn earworms [Dowd and Lagrimini 1997b, Dowd et al. 2000]. Field tests on one of the most consistently resistant tobacco lines were also run. Significantly enhanced resistance was noted to caterpillars, grasshoppers, whiteflies, and especially aphids [Dowd and Lagrimini 2006]. Fewer live aphids were present, and significantly higher percentages of aphids present were dead, on the high peroxidase compared to wild type plants on multiple sample dates. Predator presence (primarily stilt bugs - Hemiptera: Berytidae, and syrphid larvae - Diptera: Syrphidae) were not adversely affected. Significant associations were noted between aphid numbers and numbers of syrphid larvae on both wild type and high peroxidase plants even though live aphid numbers were significantly lower on high peroxidase plants. Laboratory assays have also noted no significant negative effect of *Af*MNPV virus efficacy on corn earworms fed high peroxidase tobacco or tomato compared to wild type plants [Behle et al. 2002].

This gene was also introduced into a commercial tomato variety, resulting in very high levels of total peroxidase activity (over 100 fold for leaves and fruit compared to wild type plants). Again, leaves were generally significantly more resistant to feeding by first instar corn earworms, as well as tomato hornworms (*Manduca sexta*) compared to wild type plants, although larger larvae fed equally well on both types of leaves [Dowd et al. 1998b]. Significantly higher numbers of corn earworm larvae died when fed on TOBAP compared to wild type tomato fruit. Significantly higher mortality of some first instar corn earworms, and dusky sap beetle adults was noted when fed on stem sections of high peroxidase compared to wild type plants in some cases, and the percentage of high peroxidase stems penetrated by dusky sap beetle adults was significantly lower in all cases. High peroxidase fruit was more resistant to feeding by both neonate corn earworms. Significantly higher mortality was noted for neonate corn earworm larvae that fed on fruit of high peroxidase vs. wild type plants. Stem sections were also more resistant to both insect species. Abrasion (toughness) tests run with equivalently aged ripe tomato fruit indicated that the skin of the high peroxidase tomatoes was about 8-fold more resistant to abrasive puncture compared to the wild type fruit [Dowd and Lagrimini, unpublished].

Even in a more distantly related dicot plant, sweetgum (*Liquidambar styraciflua*), insect resistance was noted in plants that expressed the TOBAP, where soluble peroxidase activity was increased about 10-fold over wild type plants in full-sized leaves [Dowd et al. 1998a].

Enhanced resistance was noted to eastern tent caterpillars (*Malacosoma americana*), fall webworms (*Hyphantria cunea*), gypsy moths (*Lymantria dispar*), cigarette beetles, and European corn borers (not a natural pest). Significantly higher mortality of fall webworms occurred and feeding damage was 30-fold less on TOBAP expressing compared to wild type sweet gum leaves; feeding damage was significantly negatively correlated with the level of peroxidase activity. Gypsy moth larvae were significantly smaller and ate significantly less leaf material when fed on the TOBAP expressing compared to wild type sweetgum leaves. However, high peroxidase leaves were more susceptible to feeding by corn earworm larvae (not a natural pest)[Dowd et al. 1998b].

The TOBAP gene was also introduced into a monocot plant. When expressed in maize, some TOBAP expressing initial transformants killed 100% of corn earworm and European corn borer larvae. Transformed maize plants also produced low mortality of fall armyworms, but significantly reduced growth of larvae in laboratory studies performed with T_0 plants [Privalle et al. 1999]. Little or no activity was noted against beet armyworms (*Spodoptera exigua*) or black cutworms (*Agrotis ipsilon*) fed T_0 plant material [Privalle et al. 1999]. Field tests of progeny of the initial transformants that were inoculated with 300 neonate European corn borers per plant had little or no damage when the insects were applied at whorl or tassel stage, whereas corresponding plants that did not express the tobacco anionic peroxidase were damaged so heavily that they all died [Privalle et al. 1999].

All of the work with TOBAP, especially that involving tobacco itself, suggests that manipulation of source-plant expression of appropriate resistance genes (which naturally occurs in resistant varieties as constitutive or inducible resistance responses), may be the most predictably effective and environmentally benign means to enhance plant resistance to insects or pathogens. We have begun to investigate some maize peroxidases, and there are indications that some of these peroxidases are involved in insect and disease resistance, including to *Aspergillus flavus* [Dowd and Johnson 2005, Dowd et al. 2007b].

Molecular Optimization of Resistance Proteins

Until sequencing and expression studies are more complete and until 3-D predictive folding and substrate/target site models are developed for the many potential insect resistance proteins, we will remain relatively uncertain of the amino acid sequence requirements for activity. One method that has been utilized to develop more active forms of proteins has been referred to as "molecular evolution". Past strategies used in molecular evolution have involved a modified PCR methodology that is relatively slow, costly, and non-specific (referred to as "error prone" PCR) [Maynard et al. 2002, Hughes et al. 2006]; and "gene shuffling", which works best for families of closely related genes where longer polypeptides can be used [Stemmer 1995]. Recent advances at our facility have involved the creation of a robotic screening system that can be used to exhaustively modify all, or a subset, of the nucleic acids in the sequence, manufacture the expression vector, introduce the vector into an expression system, and then utilize the source organism or the expressed protein to evaluate efficacy [Hughes et al. 2007].

Bioactive peptides are attractive molecules for testing different molecular evolutionary strategies due to their small size. Plants can produce bioactive peptides such as thionins [Oard and Enright 2006] and some protease inhibitors [Reeck et al. 1997] that are active against different organisms, but they are not as widespread as from other sources. Predatory and parasitic arthropods often produce neuroactive peptides, several of which are very selective toward insects. We selected an insect-specific 25 amino acid peptide produced by the harmless Carolina wolf spider, *Lycosa carolinensis*, referred to as lycotoxin-1, for studies in molecular evolution. This peptide acts by imbedding itself in cell membranes and forming pores that dissipate ion and voltage gradients, resulting in toxicity to insects such as the house fly *Musca domestica*, and bacteria such as *Escherichia coli* [Yan and Adams 1998]. When this happens to insect muscles, nerve cells depolarize and the insect becomes paralyzed [Yan and Adams 1998].

After evaluating a number of bioassay methods, it was found that feeding the peptide-expressing yeast to the larvae as point sources in plates gave a reliable and sensitive means to distinguish between activities of different forms of the peptide [Hughes et al. 2007](although presently unavailable at NCAUR, automated larval dispensing units utilized in insecticide-biotechnology companies could be readily adapted to the system for a total automation on the platform). Initially a number of peptide-expressing lines were combined. Active combinations were then subdivided into individual strains in order to identify active vs. inactive sequences. Additionally, the sequence for the polyHIS tag used to assist in purification of the peptide was separated from the active peptide sequence using a trypsin-labile sequence so that the tag would not interfere with biological activity (which has been noted with RIP proteins in our laboratory) and the peptide would not be cleaved off until it reached the insect gut. A series of orally active peptides was found, which had changes that were still consistent with the pharmacokinetic properties needed to form pores in targeted cells [Hughes et al. 2008]. One of these sequences was selected for expression studies in tobacco, and introduced using Ti plasmid. Some plants which contain the gene and appear to express the protein have caused significant mortality/feeding reductions with corn earworms, and there may be some activity against cigarette beetle larvae as well [Johnson et al. unpublished].

Effective Protein Combinations

Many plant resistance proteins do not cause 100% mortality of insects at naturally occurring concentrations. As discussed earlier, the analogy with insecticide use suggests that plant proteins can be combined to gain this level of mortality, while at the same time producing a more stable resistance gene package compared to using highly active individual genes. Again, this appears to be what the plants do naturally, as multiple resistance loci for resistance to individual organisms typically occur, and fortunately these loci often overlap for different species of organisms [e.g. McMullen and Simcox 1995]. The difficult part is determining which gene combinations to use so that particular impediments can be most effectively overcome and the fewest number of genes will be needed. Different types of protease inhibitors have been combined and produced synergistic activity against the cowpea weevil (*Callosobruchus maculatus*)[Amirhusin et al. 2007]. In other cases, highly effective "intracellular target site" proteins may be available from individual genes, but if the peritrophic membrane is the limiting factor for penetration, use of a second protein to eliminate the barrier in combination with an intracellular active protein may result in more effective control. This is in fact one of the types of combinations that has been proven most useful with microbial genes so far. There are several examples where an insect active microbial chitinase has been combined with another active protein, such as Bt [Kramer et al. 1997] or others [Dowd 2002b, Dowd et al. 2006b] to yield synergistic activity. In one study, combination of a chitinase with a Bt protein also greatly delayed the time it took to develop resistance to the Bt protein [Zhao et al. 1999]. In addition, the proteins must not interfere with one another's activity. Obviously, if a proteolytic enzyme is used, it should be specific enough so that it does not degrade another resistance protein that is also being used in combination. Of course, when new genes are introduced, their products are potentially interacting with other resistance proteins already being naturally produced by the plant. There may be a greater advantage in not just using proteins, but in using directly active proteins plus proteins that code for active secondary metabolites. However, interference between the insect-active plant secondary metabolite group referred to as condensed tannins, and the Bt protein has been reported [Navon et al. 1993], so this strategy also has to be examined carefully. Hydrolytic enzymes may act on some unexpected substrates when expressed at high enough levels [Dowd and Sparks 1988].

Our initial *in vitro* assays indicated that there were indeed some nonobvious combinations of proteins that needed to be avoided [Dowd 2002b]. Because of the wide range of potential bioactive products produced by peroxidases, we had some reluctance in trying to combine it with other genes, despite information that indicated effects against insects were most likely due to the active quinones or other oxidized products produced (as discussed above). However, when TOBAP overexpressing tobacco plants were crossed with maize RIP tobacco plants, we found significant feeding reductions, and in some cases, significantly enhanced mortality to both corn earworms and cigarette beetles compared to wild type plants [Dowd et al. 2006a]. The degree of feeding reduction was significantly negatively correlated with levels of RIP or peroxidase activity individually (indicating no synergistic activity). Because gene levels were halved compared to homozygous parents, activity was thought to be additive as it was comparable to that noted with the individual plants in the past [Dowd et al. 2006a]. Our recent work has involved combining genes based on the functional categories described above [Dowd 2002b], and we have produced two gene combination plants that appear to have proteins that interact synergistically in enhancing insect resistance, as activity is much greater than in plants expressing the individual genes [Dowd et. al. 2006b].

Again, some of the best examples of effective gene combinations have been in transgenic plant studies involving plant pathogens such as *Rhizoctonia solani* [Jach et al. 1995] and *Venturia inaequalis* [Bolar et al. 2001]. Through utilization of genes that produce proteins with activity against both insects and plant pathogens, we can potentially have the best of both worlds.

FUTURE DIRECTIONS

Chimeric proteins that bind to tissues can allow for penetration of attached active proteins. Practical plant examples of this include chitinases that have lectin-domains that facilitate binding, combining the nontoxic beta chain of ricin with other proteins [Mehlo et al.

2005] and use of bioactive lectins themselves in combination with other proteins [Zhu-Salzman et al. 2003, Fitches et al. 2004a].

Genomic sequencing allows for the potential to discover analogs of compounds or proteins active against insects that presently have unrecognized or unproven functions in plants, but that ultimately may correspond to those previously known only from microbial sources. Comparative genome work allows for the discovery of function of genes that may be present, but of unknown function, until a resistance association is found. Once genomes are known, it then also becomes possible to look for and test expressed allelar gene products for resistance activities. Relatively minor changes in leader sequence can affect cellular localization of resistance proteins [Dombrowski et al. 1993], and thus the other factors with which they may interact. Recent information indicates that even "silent" changes in nucleic acid sequence can lead to folding, and ultimately functional, changes [Kimchi-Safarty et al. 2007]. This information suggests that unexpected opportunities to discover more effective resistance alleles may exist. High-throughput methodology to examine and optimize gene sequences provides an alternative to enhancing gene product activity if natural variants of gene alleles are not effective enough.

There is opportunity to discover gene forms that are active against common targets of both insects and the pathogens that cause mycotoxins. Examples already available include peroxidases [Dowd and Johnson 2005] and RIPs [Dowd et al. 1998c, 2003, Nielsen et al. 2001]. When combined through crossing two lines of transgenic tobacco expressing each gene individually, resistance appeared to be additively enhanced [Dowd et al. 2006a]. Further challenges will include determining which genes to combine in the fewest number (for regulatory purposes), but the most effective manner to match the benchmark set for European corn borer control in maize using the CryIAb forms commercialized such as events Mon810 and Bt11. It is not inconceivable that these genes will all be discovered from the target plant for insect resistance, but the challenge will be discovering the most effective alleles.

Conclusion

A variety of genes encoding plant proteins that confer resistance to various insect species have been introduced into plants. Some of these proteins are active against multiple orders of insects. In order to promote highly effective resistance, it is likely that genes coding for different resistance proteins will need to be combined, and the challenge will be to determine which combinations should be used. Combining genes from different functional classes may be one solution to this quandry. Genome sequencing and array technology are likely to assist in the discovery of suitable insect resistance genes, and molecular evolution techniques can be used to further selectively modify resistance genes. Based on available information, in the near future it should be possible to select genes that code for resistance to both insects and fungi associated with mycotoxin problems, thereby ultimately resulting in higher yields and safer foods for humans and animals.

Acknowledgements

We thank Fernando E. Vega and Arthur K. Weissinger for critical reviews of earlier drafts of the manuscript; and our various collaborators over many years, without whose help our results would not have been possible.

Disclaimer

Names are necessary to report factually on available data; however, the USDA neither guarantees nor warrants the standard of the product, and the use of the name by the USDA implies no approval of the product to the exclusion of others that may be suitable.

References

Adams, J.R., and Bonami, J.R. (1991). *Atlas of Invertebrate Viruses*. Boca Raton, FL: CRC Press.

Amirhusin, B., Shade, R.E., Koiwa, H. Hasegawa, P.M., Bressan, R.A., Murdock, L.L. and Zhu-Salzman, K. (2007). Protease inhibitors form several classes work synergistically against *Callosobruchus maculatus*. *J. Insect Physiol.*, *53*, 734-740.

Balasubramaniam, N.K., Czapla, T.H., and Guruaj-Rao, A. (1991). Structural and functional changes associated with cyanogen bromide treatment of wheat germ agglutinin. *Plant Physiol. Biochem.*, *288*, 374-379.

Balconi, C., Lanzanova, C., Conti, E., Truilzi, T., Forlani, F., Cattaneo, M., and Lupotto, E. (2007). *Fusarium* head blight evaluation in wheat transgenic plants expressing the maize b-32 antifungal gene. *Eur. J. Plant Pathol.*, *117*, 129-140.

Barbieri, L., Batelli, M.G. and Stirpe, F. (1993). Ribosome-inactivating proteins from plants. *Biochem. Biophys. Acta.*, *1154*, 237-282.

Bass, H.W., Webster, C., O'Brian, G.R., Roberts, J.K.M. and Boston, R.S. (1992). A maize-ribosome-inactivating protein is controlled by the transcriptional activator *Opaque-2*. *Plant Cell*, *4*, 225-234.

Behle, R.W., Dowd, P.F., Tamez-Guerra, P. and Lagrimini, L.M. (2002). Effect of transgenic plants expressing high levels of a tobacco anionic peroxidase on the toxicity of *Anagrapha falcifera* nucleopolyhedrovirus to *Helicoverpa zea* (Lepidoptera: Noctuidae). *J. Econ. Entomol.*, *95*, 81-88.

Bolar, J.P., Norelli, J.L., Harman, G.E., Brown,S.K., and Aldwinckle, H.S. (2001). Synergistic activity of endochitinase and exochitinase from *Trichoderma atroviride* (*T. harzianum*) against the pathogenic fungus (*Venturia inaequalis*) in transgenic apple plants. *Transgenic Res.*, *10*, 533-543.

Bolar, J.P., Norelli, J.L., Wong, K.-W., Hayes, C.K., Harman, G.E., and Adwinckle, H.S. (2000). Plant growth and development influenced by transgenic insertion of bacterial chitinolytic enzymes. *Mol. Breed.*, *9*, 72-77.

Brandhorst, T.T., Dowd, P.F., and Kenealy, W.R. (1996). The ribosome-inactivating protein restrictocin deters insect feeding on *Aspergillus restrictus. Microbiology, 142*, 1551-1556.

Brandhorst, T.T., Dowd, P.F. and Kenealy, W.R. (2000). The effect of fungal ribosome inactivating proteins upon feeding choice in *C. freemani*, and indications of a mutualistic relationship with *A. restrictus. Mycopathologia, 152*, 155-158.

Brownleader, M.D. Ahmed, N., Trevan, M., Chapplin, M.F. and Dey, P.M. (1995). Purification and partial characterization of tomato extensin peroxidase. *Plant Physiol., 109*, 1115-1123.

Burand, J.P., Kawanshi, C.Y., and Huang, Y.S. (1986). Persistent baculovirus infections. In R.R. Granados and B.A. Federici (Eds.) *The Biology of Baculoviruses: Volume I Biological Properties and Molecular Biology* (pp. 160-173). Boca Raton, FL: CRC Press.

Casartelli, M., Corti, P., Leonardi, M.G., Fiandra, L., Burlini, N., Pennacchio, F., and Giordana, B. (2005). Absorption of albumin by the midgut of lepidoptera larva. *J. Insect Physiol., 51*, 933-940.

CAST (Council Agricultural Science and Technology). (2003). *Mycotoxins: Risks in Plant and Animal Systems. Task force Report No. 38.* Ames, IA: CAST.

Chen, H., Wilkerson, C.G., Kuchar, J.A. Phinney, B.S., and Howe, G.A. (2005). Jasmonate-inducible plant enzymes degrade essential amino acids in the herbivore midgut. *Proc. Natl. Acad. Sci. USA, 102*, 19237-19242.

Christou, P., Capell, T., Kohli, A., Gatehouse, J.A. and Gatehouse, A.M.R. (2006). Recent developments and future prospects in insect pest control in transgenic crops. *Trends Plant Sci., 11*, 302-308.

Christensen, A.H. and Quail, P.H. (1996). Ubiquitin promotor-based vectors for high-level expression of selectable and/or screenable marker genes in monocotyledonous plants. *Transg. Res., 5*, 213-218.

Clem, R.J. (2005). The role of apoptosis in defense against baculovirus infection in insects. *CTMI, 289*, 113-130.

Collinge, D.B., Kragh, K.M., Mikkelsen, J.D., Nielsen, K.K., Rasmussen, U., and Vad, K. (1993). Plant chitinases. *Plant. J., 3*, 31-40.

Czapla, T.H. (1997). Plant lectins as insect control proteins in transgenic plants. In N. Carozzi and M. Koziel (Eds.). *Advances in Insect Control: The Role of Transgenic Plants* (pp. 123-138). Bristol, PA: Taylor and Francis.

Dahiya, N., Tewari, R., and Hoondal, G.S. (2006). Biotechnological aspects of chitinolytic enzymes: A review. *Appl. Microbiol. Biotechnol., 71*, 773-782.

Desmyter, S., Vandenbussche, F., Hao, Q., Proost, P., Peumans, W.J., and vanDamme, E.J.M. (2003). Type-1 ribosome-inactivating protein from iris bulbs: a useful agronomic tool to engineer virus resistance? *Plant Mol. Biol., 51*, 567-576.

Ding, X., Gopalakrisnan, B., Johnson, L.B., White, F.F., Wang, X., Morgan, T.D., Kramer, K.J. and Muthukrishnan, S. (1998). Insect resistance of transgenic tobacco expressing an insect chitinase gene. *Transgenic Res., 7*, 77-84.

Dombrowski, J.E., Schroeder, M.R., Bednarek, S.Y., and Raikhel, N.V. (1993). Determination of the functional elements within the vacuolar targeting signal of barley lectin. *Plant Cell., 5*, 587-596.

Dowd, P.F. (1994). Enhanced maize (*Zea mays* L.) pericarp browning: Associations with insect resistance and involvement of oxidizing enzymes. *J. Chem. Ecol., 20*, 2777-2803.

Dowd, P.F. (1998). Involvement of arthropods in the establishment of mycotoxigenic fungi under field conditions. In K.K. Sinha and D. Bhatnagar (Eds.). *Mycotoxins in Agriculture and Food Safety* (pp. 307-350). New York, NY: Marcel Dekker.

Dowd, P.F. (2002a). Antiinsectan compounds derived from microorganisms. In O. Koul and G.S. Dhaliwal (Eds.) *Advances in Biopesticide Research* (pp. 13-116). Amsterdam: Harwood Academic Publishers.

Dowd, P.F. (2002b). A functional approach to more effective multigenic host plant pest resistance. *Mycopathologia. 155*, 46.

Dowd, P.F., Duvick, J.P. and Rood, T. (1997). Comparative toxicity of allelochemicals and their enzymatic oxidation products to maize fungal pathogens, emphasizing *Fusarium graminearum*. *Nat. Toxins*, *5*, 180-185.

Dowd, P.F., Herms, D.A., Berhow, M.A. and Lagrimini, L.M. (2000). Mechanisms of insect resistance in transgenic plants (over)expressing a tobacco anionic peroxidase. 1999. International Symposium for Plant Peroxidases. *Plant Peroxidase Newsl.*, *14*, 93-101.

Dowd, P.F., Holmes, R.A., Pinkerton, T.S., Johnson, E.T., Lagrimini, L.M. and Boston, R.S. (2006a). Relative activity of a tobacco hybrid expressing high levels of a tobacco anionic peroxidase and maize ribosome-inactivating protein against *Helicoverpa zea* and *Lasioderma serricorne*. *J. Agric. Food Chem.*, *54*, 2629-2634.

Dowd, P.F. and Johnson, E.T. (2005). Association of a specific cationic peroxidase isozyme with maize stress and disease resistance responses, genetic identification, and identification of a cDNA coding for the isozyme. *J. Agric. Food Chem.*, *53*, 4464-4470.

Dowd, P.F., Johnson, E.T., and Pinkerton, T.S. (2007a). Oral toxicity of β-*N*-acetyl hexosaminidase to insects. *J. Agric. Food Chem.*, *55*, 3421-3428.

Dowd, P.F., Johnson, E.T. and Pinkerton, T.S. (2007b). Allelar variants and substrate specificities of a maize anionic peroxidase potentially involved in insect resistance. Proceedings, Entomological Society of America Annual Meeting #D0672.

Dowd, P.F., Johnson, E.T., Pinkerton, T.S., Behle, R.W., and Berhow, M.A. (2006b). Examination of genes newly introduced into plants for control of corn insects associated with mycotoxin problems. Proceedings, Multicrop Aflatoxin, Fumonisin and Fungal Genomics Workshop. J.F. Robens, ed. USDA, Agricultural Research Service, Beltsville, MD. p.89

Dowd, P.F., Johnson, E.T. and Williams, W.P. (2005). Strategies for insect management targeted toward mycotoxin management. In H.K. Abbas (Ed.) *Aflatoxin and Food Safety* (pp. 517-541). Boca Raton, FL: Taylor and Francis.

Dowd,P.F. and Lagrimini, L.M. (1997a). Examination of different tobacco (*Nicotiana* spp.) types under- and overexpressing tobacco anionic peroxidase for their leaf resistance to *Helicoverpa zea*. *J. Chem. Ecol.*, *23*, 2357-2370.

Dowd, P.F. and Lagrimini, L.M. (1997b). The role of peroxidase in host insect defenses. In N. Carozzi and M. Koziel (Eds.). *Advances in Insect Control: The Role of Transgenic Plants* (pp. 195-223). Bristol, PA: Taylor and Francis.

Dowd, P.F. and Lagrimini, L.M. (2006). Examination of the biological effects of high anionic peroxidase production in tobacco plants grown under field conditions. I. Insect pest damage. *Transgenic Res.*, *15*, 197-204.

Dowd, P.F., Lagrimini, L.M. and Herms, D.A. (1998a). Differential leaf resistance to insects of transgenic sweetgum (*Liquidambar styraciflua*) expressing tobacco anionic peroxidase. *Cell. Mol. Life Sci.*, *54*, 712-720.

Dowd, P.F., Lagrimini, L.M. and Nelsen, T.C. (1998b). Relative resistance of transgenic tomato tissues expressing high levels of tobacco anionic peroxidase to different insect species. *Nat. Toxins*, *6*, 241-249.

Dowd, P.F., Mehta, A.D., and Boston, R.S. (1998c). Relative toxicity of the maize endosperm ribosome-inactivating protein to insects. *J. Agric. Food Chem.*, *46*, 3775-3779.

Dowd, P.F., and Sparks, T.C. (1988). A comparison of properties of *trans*-permethrin hydrolase and leucine aminopeptidase from the midgut of *Pseudoplusia includens* (Walker). *Pestic. Biochem. Physiol. 31*, 195-202.

Dowd, P.F. and Vega, F.E. (1996). Enzymatic oxidation products of allelochemicals as a basis for resistance against insects: Effects on the corn leafhopper, *Dalbulus maidis*. *Nat. Toxins*, *4*, 85-91.

Dowd, P.F., Zuo, W.-Z., Gillikin, J.W., Johnson, E.T. and Boston, R.S. (2003). Enhanced resistance to *Helicoverpa zea* in tobacco expressing an activated form of maize ribosome-inactivating protein. *J. Agric. Food Chem.*, *51*, 3568-3574.

Felton, G.W., Donato, K.K., Del Vecchio, R.J. and Duffey, S.S. (1992). Impact of oxidized plant phenolics on the nutritional quality of dietary protein to a noctuid herbivore, *Spodoptera exigua*. *J. Insect Physiol.*, *38*, 277-285.

Felton, G.W. and Gatehouse, J.A. (1996). Antinutritive plant defence mechanisms. In M.I. Lehane and P.F. Billingsley (Eds.) *Biology of the Insect Midgut* (pp. 373-416). London: Chapman and Hall.

Ferré, J. and Van Rie, J. (2002). Biochemistry and genetics of insect resistance to *Bacillus thuringiensis*. *Annu. Rev. Entomol.*, *47*, 501-535

Ferry, N., Edwards, M.G., Gatehouse, J., Capell, T., Christou, P. and Gatehouse, A.M.R. (2006). Transgenic plants for insect control: a forward looking scientific perspective. *Transgenic Res.*, *15*, 13-19.

Fishman, L. and Zlotkin, E. (1984). A diffusional route of transport of horseradish peroxidase through the midgut of a fleshfly. *J. Exper. Zool.*, *229*, 189-195.

Fitches, E., Edwards, M.G., Mee, C., Grishin, E., Gatehouse, A.M.R., Edwards, J.P., and Gatehouse, J.A. (2004a). Fusion proteins containing insect-specific toxins as pest control agents: snowdrop lectin delivers fused insecticidal spider venom toxin to insect hemolymph following oral ingestion. *J. Insect Physiol.*, *50*, 61-71.

Fitches, E., Wilkinson, H., Bell, H., Brown, D.P., Gatehouse, J.A., and Edwards, J.P. (2004b). Cloning, expression and functional characterization of chitinase from larvae of tomato moth (*Lacanobia oleracea*): a demonstration of the insecticidal activity of insect chitinase. *Insect Biochem. Mol. Biol.*, *34*, 1037-1050.

Gaspar, T., Penel, C., Thorpe, T. and Greppin, H. (1982). *Peroxidases: 1970-1980: A Survey of their Biochemical and Physiological Roles in Higher Plants*. Geneva,Switzerland: Universite' de Geneve, Centre de Botanique.

Gatehouse, A.M.R., Barbieri, L., Stirpe, F., and Croy, R.R.D. (1990). Effects of ribosome inactivating proteins on insect development: Differences between Lepidoptera and Coleoptera. *Entomol. Exp. Appl.*, *54*, 43-51.

Gongora, C.E. and Broadway, R.M. (2002). Plant growth and development influenced by transgenic insertion of bacterial chitinolytic enzymes. *Mol. Breed.*, *9*, 123-135.

Gongora, C.E., Wang, S., Barbehenn, R.V. and Broadway, R.M. (2001). Chitinolytic enzymes from *Streptomyces albidoflavus* expressed in tomato plants: Effects on *Trichoplusia ni*. *Entomol. Exp. Appl.*, *99*, 193-204.

Gould, F. (1998). Sustainability of transgenic insecticidal cultivars: Integrating pest genetics and ecology. *Annu. Rev. Entomol.*, *43*, 701-726.

Gröner, A. (1986). Specificity and safety of baculoviruses. In R.R. Granados and B.A. Federici (Eds.) *The Biology of Baculoviruses: Volume I Biological Properties and Molecular Biology* (pp. 177-196). Boca Raton, FL: CRC Press.

Hagmann, M. (1999). EPA, critics soften stance on pesticidal plants. *Science*, *248*, 249.

Hayakawa, T., Shimojo, E., Mori, M., Kaido, M., Furusawa, I., Miyata, S., Sano, Y., Matsumoto, T., Hashimoto, Y., and Granados, R.R. (2000). Enhancement of baculovirus infection in *Spodoptera exigua* (Lepidoptera: Noctuidae) larvae with *Autographa californica* nucleopolyhedrosis or *Nicotiana tabacum* engineered with a granulovirus enhancin gene. *Appl. Entomol. Zool.*, *35*, 163-170.

Hayakawa, T. Hashimoto, Y. Mori, M., Kaido, M., Shimojo, E. Furusawa, I., and Granados, R.R. (2004). Transgenic tobacco transformed with the *Trichoplusia ni* granulovirus enhancin gene affects insect development. *Biocontrol. Sci. Technol.*, *14*, 211-214.

Heath, M.C. (2000). Hypersensitive response-related death. *Plant Mol. Biol.*, *44*, 321-334.

Heid, M.A., Tan, L., Kamyab, A., Hare, M., Schpak, E. and Kielszewski, M.J. (2004). Di-isodityrosine is the intermolecular cross-link of isodityrosine-rich extensin analogs cross-linked *in vitro*. *J. Biol. Chem.*, *279*, 55474-55482.

Hepburn, H.R. (1985). Structure of the integument. In: G.A. Kerkut, and L.I. Gilbert (Eds.). *Comprehensive Insect Physiology, Biochemistry and Pharmacology: Volume 2. Postembryonic Development* (pp. 1-58). Oxford, UK: Pergamon Press.

Hopkins, T.L., and Harper, M.S. (2001). Lepidopteran peritrophic membranes and effects of dietary wheat germ agglutinin on their formation and structure. *Arch. Insect Biochem. Physiol.*, *47*, 100-109.

Hughes, S.R., Dowd, P.F., Hector, R.E., Panavas, T., Sterner, D.E., Qureshi, N., Bischoff, K.M., Bang, S.S., Mertens, J.A., Johnson, E.T., Li, X.L., Jackson, Jr., J.R., Caughey, R.J., Riedmuller, S.B., Bartolett, S., Liu, S., Rich, J.O., Farrelly, P.J., Butt, T.R., Labaer, J. and Cotta, M.A. 2008. Lycotoxin-1 insecticidal peptide optimized by amino acid scanning mutagenesis and expressed as a coproduct in an ethanologenic *Saccharomyces cerevisiae* strain. *J. Peptide Sci.*, in press.

Hughes, S.R., Dowd, P.F., Hector, R. Riedmuller, S.B., Bartolett, S., Mertens, J.A., Qureshi, N., Liu, S., Bischoff, K.M., Li, X.-L., Jackson, J.S. Jr., Sterner, D., Panavas, T., Rich, J.O., Farrelly, P.J., Butt, T. and Cotta, M.A. (2007). Cost-effective high-throughput fully automated construction of a multiplex library of mutagenized open reading frames for an insecticidal peptide using a plasmid-based functional proteomic robotic workcell with an improved vacuum system. *J. Assoc. Lab. Automat.*, 12, 202-212.

Hughes, S.R., Riedmuller, S.B., Mertens, J.A., Li, X.-L., Bischoff, K.M., Qureshi, N., Cotta, M.A. and Farrelly, P.J. (2006). High-throughput screening of cellulase F mutants from multiplexed plasmid sets using an automated plate assay on a functional proteomic robotic workcell. *Proteome Sci.*, *4*, 1-14.

Jach, G., Gornhardt, B., Mundy, J., Logemann, J., Pinstorf, E. Leah, R., Schell, J. and Maas C. (1995). Enhanced quantitative resistance against fungal disease by different barley antifungal proteins in transgenic tobacco. *Plant J.*, *8*, 97-109.

Johnson, E.T., Dowd, P.F., Pinkerton, T.S. (2008). Altering plant secondary metabolism to achieve broad spectrum insect control and reduce mycotoxins. In: T. Wolf, and J. Koch (Eds.). *Genetically Modified Plants*. Hauppauge, NY: Nova Science Publishers.

Kanrar, S., Venkateswari, J., Kirti, P.B., and Chopra, V.L. (2002). Transgenic Indian mustard (*Brassica juncea*) with resistance to the mustard aphid (*Lipaphis erysimi* Kalt. *Plant Cell Rept.*, *20*, 976-981.

Kieliszewski, M.J., Kamyab, A., Leykham, J.F. and Lamport, D.T.A. (1992). A histidine-rich extensin from *Zea mays* is an arabinogalactan protein. *Plant Physiol.*, *99*, 538-547.

Kieliszewski, M.J., Leykam, J.F. and Lamport, D.T.A. (1990). Structure of the threonine-rich extensin from *Zea mays*. *Plant Physiol.*, *92*, 316-326.

Kimchi-Safarty, C., Oh, J.M., Kim, I.-W., Sauna, Z.E., Calgano, A.M., Ambudkar, S.V., and Gottesman, M.M.A. (2007). 'Silent' polymorphism in the *MDR1* gene changes substrate specificity. *Science.*, *315*, 525-528.

Kramer, K.J., Muthukrisnan, S., Johnson, L. and White, F. (1997). Chitinases for insect control. In N. Carozzi and M. Koziel (Eds.) *Advances in Insect Control: The Role of Transgenic Plants* (pp. 185-93). Bristol, PA: Taylor and Francis.

Krawetz, J. and Boston, R.S. (2000). Substrate specificity of a maize ribosome-inactivating protein differs across diverse taxa. *Eur. J. Biochem.*, *267*, 1966-1974.

Laroque, A.M., and Houseman, J.G. (1990). Effect of ingested soybean, ovomucoid and corn protease inhibitors on digestive processes of the European corn borer, *Ostrinia nubilalis* (Lepidoptera: Pyralidae). *J. Insect Physiol.*, *36*, 691-697.

Lawrence, S.D. and Novak, N.G. (2006). Expression of poplar chitinase in tomato leads to inhibition of development in Colorado potato beetle. *Biotechnol. Lett.*, *28*, 593-599.

Lincoln, J.E., Richael, C., Overduin, B., Smith, K., Bostock, R. and Gilchrist, D.G. (2002). Expression of the antiapoptotic baculovirus p35 gene in tomato blocks programmed cell death and provides broad-spectrum resistance to disease. *Proc. Nat. Acad. Sci. USA*, *99*, 15217-15221.

Loc, N.T., Tinjuangnun, P., Gatehouse, A.M.R., Christou, P. and Gatehouse, J.A. (2002). Linear transgene constructs lacking vector backbone sequences generate transgenic rice plants which accumulate higher levels of proteins conferring insect resistance. *Mol. Breed.*, *9*, 231-244.

Lockshin, R.A. (1985). Programmed cell death. In: G.A. Kerkut and L.I. Gilbert (Eds.). *Comprehensive Insect Physiology, Biochemistry and Pharmacology: Volume 2. Postembryonic Development* (pp. 301-317). Oxford, UK: Pergamon Press.

Mahoney, W.C., Hermodson, M.A., Jones, B., Powers, D.D., Corfmann, R.S., and Reeck, G.R. (1984). Amino acid sequence and secondary structural analysis of the corn inhibitor of trypsin and activated Hageman Factor. *J. Biol. Chem.*, *259*, 8412-8416.

Malehorn, D.E., Borgmeyer, J.R., Smith, C.E., and Shah, D.M. (1994). Characterization and expression of an antifungal zeamatin-like protein (*Zlp*) from *Zea mays*. *Plant Physiol.*, *106*, 1471-1481.

Maynard, J., Chen, G., Georghiou, G. and Iverson, B.L. (2002). *In vitro* scanning saturation mutagenesis. *Methods Mol. Biol.*, *182*, 149-163.

McMullen, M.D. and Simcox, K.K. (1995). Genomic organization of disease and insect resistance genes in maize. *Mol. Plant-Microbe Interact.* 8: 811-815.

Medeiros, A.H., Delaibera, I., Jr. and Tingey, W.M. (2005). Aspects of potato leafhopper (Homoptera: Cicadellidae) biology on *Solanum berthaultii* and other potato genotypes. *J. Econ. Entomol.*, *98*, 1704-1709.

Mehlo, L., Gahakwa, D., Nghia, P.T., Loc, N.T., Capell, T., Gatehouse, J.A., Gatehouse, A.M.R. and Christou, P. (2005). An alternative strategy for sustainable pest resistance in genetically enhanced crops. *Proc. Nat. Acad. Sci. USA*, *102*, 7812-7816.

Mitsuahsi, W. and Oaks, A. (1994). Development of endopeptidase activities in maize *Zea mays* L.) endosperms. *Plant Physiol.*, *104*, 401-407.

Moore, K.G., Price, M.S., Boston, R.S., Weissinger, A.K. and Payne, G.A. (2004). A chitinase from Tex6 maize kernels inhibits growth of *Aspergillus flavus*. *Phytopathology.*, *94*, 82-87.

Murdock, L.L., Brookhart, G., Dunn, P.E., Foard, D.E., Kelley, S., Kitch, L., Shade, R.E., Shukle, R.H., and Wolfson, J.L. (1987). Cysteine digestive proteinases in Coleoptera. *Comp. Biochem. Physiol.*, *87B*, 783-787.

Murphy, J.M. and Hood, E.E. (1993). Molecular basis for extensin size heterogeneity in two maize varieties. *Plant Mol. Biol.*, *21*, 885-893.

Narayanan, K. (2002). Microbial control of insect pests: role of genetic engineering and tissue culture. In O. Koul and G.S. Dhaliwal, (Eds.). *Microbial Biopesticides* (pp. 117-179). New York, NY: Taylor and Francis.

National Research Council, Committee on Genetically Modified Pest Protected Plants, Board on Agricultural and Natural Resources. (2000). *Genetically Modified Pest Protected Plants: Science and Regulation*. Washington, D.C.: National Academy Press.

Navon, A., Hare, J.D., and Federici, B. (1993). Interactions among *Heliothis virescens* larvae, cotton condensed tannin and the CryIA(c) delta endotoxin of *Bacillus thuringiensis*. *J. Chem. Ecol.*, *19*, 2485-2499.

Nielsen, K., and Boston, R.S. (2001). Ribosome-inactivating proteins: a plant perspective. *Annu. Rev. Plant Physiol. Mol. Biol.*, *52*, 785-816.

Nielsen, K., Payne, G.A., and Boston, R.S. (2001). Maize ribosome-inactivating protein inhibits normal development of *Aspergillus nidulans* and *Aspergillus flavus*. *Mol. Plant Microbe Interact.*, *14*, 164-172.

Nucere, J.N., Cleveland, T.E., and Dishinger, C. (1991). Existence of chitinase activity in mature corn kernels (*Zea mays* L.). *J. Agric. Food Chem.*, *39*, 1326-1328.

Oard, S.V., and Enright, F.M. (2006). Expression of the antimicrobial peptides in plants to control phytopathogenic bacteria and fungi. *Plant Cell Rept.*, *25*, 561-572.

Pechan, T., Cohen, A., Williams, W.P., and Luthe, D.S. (2002). Insect feeding mobilizes a unique plant defense protease that disrupts the peritrophic matrix of caterpillars. *Proc. Nat. Acad. Sci. USA*. *99*, 13319-13323.

Peterbauer, C.K., Lorito, M., Hayes, C.K., Harman, G.E. and Kubicek, C.P. (1996). Molecular cloning and expression of the *nag1* gene (*N*-acetyl-B-D-glucosaminidase-encoding gene) from *Trichoderma harzianum* P1. *Curr. Gen.*, *30*, 325-331.

Pinkerton, T.S., Dowd, P.F. and Behle, R.W. (2008). Effects of oral ingestion of maize expressed baculovirus antiapoptotic protein p35 on feeding and nucleopolyhedron virus lethality in caterpillars. (submitted).

del Pozo, O., and Lam, E. (2003). Expression of the baculovirus p35 protein in tobacco affects cell death progression and compromises N gene-mediated disease resistance response to tobacco mosaic virus. *Mol. Plant Microbe Interact.*, *16*, 485-494.

Privalle, L., Estruch, J., Wright, M., Hill, M.B., Dowd, P.F., and Lagrimini, L.M. (1999). Methods for conferring insect resistance to a monocot using a peroxidase coding sequence. U.S. Patent #6,002,068.

Purcelle, J.P., Greenplate, J.T., and Sammons, R.D. (1992). Examination of midgut luminal proteinase activities of six economically important insects. *Insect Biochem. Mol. Biol.*, *22*, 41-47.

Pusztai, A. (1991). *Plant Lectins*. Cambridge, UK: Cambridge University Press.

Quilis, J., Meynard, D., Vila, L., Avilobs, F.X., Guiderdoni, E., and San-Segundo, B. (2007). A potato carboxypeptidase inhibitor gene provides pathogen resistance in transgenic rice. *Plant Biotechnol. J.*, *5*, 537-553.

Reeck, G.R., Kramer, K.J., Baker, J.E., Kanost, M.R., Fabrick, J.A. and Behnke, C.A. (1997). Proteinase inhibitors and resistance of transgenic plants to insects. In N. Carozzi and M. Koziel (Eds.) *Advances in Insect Control: The Role of Transgenic Plants* (pp. 157-183). Bristol, PA: Taylor and Francis.

Sadasivam, S. and Thayumanavan, B. (2003). *Molecular Host Plant Resistance to Insects.* New York, NY: Marcel Dekker.

Scandiucci de Freitas, D., Felipe Coelho, M.C., Souza, J.R., Marques, A. and Moaris Ribeiro, B. (2007). Introduction of the anti-apoptotic baculovirus p35 gene in passion fruit induces herbicide tolerance, reduced bacterial lesions, but does not inhibit passion fruit woodiness disease progress induced by cowpea aphid-borne mosaic virus (CABMV). *Biotechnol. Lett.*, *29*, 79-87.

Schnabelrauch, L.S., Kielszewski, M., Upham, B.L., Alizedeh, H. and Lamport, D.T.A. (1996). Isolation of pI 4.6 extensin peroxidase from tomato cell suspension cultures and identification of Val-Tyr-Lys as putative intermolecular cross-link site. *Plant J. Cell Mol. Biol.*, *9*, 477-489.

Shelton, A.M., Zhao, J.-Z., and Roush, R.T. (2002). Economic, ecological, food safety, and social consequences of the deployment of Bt transgenic plants. *Annu. Rev. Entomol.*, *47*, 501-535.

Siegfried, B.D., Spencer, T., Crespo, A., Storer, N.P., Head, G.P., Owens, E.D. and Guyer, D. (2007). Ten years of Bt resistance monitoring in the European corn borer: What we know, what we don't know, and what we can do better. *Amer. Entomol.*, *53*, 208-214.

Sobero'n, M., Pardo-Lo'pez, L., Lo'pez, I., Go'mez, I., Tabashnik, B.E. and Bravo, A. (2007). Engineering modified Bt toxins to counter insect resistance. *Science.*, *318*, 1640-1642.

Srinivasan, A., Chougule, N.P., Giri, A.P., Gatehouse, J.A., and Gupta, V.S. (2005). Podborer (*Helicoverpa armigera* Hubn.) does not show specific adaptations in gut proteinases to dietary *Cicer arietinum* Kunitz proteinase inhibitor. *J. Insect Physiol.*, *51*, 1268-1276.

Stemmer, W.P.C. (1995). Searching sequence space. *Biotechnology*, *13*, 549-553.

Stennicke, H.R., Ryan, C.A., and Salvesen, G.S. (2002). Reprieval from execution: the molecular basis of capsase inhibition. *TRENDS Biochem. Sci.*, *27*, 94-101.

Stirpe, F., Barbieri, L., Batelli, M.G., Isoria, M., and Lappi, D. (1992). Ribosome inactivating proteins from plants: Present status and future prospects. *Bio/Technology*, *10*, 405-412.

Walsh, T.A., Morgan, A.E., and Hey, T.D. (1991). Characterization and molecular cloning of a proenzyme form of a ribosome inactivating protein from maize. *J. Biol. Chem.*, *266*, 23422-23427.

Wang, J.-S. and Tang, L. (2005). Epidemiology of aflatoxin exposure and human liver cancer. In H.K. Abbas (Ed.) *Aflatoxin and Food Safety* (pp. 195-211). Boca Raton, FL: Taylor and Francis.

Wei, G. and Shirsat, A.H. (2006). Extensin over-expression in *Arabidopsis* limits pathogen invasiveness. *Mol. Plant Pathol.*, *7*, 579-592.

Wilson, L.G. and Fry, J.C. (1986). Extensin, a major cell wall glycoprotein. *Plant Cell Environ.* 9: 239-260.

Wolfson, J.L., and Murdock, L.L. (1990). Diversity in digestive proteinase activity among insects. *J. Chem. Ecol.*, *16*, 1089-1102.

Wu, F., Miller, J.D. and Casman, E.A. (2005). Bt corn and mycotoxin reduction: an economic perspective. In H.K. Abbas (Ed.) *Aflatoxin and Food Safety* (pp. 459-482). Boca Raton, FL: Taylor and Francis.

Wu, S., Kriz, A.L., and Windholm, J.M. (1994). Molecular analysis of two cDNA clones encoding acidic class I chitinase in maize. *Plant Physiol.*, *105*, 1097-1105.

Yan, L. and Adams, M.E. (1998). Lycotoxins, antimicrobial peptides from venom of the wolf spider *Lycosa carolinensis*. *J. Biol. Chem.*, *273*, 2059-2066.

Zhao, J.Z., Fan, Y.L., Fan, X.L., Shi, X.P. and Lu, M.G. (1999). Evaluation of transgenic tobacco expressing two insecticidal genes to delay resistance development of *Helicoverpa armigera*. *Chin Sci. Bull.*, *44*, 1871-1873.

Zhe-Fu, L., Danqui, W., Annuo, L., and Weiqin, Z. (1992). Chitinases from seeds of *Zea mays* and *Coix lachryma-jobi* L. Purification and some properties. *Proc. Biochem.*, *27*, 83-88.

Zhu-Salzman, K., Ahn, J.E., Salzman, R.A., Koiwa, H., Shade, R.E., and Balfe, S. (2003). Fusion of a soybean cysteine protease inhibitor and a legume lectin enhances anti-insect activity synergistically. *Agric. Forest Entomol.*, *5*, 317-323.

In: Genetically Modified Plants
Editors: Timm Wolf and Jonas Koch

ISBN 978-1-60456-696-3

Chapter 7

ALTERING PLANT SECONDARY METABOLISM TO ACHIEVE BROAD SPECTRUM INSECT CONTROL AND REDUCE MYCOTOXINS

Eric T. Johnson, Patrick F. Dowd and T. Scott Pinkerton
Crop Bioprotection Research Unit, USDA-Agricultural Research Service, 1815 N. University St., Peoria, IL 61604, USA

ABSTRACT

Mycotoxins are toxic or carcinogenic compounds produced by fungi, including *Aspergilllus* and *Fusarium* molds that colonize seeds of maize, cotton, peanuts or tree nuts. Insect herbivory of plant tissue often enhances mold infection. Transgenic maize expressing the *Bacillus thuringiensis* (Bt) toxin produce ears with lower mycotoxin levels when the target insect pest is effectively controlled. However, Bt toxins are generally species specific and therefore new transgenic strategies for broad insect control need to be developed. The vast diversity of plant secondary biochemicals reflects the strategy of non-motile plants to defend themselves from pathogens and animal herbivores. Research in recent years revealed that many of these plant biochemicals are synthesized by a number of enzymes that are coordinately regulated at the gene level by transcription factors. Altering the expression of these transcription factors by genetic engineering opens the possibility that these secondary biochemicals may be synthesized in specific tissues to combat insect herbivory and reduce mycotoxin contamination. Alternatively, well studied secondary biochemical pathways may be modified or transferred from one plant species to another and tested for effective insect resistance. Transgenic manipulation of secondary metabolism must be carefully considered because redirection of plant energy reserves may hinder plant yield. The advent of genetic engineering and advances in plant biochemistry has opened up the exciting possibilities of utilizing the diversity of secondary biochemicals for protecting valuable crop commodities.

INTRODUCTION

Plants synthesize primary metabolites (carbohydrates, lipids and amino acids) which are crucial for their day-to-day survival and serve as chemical building blocks for higher organisms. Plants also produce low molecular weight chemicals, the secondary metabolites, which are also extremely important for the plant's continued existence. Plants have developed chemical defense measures, which generally include secondary metabolites, against pathogens, competitors or herbivores. Nearly 100,000 secondary metabolites have been discovered from the plant kingdom but only half of the structures are known [91]. Based on their biosynthetic pathway, secondary metabolites can be classified into three major groups: the alkaloids, the terpenoids, and the phenylpropanoids and allied phenolic compounds [18]. Terpenoids are derived from the five-carbon precursor isopentenyl diphosphate while the alkaloids, which contain one or more nitrogen atoms, are primarily synthesized from amino acids [18]. The phenolic compounds come from either the shikimic acid pathway or the malonate-acetate pathway [18].

Biosynthesis of secondary metabolites from primary metabolites (such as amino acids) usually requires a number of enzymatic steps. Each of the biosynthetic enzymes is expressed in the cell to complete synthesis of the metabolite. In recent years, studies have documented the presence of metabolic channels in cells where each biosynthetic enzyme in a particular pathway is in close proximity to each other and biosynthetic intermediates can be passed between from one enzyme to the other [56]. Coordinated biosynthesis of secondary metabolite requires a constant level of each enzyme, which is regulated in part at the gene level. Expressing a master regulator (i.e. inducer) of all the genes coding for enzyme of a biosynthetic pathway would ensure that sufficient enzyme is present for secondary metabolite production, and this regulator has been identified for a number of plant secondary metabolite pathways [63;81;118]. However, it is also possible that the biosynthetic pathway is regulated by a rate-limiting enzyme which itself is regulated at the gene or protein (e.g. phosphorylation) level. Genetic engineering of a secondary metabolite is clearly more straightforward with a known master regulator that controls the entire pathway, but modulation of a key enzyme(s) that influences the levels or types of secondary metabolites has been accomplished [22;77;103]. The greatest hindrance to the engineering of secondary metabolites in plants is the limited knowledge of their biosynthetic pathway and incomplete genetic characterization. Others have discussed the recent advances in secondary metabolite engineering [39;118;125] but this review will focus on describing the few studies that have engineered secondary metabolites for effective insect resistance. In addition, the potential for engineering pathways for well characterized secondary metabolites for insect resistance will be discussed.

ANTHOCYANINS AND FLAVONOIDS

While it is generally agreed that anthocyanins are synthesized in floral tissues to attract pollinators [63], there is data to suggest that these pigments have other roles. Some have postulated that the anthocyanins may protect plant tissues from light stress, or serve as antioxidants or function as osmoregulators [16;41;113]. However, others have argued that anthocyanins may be part of a defense strategy against potential herbivores [78;104]. Studies of 2 tree species found that leaf anthocyanins possibly play a role in herbivore protection [58;105]. It may be possible that anthocyanins inhibit larval growth. Some anthocyanins from cotton or petunia flowers can inhibit the development of insect larvae [46;52;53]. Transforming anthocyanin transcription factors into a variety of plants can potentially determine whether they do serve as anti-herbivore molecules. Although the *pap1-D* mutant overexpresses other molecules besides anthocyanins, the growth of first instar fall amyworms was inhibited while eating *pap1-D* leaves compared to wild type leaves [55].

The anthocyanin pathway has been well characterized both at the enzyme and gene level [63]. The studies have demonstrated that anthocyanin and flavonoid biosynthesis is controlled by a small number of master regulators that control multiple genes. For example, in maize, the combination of a basic helix-loop-helix (BHLH) protein (R) and MYB (DNA-binding protein encoded by avian myeloblastosis virus) like protein (C1) control transcriptional activation of anthocyanin biosynthesis [38;76]. This combination of BHLH and MYB proteins for control is not limited to a single plant genus, as the maize R and C1 combination also functions in other plants to induce anthocyanin production [73]. In *Arabidopsis thalliana* (Arabidopsis), a single MYB-like protein (PAP1) induces the production of anthocyanins, lignins and flavonoids in all tissues [12].

The maize MYB protein P1 induces a secondary metabolite pathway distinct from anthocyanins without the need for a BHLH protein [42]. The maize *p1* transcription factor regulates production of 3-deoxy anthocyanins, phlobaphenes (polymers of flavan-4-ols), and C-glycosyl flavones [42]. A number of the biosynthetic steps regulated by *p1* are also utilized by anthocyanin biosynthesis [42]. In silks, *p1* induces the biosynthesis of maysin (Figure 1), a C-glycosyl flavone that inhibits the growth of corn earworm larvae [14]. Corn lines that have high levels of maysin in silks usually have red kernels colored by phlobaphenes, which are generally not preferred by consumers. Transgenic expression of *p1* in silks by a putative silk-specific promoter induced maysin synthesis and increased resistance to first instar corn earworm larvae [54]. However, browning of the transgenic kernels was observed, which indicated that the gene promoter utilized was not tissue specific [54]. This study demonstrated that engineering a transcription factor to produce secondary metabolites (maysin) can increase insect resistance, but tissue specific expression is dependent on effective selective gene promoters.

Isoflavonoids are a group of secondary metabolites produced primarily in legumes, where they are responsible for interactions with microbes [5;101]. Studies have demonstrated that some isoflavonoids act as deterrents, toxins or growth inhibitors to a variety of insects [65;95;107;126;131]. Some isoflavonoids are synthesized by isoflavone synthase from naringenin, which is also a precursor molecule of anthocyanins [72]. The cloning of isoflavone synthase has enabled the metabolic engineering of isoflavones in both legume and other plant species [20;72;111;117]. Induction of the phenylpropanoid pathway in soybean by maize R and C1 transcription factors and concurrent suppression of flavonone 3-hydroxylase (F3H) activity (by expressing the F3H gene in between synthetic inverted repeat sequences) effectively blocked anthocyanin production and increased isoflavone levels [132]. The success of these transgenic plant studies indicate that experiments can be performed to test the effects of elevated levels of isoflavonoids on insect larvae. The non-legume transgenic plants (tobacco, petunia and lettuce) engineered with soybean isoflavone synthase all produced genistein. Genistin (Figure 1), a glucoside of genistein, was implicated in resistance to *Anticarsia gemmatalis*, a pest of soybean [95].

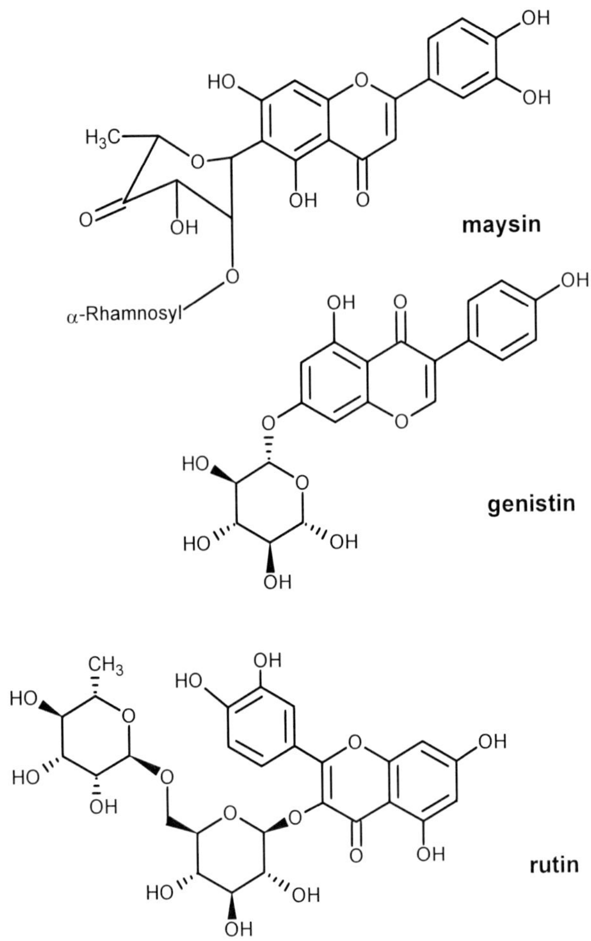

Figure 1. Representative phenolics.

Rutin (quercetin 3-rutinoside, Figure 1), a flavonol glycoside, has deleterious effects on a number of insect species [10;27;49;50;112] but does not appear to affect *Manduca sexta* or *Heliothis virescens* [7]. Experiments have demonstrated that rutin oxidized by peroxidases or tyrosinases (which may occur upon herbivore attack) could be the insect toxic molecule [25]. Chalcone isomerase (CHI) catalyzes the reaction of chalcone to (2S)-naringenin [70]. Levels of rutin in transgenic tomato fruit peel were improved up to 78 fold over control fruit peels by the overexpression of petunia CHI [83]. Overexpression of *Saussurea medusa* (a Chinese medicinal plant) CHI in tobacco plants increased leaf rutin levels [70]. CHI overexpression enhanced floral levels of flavonoids (which includes rutin), but not anthocyanin content, which indicates the presence of an unusual regulatory mechanism [70]. No other flavonols or isoflavones were measured in the CHI overexpressing leaves [70]. Unfortunately, none of the CHI overexpressing tissues mentioned above have been tested for insect resistance.

Alkaloids

Alkaloids encompass a wide variety of secondary metabolites in 20% of all plants [101]. The current definition includes all nitrogen-containing compounds that are not classified as non-protein amino acids, amines, cyanogenic glycosides, glucosinolates, peptides, cofactors, phytohormones, purines or pyrimidines [101]. Alkaloids are classified into 12 groups based on their chemical structure, and many of them have activity against insects [101]. Unlike phenolic compounds, there is no general or common alkaloid biosynthetic pathway [101]. Thus the entire or partial biosynthetic pathway of an alkaloid with insect bioactivity would need to be engineered into crops depending on whether the crop of interest is capable of synthesizing some types of alkaloids. Biosynthetic pathways of indole (vinblastine, vincristine and captothecin), tropane (atropine, cocaine), pyridine (nicotine), benzylquinoline (morphine, codeine) and benzoxazinoid alkaloids have been fairly well elucidated, but some enzymatic steps are still unknown and very few transcription factors have been identified [39]. In addition, studies determined that biochemical regulation of the indole alkaloid pathway was complex, with intermediates being transported among membrane components, cells and tissues [93]. This makes the prospect of transfer of entire alkaloid pathways to crops that do not produce alkaloids difficult with current knowledge and technology.

The existing understanding of alkaloid biosynthesis and availability of genes can still be exploited for use in crop resistance. Two studies examined the transgenic production of the terpenoid indole alkaloid (TIA) precursor tryptamine (Figure 2) in tobacco and poplar [35;36]. Tryptamine is synthesized from tryptophan by tryptophan decarboxylase [36]. Products of this single gene addition significantly inhibited the growth of two different insect larvae [36]. Though not tested for insect resistance, transgenic petunia plants with elevated tryptamine levels have also been made [116]. These studies demonstrate that elevated levels of some alkaloid biosynthetic intermediates can have detrimental effects on insect growth. Experiments manipulating levels of different alkaloid biosynthetic intermediates for insect resistance may be worth pursuing provided levels of primary metabolites are not significantly compromised.

The benzoxazinoids (Bx) are alkaloids produced in grasses (rye, wheat and maize) that act as defensive compounds against pathogens and insects [86]. 2,4-dihydroxy-7-methoxyl-1,4-benzoxazin-3-one (DIMBOA, Figure 2) is the predominant Bx of maize and is stored as a glucoside in the vacuole [119]. DIMBOA has a broad range of activity, affecting bacteria, fungi and insects such as the corn borer and aphid [80]. When plant tissue is ruptured, DIMBOA glucosides are hydrolyzed by resident β-glucosidases [90] producing DIMBOA aglycones, which have lethal and antifeedant effects on insects [62;87]. A number of studies determined that DIMBOA levels are highest in young seedlings but then decline as the plant ages [8;15;61]. The decline in DIMBOA levels parallels changes in resistance to some maize feeding insects; most strains of maize are resistant to first brood European corn borer larvae as seedlings but most strains lose this resistance as the plant matures [61]. Significant correlations were found between first brood European corn borer larvae resistance and Bx levels in maize [60]. The pathway of DIMBOA biosynthesis has been elucidated and all of the structural genes (*Bx1* through *Bx6*) of the pathway have been cloned [30;31]. It would be technically challenging to transform the entire pathway into a crop using robust promoters for each biosynthetic gene. A more promising approach may utilize genomics to identify possible transcription factors that regulate the entire pathway.

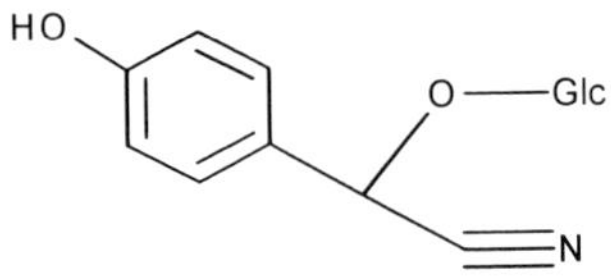

Dhurrin

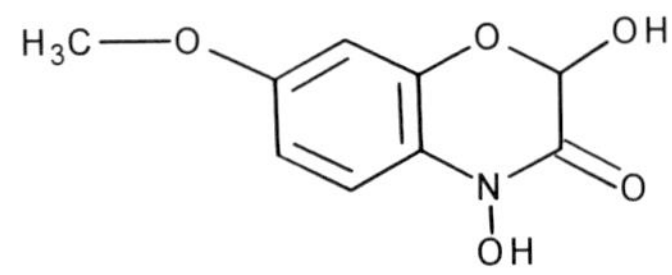

DIMBOA

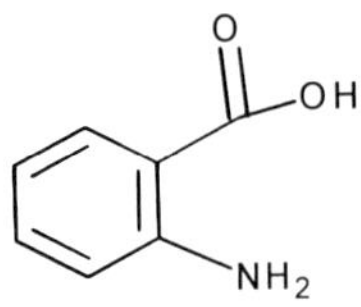

Anthranilate

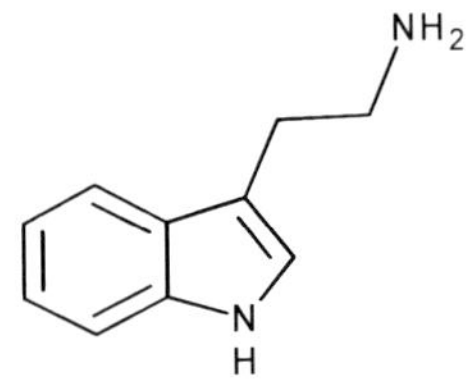

Tryptamine

Figure 2. Representative alkaloids, alkaloid precursor, and a cyanogenic glucoside.

Tryptophan plays an important role in plants not only as an amino acid for protein production, but also as a precursor for indole and quinoline alkaloids [101]. Tryptophan biosynthesis plays a key role in the synthesis of DIMBOA [80]. A key point of metabolic control of tryptophan and related compounds is the enzyme anthranilate synthase (EC 4.1.3.27), which converts chorismate to anthranilate [134]. Anthranilate (Figure 2) is a precursor for acridone and furoquinoline alkaloids that are found mainly in the Rutaceae family [11;99]. Anthranilate synthase enzymatic activity is allosterically controlled by tryptophan levels. Two genes have been identified in *Arabidopsis* as encoding anthranilate synthase and one of those genes, *ASA1*, has been shown to be induced by wounding and bacterial pathogen infiltration [88]. Mutations in anthranilate synthase that can abolish allosteric control can lead to an increased amount of tryptophan as well as the accumulation of derivatives of anthranilate. Several plant antranilate synthase mutants have been identified [89;128;129]. Among the phenotypes associated with anthranilate synthase mutations is blue fluorescence due to the accumulation of anthranilate and related compounds. Maize seeds of an L289xI205 cross were exposed to radiation from a nuclear test in 1946 and several mutants were described by Teas and Anderson, including two with a blue fluorescence phenotype [115]. One of these mutants, *bf-1*, is characterized by blue fluorescence in seedlings and anthers under UV illumination, and by volatiles producing a characteristic grape-like odor due to the accumulation of anthranilate derived compounds [84]. The mutant is also resistant to feedback regulation of tryptophan biosynthesis by tryptophan [108]. Sequencing of cDNA from *bf-1* plants showed a point mutation in a conserved chorismate binding region of the anthranilate synthase α subunit which caused a leucine to proline shift at amino acid residue 531 [94]. This mutation is unique compared to other observed mutations in plant anthranilate synthases that convey resistance to feedback inhibition of tryptophan biosynthesis. The *bf-1* anthranilate synthase with an added c-terminal *myc* epitope tag was cloned into the plant expression vector pAHC25 and transferred into maize line *HiII* by particle bombardment [94]. The transgenic plants expressing the *bf-1* mutant anthranilate synthase did not show the strong odor phenotype associated with the *bf-1* mutant plants and only mild fluorescence was observed in some transgenic plants [94]. Callus derived from a transgenic line did show resistance to the anthranilate synthase inhibitory compound 6-methyl anthranilate [94]. Similar blue fluorescent mutants in *Arabidopsis thaliana* have been shown to be dependant on an UDPglucose:salicylic acid glucosyltransferase [97]. Currently we are working to characterize the maize version of this enzyme and another potential modification enzyme, salicylic acid carboxymethyl-transferase, which transfers a methyl group from s-adenosylmethionine to salicylic acid and related compounds (benzoic acid and anthranilic acid) to form methyl esters [67]. The action of these enzymes could explain the presence of both methyl anthranilate and anthranilate-glucose conjugate in the *bf-1* mutant. Work on these enzymes is also of interest because of the recent discovery that methyl salicylate and the salicylic acid carboxymethyltransferase are critical components of the plant systemic acquired resistance pathway in tobacco [92].

Field assays with three different lines of *bf-1* indicated reduced feeding by flea beetles compared to other inbreds and commercial hybrids of the same age growing in the same field (Dowd, unpublished results). Laboratory assays with leaf sections from 3 leaf plants indicated the *bf-1* line tested had significantly less feeding by fall armyworms compared to other lines including its original source hybrid, plus Oh43, Hi-II, I205 and L289 (Dowd, unpublished results). Assays with transformed plants using the mutant *bf-1* gene indicated feeding was significantly inversely correlated with transgenic protein expression levels for corn earworms, and the same trend was noted with fall armyworms [24]. Diet incorporation assays indicated that diet containing 1% methyl anthranilate was significantly active against corn earworms and fall armyworms, causing 100% mortality in both species, and significant feeding reductions occurred when methyl anthranilate was at 0.1% in diet (Pinkerton and Dowd, unpublished results). Although no significant mortality was noted, anthranilic acid caused significant reductions in growth for both corn earworms and fall armyworms at 1% in diet (Pinkerton and Dowd, unpublished results).

Terpenoids

Terpenoids are used for a wide range of commercial applications such as flavoring agents, pharmaceutical, perfumes, insecticides and anti-microbial agents [3]. They are synthesized by the melavonate pathway in the cytosol or the Rohmer (non-melavonate) pathway in plastids. Addition of isopentenyl diphosphate groups to dimethyl allyl diphosphate builds geranyl diphosphate (C10), farnesyl diphosphate (C15) or geranylgeranyl diphosphate (C20). Terpene synthases or cyclases then convert these building blocks to the C10 monoterpenes, C15 sesquiterpenes, or C20 diterpenes. Chloroplasts are the site of monoterpene and diterpene biosynthesis while sesquiterpene biosynthesis occurs in the cytosol [3].

Some diterpenes and sequiterpenes kill insects or inhibit their growth [101], but biosynthetic modifications to monoterpene biosynthesis are the easiest to modify by current technologies (see below). Therefore this review will briefly highlight some of the ways in which monoterpenes help plants fight insect attack. Volatile monoterpenes can act as insect repellents. Pure 1,8-cineole is a repellent of mosquitoes, confused flour beetles, and the American cockroach [101;122]. The bicyclic monoterpene isoborneol repels subterranean termites when applied to soil [9]. A number of monoterpenes (e.g. (±)-linalool (Figure 3) and (±)-camphor (Figure 3)) from *Artemisia vulgaris* successfully repelled mosquitoes [51].

Monoterpenoids can also be feeding deterrents. Limonene (Figure 3) and pulegone (Figure 3) are feeding deterrents to cat fleas (*Ctenocephalides felis*) and fall armyworm (*Spodoptera frugiperda*), respectively [101]. Verbenone suppressed the feeding of the pine weevil [71] while both S (+) and R (-) carvone had antifeedant activity against the pales weevil [102].

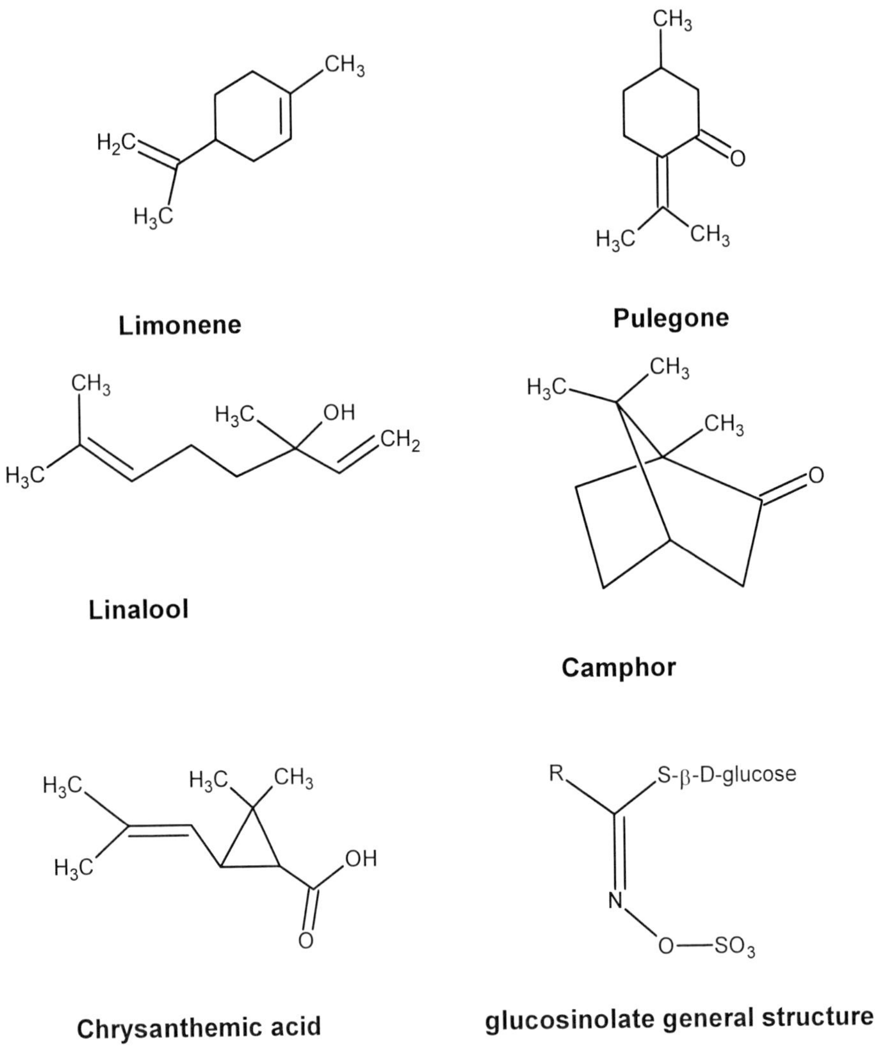

Figure 3. Representative terpenoids and a glucosinolate.

The pyrethroids from *Tanacetum* species are insect neurotoxin terpenoids that affect a wide range of insect species (including caterpillars, beetles and flies) but are rapidly metabolized and excreted in mammals [47]. Pyrethroids are unstable in air and light and thus synthetic pyrethroids (e.g. allethrin) that are more stable have been developed for commercial production [101]. The gene for the first reaction of pyrethroid biosynthesis has been cloned [98], but the genes for the remaining 2 reactions to the insect active compound chrysanthemic acid (Figure 3) have not been isolated. Monoterpenoids, especially pulegone, caused mortality of larvae of *Ostrinia nubialis* when incorporated into artificial diet or when added to the diet surface [68]. Pulegone can also inhibit the larval growth of *Spodoptera eridania* [45].

Efforts to modify terpenoid levels by genetic engineering have been most successful when monoterpene synthases were overexpressed, which increased and/or modified monoterpene profiles [3]. For example, Arabidopsis leaves emit traces of only one monoterpene, limonene [3]. When transformed with a strawberry nerolidol synthase 1 gene targeted to plastids, transgenic Arabidopsis leaves emitted high levels of linalool, in some lines reaching ~7-13 μg day^{-1} $plant^{-1}$ [2]. In addition, the monoterpene profiles of a number of other plants have been modified, including petunia, tobacco, carnation, tomato and potato [3;66;69;74;75]. In the tobacco study, three different lemon monoterpene synthases were transformed into tobacco and then combined into one line by breeding to produce a plant that had higher levels of volatile monoterpenes than wild type plants [75].

Manipulation of monoterpenes for insect resistance is attractive because many important crops, including maize, wheat and rice all synthesize terpenoids and therefore have the basic building blocks of monoterpenoids available in either the cytosol or plastids [17;37;121]. On the other hand, plants that produce high levels of monoterpenes (e.g. mint) store these compounds as oil blends in glandular trichomes [13], presumably to reduce volatility and avoid cytotoxicity. Plants that produce low levels of monoterpenes may have difficulty in storing elevated levels of these compounds. For example, some transgenic potato plants expressing high levels of the strawberry nerolidol synthase 1 gene exhibited bleaching under greenhouse conditions [3]. It remains to be tested whether elevated levels of monoterpenes that preserve plant morphology and yield will have better insect resistance.

A number of volatiles are emitted by plants after attack to attract predators or parasitoids of herbivores, which is termed “indirect defense” [21]. For example, a blend of volatiles, primarily composed of terpenes, is produced by maize after attack by *Spodoptera* larvae, which causes parasitic wasps (*Cotesia marginiventris*) to oviposit on the larvae [123;124]. The blend of maize volatiles varies among maize varieties, which has made identification of the bioactive volatile(s) difficult [40]. Rice plants also emit a blend of sesquiterpenes after treatment with methyl jasmonate, which mimics insect attack [17;79].

The identification of many terpene synthases in recent years has led to the transgenic manipulation of volatile production for indirect defense. The maize terpene synthase clone 10 (*tps10*) catalyzes the conversion of farnesyl diphosphate into a blend of (*E*)-β-farnesene, (*E*)-α-bergamotene, and 7 other minor products in a bacterial system that mirrors the terpene profile of insect damaged maize B73 plants [106]. When *tps10* was transformed into Arabidopsis, the same volatile mixture described above was produced by the rosette leaves and successfully attracted female parasitic wasps (*C. marginiventris*) that had a previous experience of finding host larva on the *tps10* transgenic leaves [106]. Another group transformed Arabidopsis with a strawberry nerolidol synthase gene which produced the homoterpenes (3*S*)-(*E*)-nerolidol and 4,8-dimethyl-1,3(*E*),7-nonatriene in some of the primary transformants [57]. These transgenic plants attracted carnivorous predatory mites (*Phytoseiulus persimilis*) which can benefit plants under herbivore attack [57]. Overexpression of *OsTPS3* (rice sesquiterpene synthase 3) in transgenic rice resulted in more (*E*)-β-caryophyllene after methyl jasmonate treatment which attracted more parasitoid wasps (*Anagrus nilaparvatae*) than wild type plants [17]. While these studies certainly indicate that induced volatile blends can be changed via transgenic methodology to attract beneficial insects, it remains to be determined if indirect defense is a viable strategy in commercial crop protection.

Glucosinolates

These compounds, produced mainly by members of the Brassicaceae, contribute to the unique flavors and aromas of cruciferous vegetables [43]. In addition, glucosinolates are involved in plant defense, auxin homeostasis, and can prevent cancer in humans [43]. The core glucosinolate structure is synthesized from certain amino acids that have a beta-thioglucosyl residue attached to the original alpha-carbon to create a sulfated ketoxime (see Figure 3). Side chain modification and elongation of the amino acid contributes to ~120 known glucosinolate structures [28]. Glucosinolates are normally stored in vacuoles, and cell rupture leads to contact with beta-thioglucosidases called myrosinases [43]. The degradation of glucosinates leads to the creation of bioactive compounds such as isothiocyanates, thiocyanates, nitriles (with β-hydroxylated side chains), and epithionitriles that protect against herbivores and pathogens [101]. For example, isothiocyanates reduce the survival and growth of *Pieris rapae*, an herbivore of plants of the Brassicaceae family, in a dose dependent manner [1]. Glucosinolates also interfere with insect oviposition and egg hatchability, act as phagostimulants for insects adapted to plants of the Brassicaceae, deter feeding of insects, and are toxic to some insects that don't feed on crucifers [101].

Glucosinolates, their biosynthetic pathway, and quantitative trait loci have been extensively studied in Arabidopsis [44;127;135]. Most of the biosynthetic steps of the core glucosinolate structure have been identified [44]. In addition, a number of gene regulators of the pathway have also been identified [32-34;48;109]. If these Arabidopsis regulators function similarly in other species in the Brassicaceae, it may be possible to increase the levels of glucosinolates in important *Brassica* crops like radish, mustard and rapeseed. Transferring the entire core glucosinolate pathway (once it is fully identified) to non-*Brassica* species in the future may be possible. However, synthesized glucosinolates would need to be properly transported to the vacuole for storage and glucosinolate vacuolar transporters have yet not been identified [43]. In addition, a suitable myrosinase gene would also need to be transformed into the non-*Brassica* species and expressed in such a manner that it would not come in contact with glucosinolates. Studies of *Brassica napus* demonstrated that myrosinases were stored in idioblasts called myrosin cells, which were found in all plant organs [4]. Localization of glucosinolates is not as well studied, but it is clear they are stored in vacuoles and not in myrosin cells [4;59]. Strategies for differential storage of glucosinolates and myrosinase need to be identified before glucosinolates can be utilized in plants outside of the Brassicaceae.

Cyanogenic Glucosides

Cyanogenic glucosides (CG; see example of dhurrin, Figure 2) are derived from amino acids that release glucose, a ketone or aldehyde, and toxic HCN when hydrolyzed by specific β-glucosidases followed by hydroxynitrile lyase [82;101]. CGs are synthesized in the cytoplasm and stored in the vacuole. To avoid initiating hydrolysis, the CGs and hydrolyzing enzyme are usually stored in different cells; in sorghum the CGs are stored in epidermal cells while the degrading enzyme is in adjacent mesophyll cells [101]. Hydrolysis can occur under conditions of herbivory, trampling, heat or frost where cellular integrity is compromised [101].

At least 75 different CG have been identified from some 2650 plants, including ferns, gymnosperms, and the angiosperms [101]. Important crop plants containing CGs include sorghum, almond, lima bean, white clover and cassava. Cassava tubers are a major food staple in developing countries. Because the CGs are heat stable, grinding or grating of cassava tubers are necessary for the CGs to come in contact with the β-glucosidase and release the toxic HCN [96]. The free HCN can then be removed by water, cooking, air-drying or sun-drying [96]. CGs are converted to thiocyanates in mammals. Chronic ingestion of CGs coupled with iodine deficiency can lead to goiter and cretinism because thiocyanates interfere with iodine uptake by the thyroid gland [100]. Future biotechnological strategies to utilize CGs for insect resistance in crop plants must consider the necessary post-harvest treatments.

CG adapted insects can sequester CGs, detoxify the released HCN using 3-cyanolalanine synthase or inhibit CG hydrolysis through insect produced β-glycosidases [101]. CGs are toxic to non-adapted herbivores, primarily through cyanide inhibition of cytochrome oxidase and other respiratory enzymes [101]. There is some evidence to suggest that hydrolyzed ketones and aldehydes of CGs can also be toxic [110]. The presence of CGs in different lines of *Trifolium repens* caused various grasshoppers, aphid, slugs, snails and deer to prefer acyanogenic lines [26]. HCN release rates from sorghum leaves depend on plant age, variety differences and other environmental conditions [130]. For example, *Locusta migratoria* will often reject young sorghum leaves but eat older leaves because of differential HCN release rates [130].

Biosynthesis of the CGs have been clearly elucidated in *Sorghum bicolor*, which produces the CG dhurrin (Figure 2) using 3 enzymes that have been genetically identified [6]. The entire dhurrin biosynthetic pathway was engineered into Arabidopsis and the transgenic plants accumulated dhurrin up to 4% dry weight [114]. The cruciferous flea beetle *P. nemorum* was significantly deterred by dhurrin accumulating plants in choice tests comparing wild type tissue [114]. In addition, most flea beetle larvae died while feeding on the dhurrin accumulating Arabidopsis [114]. The three genes for dhurrin biosynthesis were also transformed into grapevine hairy roots that produced up to 100 mg HCN kg^{-1} fresh weight [29]. However, specialist root-sucking insects (*Daktulosphaira vitifoliae*) fed on the cyanogenic transgenic roots and control roots equally well; the authors speculated that higher levels of dhurrin may be necessary for resistance, but only 1 insect species was tested [29]. The successful transformation of the dhurrin biosynthetic pathway from sorghum to two distantly related plants indicates that economically important crops could be transformed as well. The only limitation to development could be the post-harvest treatment necessary for cyanogenic plant tissue that is consumed by humans. Alternatively, the dhurrin biosynthetic pathway could potentially be expressed in tissues (by tissue-specific gene promoters) that are susceptible to insect attack but are consumed in small amounts by humans (e.g. maize husk, silk or tassel).

Secondary Metabolites and Plant Performance

Manipulation of secondary metabolites to enhance insect resistance can potentially hinder other aspects of plant growth and development. However, only a handful of studies have actually measured the metabolic costs of secondary metabolism. As mentioned above, the

Arabidopsis *pap1-D* mutant overexpresses a MYB-like transcription factor that results in the induction of anthocyanins, lignins and flavonoids in all tissues [12]. More recent and extensive analysis of around 1800 metabolites of wild type and *pap1-D* mutant plants found that there were only slight differences in the metabolic profile between wild type and the *pap1-D* mutant apart from the substantial increase in anthocyanins and flavonols [120]. The more recent analysis [120] does not indicate changes in lignin accumulation between wild type and *pap1-D* plants while the earlier analysis demonstrated the *pap1-D* mutant produces more lignin [12]. Transcriptome analysis revealed that 38 genes were induced by PAP1 transcription factor [120]. Surprisingly, there were no significant differences in the levels of 16 amino acids, or 12 sugars and anions between wild type and *pap1-D* plants that had been grown in vermiculite for 4 weeks or in sterile culture media for 3 weeks [120]. In addition, there were no changes in levels of phenylalanine between wild type and the *pap1-D* mutant, which indicates that the pools of free phenylalanine were sufficient for the enhanced anthocyanin and flavonol production [120].

Similar results were found for Arabidopsis plants engineered with 3 sorghum genes that synthesized the CG dhurrin up to 4% of the leaf dry weight [64]. The first step of dhurrin biosynthesis utilizes tyrosine, and no significant changes in the free amino acid pools were found between ~30 day old wild type and dhurrin-producing transgenic plants [64]. Analysis of small and global microarrays found very few changes in gene expression between the wild type and dhurrin-transgenic plants [64]. The small metabolic changes observed in the dhurrin transgenic plants may be due in part to the ability of the enzymes of these biosynthetic pathways to form metabolons, which are multiprotein complexes that effectively channel pathway intermediates from one enzyme to the other as discussed above. Recent studies found that the enzymes of the dhurrin biosynthetic pathway do form a metabolon [85]. Arabidopsis plants transformed with only the first 2 genes of dhurrin biosynthesis were stunted and displayed pronounced changes in the transcriptome and metabolome because of the necessary detoxification of *p*-hydroxymandelonitrile, a dhurrin pathway intermediate [64]. In the future, it may be possible to transform crop plants with multistep biosynthetic pathways from heterologous species, as achieved with the dhurrin pathway. Changes to the host transcriptome and metabolome can potentially be minimized by assuring that the donor pathway forms a biosynthetic metabolon and does not "leak" metabolites.

In our studies of the Arabidopsis *pap1-D* mutants, we grew plants with solid purple leaves as well as leaves that only had purple veins, which suggested that some genetic changes had occurred in some progeny of the original mutant [55]. We found the plants with purple-veined leaves produced similar numbers of inflorescences but significantly fewer siliques compared to wild type plants [55]. The solid purple plants produced significantly fewer inflorescences and siliques compared to wild type plants [55]. These findings suggest that the significant production of phenylpropanoid metabolites in the solid purple *pap1-D* plants lowered plant productivity. Perhaps the increased demand for protein deposition in seeds and the continued production of phenylpropanoid metabolites directed by the PAP1 transcription factor begins to deplete the free amino acid pool in late plant development and requires the catabolism of protein or more amino acid synthesis from photosynthate. Based on the number of days (~53) from sowing to flowering of the Arabidopsis Columbia ecotype [19], we presume that the metabolome-analyzed Arabidopsis (Columbia ecotype) wild type and *pap1-D* plants [120], which were metabolically similar apart from anthocyanin and flavonol production, were likely not flowering at the time of leaf analysis. Therefore it remains to be determined if flowering and subsequent seed set significantly alters metabolism in the *pap1-D* mutants. In any case, if secondary metabolism is manipulated in any tissues of crop plants for insect resistance, metabolic studies will need to be measured to ensure that primary metabolism is not compromised.

Conclusion

As demonstrated from this short review, there is wide diversity of plant secondary metabolites that can potentially serve as insect resistant molecules. The advantage to studying these secondary metabolites lies in the fact that we are simply trying to identify and utilize the chemical resources that nature has developed over the millennia. The greatest hindrance to utilizing these secondary metabolites is the lack of knowledge of the basic biochemistry of their biosynthesis. Once the biosynthetic enzymes and genes are identified, the pathways may be transformed into model species or the crops themselves. We imagine that as technology improves in the next decade that it will be easier to transform plants with entire biosynthetic pathways. Two metabolic studies of transgenic plants demonstrated that secondary metabolism could be manipulated with very little cost to primary metabolism, but neither of these studies documented the metabolome as the plants shifted their resources into reproduction. Alternatively, if secondary metabolites are identified that are highly toxic to insects and not vertebrates, it may be possible to express them in plants at levels that do not incur a heavy metabolic cost. In addition, some secondary metabolites, such as anthocyanins [133], can benefit human health while potentially reducing insect damage. Reductions in insect herbivory in crops susceptible to fungal colonization (e.g. maize expressing *Bacillus thuringiensis* crystal protein) can result in lower levels of mycotoxins that are harmful to humans [23]. We believe that the utility of secondary metabolite engineering will greatly benefit agriculture in the future.

Acknowledgements

We thank Mark Berhow reviewing a draft of the manuscript. We thank Dave Lee for technical assistance.

Disclaimer

Mention of trade names or commercial products in this article is solely for the purpose of providing specific information and does not imply recommendation or endorsement by the U.S. Department of Agriculture.

REFERENCES

[1] Agrawal, A. A. and Kurashige, N. S. (2003). A role for isothiocyanates in plant resistance against the specialist herbivore *Pieris rapae*. *J. Chem. Ecol.*, *29*, 1403-1415.

[2] Aharoni, A., Giri, A. P., Deuerline, S., Griepink, F., de-Kogel, W.-J., Verstappen, F. W. A., Verhoeven, H. A., Jongsma, M. A., Schwab, W. and Bouwmeester, H. J. (2003). Terpenoid metabolism in wild-type and transgenic Arabidopsis plants. *Plant Cell*, *15*, 2866-2884.

[3] Aharoni, A., Jongsma, M. A., Kim, T. Y., Ri, M. B., Giri, A. P., Verstappen, F. W. A., Schwab, W. and Bouwmeester, H. J. (2006). Metabolic engineering of terpenoid biosynthesis in plants. *Phytochem. Rev.*, *5*, 49-58.

[4] Andreasson, E., Bolt, J. L., Hoglund, A. S., Rask, L. and Meijer, J. (2001). Different myrosinase and idioblast distribution in Arabidopsis and *Brassica napus*. *Plant Physiol.*, *127*, 1750-1763.

[5] Aoki, T., Akashi, T. and Ayabe, S. (2000). Flavonoids of leguminous plants: structure, biological activity and biosynthesis. *J. Plant Res.*, *113*, 475-488.

[6] Bak, S., Paquette, S., Morant, M., Morant, A. V., Saito, S., Zagrobelny, M., Jørgensen, K., Osmani, S., Jørgensen, B. and Møller, B. L. (2006). Cyanogenic glucosides: A case study for evolution and application of cytochromes P450. *Phytochem. Rev.*, *5*, 309-329.

[7] Bi, J. L., Felton, G. W., Murphy, J. B., Howles, P. A., Dixon, R. A. and Lamb, C. J. (1997). Do plant phenolics confer resistance to specialist and generalist insect herbivores? *J. Agric. Food Chem.*, *45*, 4500-4504.

[8] Bing, J. W., Guthrie, W. D., Dicke, F. F. and Obrycki, J. J. (1990). Relation of corn leaf aphid (Homoptera: Aphididae) colonization to DIMBOA content in maize inbred lines. *J. Econ. Entomol.*, *83*, 1626-1632.

[9] Bläske, V. U., Hertel, H. and Forschler, B. T. (2003). Repellent effects of isoborneol on subterranean termites (Isoptera: Rhinotermitidae) in soils of different composition. *J. Econ. Entomol.*, *96*, 1267-1274.

[10] Bloem, K. A. and Duffey, S. S. (1990). Interactive effect of protein and rutin on larval *Heliothis zea* and the endoparasitoid *Hyposoter exiguae*. *Entomol. Exp. Appl.*, *54*, 149-160.

[11] Bohlmann, J., DeLuca, V., Eilert, U. and Martin, W. (1995). Purification and cDNA cloning of anthranilate synthase from *Ruta graveolens*: modes of expression and properties of native and recombinant enzymes. *Plant J.*, *7*, 491-501.

[12] Borevitz, J. O., Xia, Y., Blount, J., Dixon, R. A. and Lamb, C. (2000). Activation tagging identifies a conserved MYB regulator of phenylpropanoid biosynthesis. *Plant Cell*, *12*, 2383-2394.

[13] Broun, P., Liu, Y., Queen, E., Schwarz, Y., Abenes, M. L. and Leibman, M. (2006). Importance of transcription factors in the regulation of plant secondary metabolism and their relevance to the control of terpenoid accumulation. *Phytochem. Rev.*, *5*, 27-38.

[14] Byrne, P. F., Darrah, L. L., Snook, M. E., Wiseman, B. R., Widstrom, N. W., Moellenback, D. J. and Barry, B. D. (1996). Maize silk browning, maysin content, and antibiosis to the corn earworm, *Helicoverpa zea* (Boddie). *Maydica*, *41*, 13-18.

[15] Cambier, V., Hance, T. and de Hoffman, E. (2000). Variation of DIMBOA and related compounds content in relation to the age and plant organ in maize. *Phytochemistry, 53*, 223-229.

[16] Chalker-Scott, L. (2002). Do anthocyanins function as osmoregulators in leaf tissues? *Adv. Bot. Res., 37*, 103-127.

[17] Cheng, A.-X., Xiang, C.-Y., Li, J.-X., Yang, C.-Q., Hu, W.-L., Wang, L.-J., Lou, Y.-G. and Chen, X.-Y. (2007). The rice (*E*)-β-caryophyllene synthase (OsTPS3) accounts for the major inducible volatile sesquiterpenes. *Phytochemistry, 68*, 1632-1641.

[18] Croteau, R. B., Kutchan, T. M., and Lewis, N. G. (2000). Natural Products (Secondary Metabolites). In B. B. Buchanan, W. Gruissem, and R. L. Jones (Eds.), *Biochemistry and Molecular Biology of Plants* (pp. 1250-1318). Rockville, MD: American Society of Plant Physiologists.

[19] De Paepe, A., De Grauwe, L., Bertrand, S., Smalle, J. and Van der, S. D. (2005). The *Arabidopsis* mutant *eer2* has enhanced ethylene responses in the light. *J. Exp. Bot., 56*, 2409-2420.

[20] Deavours, B. E. and Dixon, R. A. (2005). Metabolic engineering of isoflavonoid biosynthesis in alfalfa. *Plant Physiol., 138*, 2245-2259.

[21] Dicke, M. (1999). Evolution of induced indirect defense of plants. In C. D. Harwell, and R. Trollian (Eds.), *The Ecology and Evolution of Inducible Defenses* (pp. 62-88). Princeton, NJ: Princeton University Press.

[22] Diemer, F., Caissard, J. C., Moja, S., Chalchat, J. C. and Jullien, F. (2001). Altered monoterpene composition in transgenic mint following the introduction of 4*S*-limonone synthase. *Plant Physiol. Biochem., 39*, 603-614.

[23] Dowd, P. F., Johnson, E. T., and Williams, W. P. (2005). Strategies for insect managment targeted toward mycotoxin management. In H. K. Abbas (Ed.), *Aflatoxin and Food Safety* (pp. 517-541). Boca Raton: CRC Press.

[24] Dowd, P. F., Johnson, E. T., and Pinkerton, T. S. (2007). Selectable markers with potential activity agains insects, plus other insect-oriented strategies for mycotoxin reduction in Midwest corn. *Proceedings of the Multicrop Aflatoxin, Fumonisin and Fungal Genomics Workshop.*

[25] Dowd, P. F. and Vega, F. E. (1996). Enzymatic oxidation products of allelochemicals as a basis for resistance against insects: Effects on the corn leafhopper *Dalbulus maidis*. *Natural Toxins, 4*, 85-91.

[26] Dritschilo, W., Krummel, J., Nafus, D. and Pimentel, D. (1979). Herbivorous insects colonizing cyanogenic and acyanogenic *Trifolium repens*. *Heredity, 42*, 49-56.

[27] Elliger, C. A., Wong, Y., Chan, B. G. and Waiss Jr., A. C. (1981). Growth inhibitors in tomato (*Lypersicon*) to tomato fruitworm (*Heliothis zea*). *J. Chem. Ecol., 7*, 753-758.

[28] Fahey, J. W., Zalcmann, A. T. and Talalay, P. (2001). The chemical diversity and distribution of glucosinolates and isothiocyanates among plants. *Phytochemistry, 56*, 5-51.

[29] Franks, T. K., Powell, K. S., Choimes, S., Marsh, E., Iocco, P., Sinclair, B. J., Ford, C. M. and van Heeswijck, R. (2006). Consequences of transferring three sorghum genes for secondary metabolite (cyanogenic glucoside) biosynthesis to grapevine hairy roots. *Transgenic Res., 15*, 181-195.

[30] Frey, M., Chomet, P., Glawischnig, E., Stettner, C., Grun, S., Winklmair, A., Eisenreich, W., Bacher, A., Meeley, R. B., Briggs, S. P., Simcox, K. and Gierl, A.

(1997). Analysis of a chemical plant defense mechanism in grasses. *Science*, *277*, 696-699.

[31] Frey, M., Huber, K., Park, W. J., Sicker, D., Lindberg, P., Meeley, R. B., Simmons, C. R., Yalpani, N. and Gierl, A. (2003). A 2-oxoglutarate-dependent dioxygenase is integrated in DIMBOA-biosynthesis. *Phytochemistry*, *62*, 371-376.

[32] Gigolashvili, T., Berger, B., Mock, H. P., Muller, C., Weisshaar, B. and Flugge, U. I. (2007). The transcription factor HIG1/MYB51 regulates indolic glucosinolate biosynthesis in *Arabidopsis thaliana*. *Plant J.*, *50*, 886-901.

[33] Gigolashvili, T., Engqvist, M., Yatusevich, R., Muller, C. and Flugge, U. I. (2008). HAG2/MYB76 and HAG3/MYB29 exert a specific and coordinated control on the regulation of aliphatic glucosinolate biosynthesis in *Arabidopsis thaliana*. *New Phytol.*, *177*, 627-642.

[34] Gigolashvili, T., Yatusevich, R., Berger, B., Muller, C. and Flugge, U. I. (2007). The R2R3-MYB transcription factor HAG1/MYB28 is a regulator of methionine-derived glucosinolate biosynthesis in *Arabidopsis thaliana*. *Plant J.*, *51*, 247-261.

[35] Gill, R. I. S. and Ellis, B. E. (2006). Over-expression of tryptophan decarboxylase gene in poplar and its possible role in resistance against *Malacosoma disstria*. *New Forests*, *31*, 195-209.

[36] Gill, R. I. S., Ellis, B. E. and Isman, M. B. (2003). Tryptamine-induced resistance in tryptophan decarboxylase transgenic poplar and tobacco plant against their specific herbivores. *J. Chem. Ecol.*, *29*, 779-793.

[37] Gitelson, I. I., Tikhomirov, A. A., Parshina, O. V., Ushakova, S. A. and Kalacheva, G. S. (2003). Volatile metabolites of higher plant crops as a photosynthesizing life support system component under temperature stress at different light intensities. *Adv. Space Res.*, *31*, 1781-1786.

[38] Goff, S. A., Klein, T. M., Roth, B. A., Fromm, M. E., Cone, K. C., Radicella, J. P. and Chandler, V. L. (1990). Transactivation of anthocyanin biosynthetic genes following transfer of B regulatory genes into maize tissues. *EMBO J.*, *9*, 2517-2522.

[39] Gómez-Galera, S., Pelacho, A. M., Gené, A., Capell, T. and Christou, P. (2007). The genetic manipulation of medicinal and aromatic plants. *Plant Cell Rep.*, *26*, 1689-1715.

[40] Gouinguene, S., Degen, T. and Turlings, T. C. J. (2001). Variability in herbivore-induced odor emissions among maize cultivars and their wild ancestors. *Chemoecology*, *11*, 9-16.

[41] Gould, K. S., McKelvie, J. and Markham, K. R. (2002). Do anthocyanins function as antioxidants in leaves? Imaging of H_2O_2 in red and green leaves after mechanical injury. *Plant, Cell Environ.*, *25*, 1261-1269.

[42] Grotewold, E., Drummond, B. J., Bowen, B. and Peterson, T. (1994). The myb-homologous P gene controls phlobaphene pigmentation in maize floral organs by directly activating a flavonoid biosynthetic gene subset. *Cell*, *76*, 543-553.

[43] Grubb, C. D. and Abel, S. (2006). Glucosinolate metabolism and its control. *Trends Plant Sci.*, *11*, 89-100.

[44] Halkier, B. A. and Gershenzon, J. (2006). Biology and biochemistry of glucosinolates. *Annu. Rev. Plant Biol.*, *57*, 303-333.

[45] Hassanali, A. and Lwande, W. (1989). Antipest secondary metabolites from African plants. *ACS Symp. Series*, *387*, 78-94.

[46] Hedin, P. A., Jenkins, J. N., Collum, D. H., White, W. H., Parrott, W. L. and MacGown, M. W. (1983). Cyanidin-3-β-glucoside, a newly recognized basis for resistance in cotton to the tobacco budworm *Heliothis virescens* (Fab.)(Lepidoptera: Noctuidae). *Experientia, 39*, 799-801.

[47] Henrick, C. A. (1995). Pyrethroids. In C. R. A. Godfrey (Ed.), *Agrochemicals from Natural Products* (pp. 63-145). New York: Marcel Dekker.

[48] Hirai, M. Y., Sugiyama, K., Sawada, Y., Tohge, T., Obayashi, T., Suzuki, A., Araki, R., Sakurai, N., Suzuki, H., Aoki, K., Goda, H., Nishizawa, O. I., Shibata, D. and Saito, K. (2007). Omics-based identification of *Arabidopsis* Myb transcription factors regulating aliphatic glucosinolate biosynthesis. *Proc. Natl. Acad. Sci. U. S. A., 104*, 6478-6483.

[49] Hoffmann-Campo, C. B., Harborne, J. B. and McCaffery, A. R. (2001). Pre-ingestive and post-ingestive effects of soya bean extracts and rutin on *Trichoplusia ni* growth. *Entomol. Exp. Appl., 98*, 181-194.

[50] Hoffmann-Campo, C. B., Neto, J. A. R., de Oliveira, M. C. and Oliveira, L. J. (2006). Detrimental effect of rutin on *Anticarsia gemmatalis*. *Pesqui. Agropecu. Bras., 41*, 1453-1459.

[51] Hwang, Y.-S., Wu, K.-H., Kumamoto, J., Axelrod, H. and Mulla, M. S. (1985). Isolation and identification of mosquito repellents in *Artemisia vulgaris*. *J. Chem. Ecol., 11*, 1297-1306.

[52] Jenkins, J. N., Hedin, P. A., Parrott, W. L., McCarty, J. C. and White, W. H. (1983). Cotton allelochemics and growth of tobacco budworm larvae. *Crop Sci., 23*, 1195-1198.

[53] Johnson, E. T., Berhow, M. A. and Dowd, P. F. (2008). Colored and white sectors from star-patterned petunia flowers display differential resistance to corn earworm and cabbage looper larvae. *J. Chem. Ecol.,* in press.

[54] Johnson, E. T., Berhow, M. A. and Dowd, P. F. (2007). Expression of a maize *Myb* transcription factor driven by a putative silk-specific promoter significantly enhances resistance to *Helicoverpa zea* in transgenic maize. *J. Agric. Food Chem., 55*, 2998-3003.

[55] Johnson, E. T. and Dowd, P. F. (2004). Differentially enhanced insect resistance, at a cost, in *Arabidopsis thaliana* constitutively expressing a transcription factor of defensive metabolites. *J. Agric. Food Chem., 52*, 5135-5138.

[56] Jørgensen, K., Rasmussen, A. V., Morant, M., Nielsen, A. H., Bjarnholt, N., Zagrobelny, M., Bak, S. and Møller, B. L. (2005). Metabolon formation and metabolic channelling in the biosynthesis of plant natural products. *Curr. Opin. Plant Biol., 8*, 280-291.

[57] Kappers, I. F., Aharoni, A., van Herpen, T. W., Luckerhoff, L. L., Dicke, M. and Bouwmeester, H. J. (2005). Genetic engineering of terpenoid metabolism attracts bodyguards to *Arabidopsis*. *Science, 309*, 2070-2072.

[58] Karageorgou, P. and Manetas, Y. (2006). The importance of being red when young: Anthocyanins and the protection of young leaves of *Quercus coccifera* from insect herbivory and excess light. *Tree Physiol., 26*, 613-621.

[59] Kelly, P. J., Bones, A. and Rossiter, J. T. (1998). Sub-cellular immunolocalization of the glucosinolate sinigrin in seedlings of *Brassica juncea*. *Planta, 206*, 370-377.

[60] Klun, J. A., Guthrie, W. D., Hallauer, A. R. and Russell, W. A. (1970). Genetic nature of the concentration of 2,4-dihydroxy-7-methoxy (2*H*)-1,4-benzoxazin-3-(4*H*)-one and

resistance to the European corn borer in a diallel set of eleven maize inbreds. *Crop Sci.*, *10*, 87-90.

[61] Klun, J. A. and Robinson, J. F. (1969). Concentration of two 1,4-benzoxazinones in dent corn at various stages of development of the plant and its relation to resistance of the host plant to the European corn borer. *J. Econ. Entomol.*, *62*, 214-220.

[62] Klun, J. A., Tipton, C. L. and Brindley, T. A. (1967). 2,4-dihydroxy-7-methoxy-1,4-benzoxasin-3-one (DIMBOA), an active agent in the resistance of maize to the European corn borer. *J. Econ. Entomol.*, *60*, 1529-1533.

[63] Koes, R., Verweij, W. and Quattrocchio, F. (2005). Flavonoids: a colorful model for the regulation and evolution of biochemical pathways. *Trends Plant Sci.*, *10*, 236-242.

[64] Kristensen, C., Morant, M., Olsen, C. E., Ekstrom, C. T., Galbraith, D. W., Moller, B. L. and Bak, S. (2005). Metabolic engineering of dhurrin in transgenic *Arabidopsis* plants with marginal inadvertent effects on the metabolome and transcriptome. *Proc. Natl. Acad. Sci. U. S. A.*, *102*, 1779-1784.

[65] Lane, G. A., Biggs, D. R., Russell, G. B., Sutherland, O. R. W., Williams, E. M., Maindonald, J. H. and Donnell, D. J. (1985). Isoflavonoid feeding deterrents for *Costelytra zealandica* structure-activity relationships. *J. Chem. Ecol.*, *11*, 1713-1735.

[66] Lavy, M., Zuker, A., Lewinsohn, E., Larkov, O., Ravid, U., Vainstein, A. and Weiss, D. (2002). Linalool and linallol oxide production in transgenic carnation flowers expressing the *Clarkia breweri* linalool synthase gene. *Mol. Breeding*, *9*, 103-111.

[67] Lee, H. I., Leon, J. and Raskin, I. (1995). Biosynthesis and metabolism of salicylic acid. *Proc. Natl. Acad. Sci. U. S. A.*, *92*, 4076-4079.

[68] Lee, S., Tsao, R. and Coats, J. R. (1999). Influence of dietary applied monoterpenoids and derivatives on survival and growth of the European corn borer (Lepidoptera: Pyralidae). *J. Econ. Entomol.*, *92*, 56-67.

[69] Lewinsohn, E., Schalechet, F., Wilkinson, J., Matsui, K., Tadmor, Y., Nam, K. H., Amar, O., Lastochkin, E., Larkov, O., Ravid, U., Hiatt, W., Gepstein, S. and Pichersky, E. (2001). Enhanced levels of the aroma and flavor compound *S*-linalool by metabolic engineering of the terpenoid pathway in tomato fruits. *Plant Physiol.*, *127*, 1256-1265.

[70] Li, F., Jin, Z., Qu, W., Zhao, D. and Ma, F. (2006). Cloning of a cDNA encoding the *Saussurea medusa* chalcone isomerase and its expression in transgenic tobacco. *Plant Physiol. Biochem.*, *44*, 455-461.

[71] Lindgren, B. S., Nordlander, G. and Birgersson, G. (1996). Feeding deterrence of verbenone to the pine weevil, *Hylobius abietis* (L.)(Col., Curculionidae). *J. Appl. Entomol.*, *120*, 397-403.

[72] Liu, R., Hu, Y., Li, J. and Lin, Z. (2007). Production of soybean isoflavone genistein in non-legume plants via genetically modified secondary metabolism pathway. *Metab. Eng.*, *9*, 1-7.

[73] Llyod, A. M., Walbot, V. and Davis, R. W. (1992). *Arabidopsis* and *Nicotiana* anthocyanin production activated by maize regulators *R* and *C1*. *Science*, *258*, 1773-1775.

[74] Lücker, J., Bouwmeester, H. J., Schwab, W., Blaas, J., van der Plas, L. H. and Verhoeven, H. A. (2001). Expression of *Clarkia S*-linalool synthase in transgenic petunia plants results in the accumulation of *S*-linalyl-β-D-glucopyranoside. *Plant J.*, *27*, 315-324.

[75] Lücker, J., Schwab, W., van Hautum, B., Blaas, J., van der Plas, L. H., Bouwmeester, H. J. and Verhoeven, H. A. (2004). Increased and altered fragrance of tobacco plants after metabolic engineering using three monoterpene synthases from lemon. *Plant Physiol.*, *134*, 510-519.

[76] Ludwig, S. R., Bowen, B., Beach, L. and Wessler, S. R. (1990). A regulatory gene as a novel visible marker for maize transformation. *Science*, *247*, 449-450.

[77] Mahmoud, S. S. and Croteau, R. B. (2001). Metabolic engineering of essential oil yield and composition in mint by altering expression of deoxyxylulose phosphate reductoisomerase and menthofuran synthase. *Proc. Natl. Acad. Sci. U. S. A.*, *98*, 8915-8920.

[78] Manetas, Y. (2006). Why some leaves are anthocyanic and why most anthocyanic leaves are red? *Flora: Morph. Dist. Func. Ecol. Plants*, *201*, 163-177.

[79] Martin, D. M., Gershenzon, J. and Bohlmann, J. (2003). Induction of volatile terpene biosynthesis and diurnal emission by methyl jasmonate in foliage of Norway Spruce. *Plant Physiol.*, *132*, 1586-1599.

[80] Melanson, D., Chilton, M. D., Masters-Moore, D. and Chilton, W. S. (1997). A deletion in an indole synthase gene is responsible for the DIMBOA-deficient phenotype of *bxbx* maize. *Proc. Natl. Acad. Sci. U. S. A.*, *94*, 13345-13350.

[81] Memelink, J. and Gantet, P. (2007). Transcription factors involved in terpenoid indole alkaloid biosynthesis in *Catharanthus roseus*. *Phytochem. Rev.*, *6*, 353-362.

[82] Morant, A. V., Jørgensen, K., Jørgensen, B., Dam, W., Olsen, C. E., Møller, B. L. and Bak, S. (2007). Lessons learned from metabolic engineering of cyanogenic glucosides. *Metabolomics*, *3*, 383-398.

[83] Muir, S. R., Collins, G. J., Robinson, S., Hughes, S., Bovy, A., Ric, D., V, van Tunen, A. J. and Verhoeyen, M. E. (2001). Overexpression of petunia chalcone isomerase in tomato results in fruit containing increased levels of flavonols. *Nat. Biotechnol.*, *19*, 470-474.

[84] Neuffer, M. G., Coe, E. H., Jr., and Wessler, S. R. (1997). *Mutants of Maize*. Cold Spring Harbor, NY: Cold Spring Harbor Press.

[85] Nielsen, K. A., Tattersall, D. B., Jones, P. R. and Moller, B. L. (2008). Metabolon formation in dhurrin biosynthesis. *Phytochemistry*, *69*, 88-98.

[86] Niemeyer, H. M. (1988). Hydroxamic acids (4-hydroxy-1,4-benzoxazin-3-ones), defence chemicals in the Gramineae. *Phytochemistry*, *27*, 3349-3358.

[87] Niemeyer, H. M., Pesel, E., Franke, S. and Francke, W. (1989). Ingestion of the benzoxazinone DIMBOA from wheat plants by aphids. *Phytochemistry*, *28*, 2307-2310.

[88] Niyogi, K. K. and Fink, G. R. (1992). Two anthranilate synthase genes in Arabidopsis: defense-related regulation of the tryptophan pathway. *Plant Cell*, *4*, 721-733.

[89] Niyogi, K. K., Last, R. L., Fink, G. R. and Keith, B. (1993). Suppressors of *trp1* fluorescence identify a new Arabidopsis gene, *TRP4*, encoding the anthranilate synthase beta subunit. *Plant Cell*, *5*, 1011-1027.

[90] Oikawa, A., Ebisui, K., Sue, M., Ishihara, A. and Iwamura, H. (1999). Purification and characterization of a β-glucosidase specific for 2,4-dihydroxy-7-methoxy-1,4-benzoxazin-3-one (DIMBOA) glucoside in maize. *Z. Naturforsh.*, *54c*, 181-185.

[91] Oksman-Caldentey, K. M. and Inzé, D. (2004). Plant cell factories in the post-genomic era: new ways to produce designer secondary metabolites. *Trends Plant Sci.*, *9*, 433-440.

[92] Park, S. W., Kaimoyo, E., Kumar, D., Mosher, S. and Klessig, D. F. (2007). Methyl salicylate is a critical mobile signal for plant systemic acquired resistance. *Science*, *318*, 113-116.

[93] Pasquali, G., Porto, D. D. and Fett-Netto, A. G. (2006). Metabolic engineering of cell cultures versus whole plant complexity in production of bioactive monoterpene indole alkaloids: recent progress related to old dilemma. *J. Biosci. Bioeng.*, *101*, 287-296.

[94] Pinkerton, T. S. and Dowd, P. F. (2007). Expression of an anthranilate synthase from maize mutant *bf-1* in maize line Hi II. *Final Program of the Annual Meeting of the American Society of Plant Biologists*, Abstract P45021.

[95] Piubelli, G. C., Hoffmann-Campo, C. B., Moscardi, F., Miyakubo, S. H. and de Oliveira, M. C. (2005). Are chemical compounds important for soybean resistance to *Anticarsia gemmatalis*? *J. Chem. Ecol.*, *31*, 1509-1525.

[96] Poulton, J. E. (1989). Toxic compounds in plant foodstuffs: cyanogens. In J. E. Kinsella, and W. G. Soucie (Eds.), *Food Proteins* (pp. 381-401). Champaign, IL: American Oil Chemists' Society.

[97] Quiel, J. A. and Bender, J. (2003). Glucose conjugation of anthranilate by the *Arabidopsis* UGT74F2 glucosyltransferase is required for tryptophan mutant blue fluorescence. *J. Biol Chem.*, *278*, 6275-6281.

[98] Rivera, S. B., Swedlund, B. D., King, G. J., Bell, R. N., Hussey, C. E., Jr., Shattuck-Eidens, D. M., Wrobel, W. M., Peiser, G. D. and Poulter, C. D. (2001). Chrysanthemyl diphosphate synthase: isolation of the gene and characterization of the recombinant non-head-to-tail monoterpene synthase from *Chrysanthemum cinerariaefolium*. *Proc. Natl. Acad. Sci. U. S. A.*, *98*, 4373-4378.

[99] Rohde, B., Hans, J., Martens, S., Baumert, A., Hunziker, P. and Matern, U. (2008). Anthranilate *N*-methyltransferase, a branch-point enzyme of acridone biosynthesis. *Plant J.*, *53*, 541-553.

[100] Rosling, H. (1987). *Cassava Toxicity and Food Security* (p. 16). Uppsula, Sweden: Tryck kontact.

[101] Sadasivam, S., and Thayumanavan, B. (2003). *Molecular Host Plant Resistance to Pests*. New York: Marcel Dekker, Inc.

[102] Salom, S. M., Gray, J. A., Randall Alford, A., Mulesky, M., Fettig, C. J. and Woods, S. A. (1996). Evaluation of natural products as antifeedants for the pales weevil (Coleoptera: Curculionidae) and as fungitoxins for *Leptographium procerum*. *J. Entomol. Sci.*, *31*, 453-465.

[103] Sato, F., Hashimoto, T., Hachiya, A., Tamura, K., Choi, K. B., Morishige, T., Fujimoto, H. and Yamada, Y. (2001). Metabolic engineering of plant alkaloid biosynthesis. *Proc. Natl. Acad. Sci. U. S. A.*, *98*, 367-372.

[104] Schaefer, H. M. and Rolshausen, G. (2006). Plants on red alert: do insects pay attention? *BioEssays*, *28*, 65-71.

[105] Schlindwein, C. C. D., Feet-Neto, A. G. and Dillenburg, L. R. (2006). Chemical and mechanical changes during leaf expansion of four woody species of a dry Restinga woodland. *Pl. Biol.*, *8*, 430-438.

[106] Schnee, C., Kollner, T. G., Held, M., Turlings, T. C., Gershenzon, J. and Degenhardt, J. (2006). The products of a single maize sesquiterpene synthase form a volatile defense signal that attracts natural enemies of maize herbivores. *Proc. Natl. Acad. Sci. U. S. A.*, *103*, 1129-1134.

[107] Simmonds, M. S. and Stevenson, P. C. (2001). Effects of isoflavonoids from *Cicer* on larvae of *Heliocoverpa armigera. J. Chem. Ecol.*, *27*, 965-977.

[108] Singh, M. and Widholm, J. M. (1975). Study of a corn (*Zea mays* L.) mutant (*blue fluorescent-1*) which accumulates anthranilic acid and its beta-glucoside. *Biochem. Genet.*, *13*, 357-367.

[109] Skirycz, A., Reichelt, M., Burow, M., Birkemeyer, C., Rolcik, J., Kopka, J., Zanor, M. I., Gershenzon, J., Strnad, M., Szopa, J., Mueller-Roeber, B. and Witt, I. (2006). DOF transcription factor AtDof1.1 (OBP2) is part of a regulatory network controlling glucosinolate biosynthesis in Arabidopsis. *Plant J.*, *47*, 10-24.

[110] Spencer, K. (1988). K. Spencer (Ed.), *Chemical Mediation of Coevolution* (pp. 167-240). Orlando, FL: Academic Press.

[111] Sreevidya, V. S., Srinivasa, R. C., Sullia, S. B., Ladha, J. K. and Reddy, P. M. (2006). Metabolic engineering of rice with soybean isoflavone synthase for promoting nodulation gene expression in rhizobia. *J. Exp. Bot.*, *57*, 1957-1969.

[112] Stamp, N. E. (1994). Interactive effects of rutin and constant versus alternating temperatures on performance of *Manduca sexta* caterpillars. *Entomol. Exp. Appl.*, *72*, 125-133.

[113] Steyn, W. J., Wand, S. J. E., Holcroft, D. M. and Jacobs, G. (2002). Anthocyanins in vegetative tissues: A proposed unified function in photoprotection. *New Phytol.*, *155*, 349-361.

[114] Tattersall, D. B., Bak, S., Jones, P. R., Olsen, C. E., Nielsen, J. K., Hansen, M. L., Hoj, P. B. and Moller, B. L. (2001). Resistance to an herbivore through engineered cyanogenic glucoside synthesis. *Science*, *293*, 1826-1828.

[115] Teas, H. J. and Anderson, E. G. (1950). Blue fluorescent, a new mutant in maize. *Genetics*, *35*, 696-697.

[116] Thomas, J. C., Akroush, A. M. and Adamus, G. (1999). The indole alkaloid tryptamine produced in transgenic *Petunia hybrida. Plant Physiol. Biochem.*, *37*, 665-670.

[117] Tian, L. and Dixon, R. A. (2006). Engineering isoflavone metabolism with an artificial bifunctional enzyme. *Planta*, *224*, 496-507.

[118] Tian, L., Pang, Y. and Dixon, R. A. (2008). Biosynthesis and genetic engineering of proanthocyanidins and (iso)flavonoids. *Phytochem. Rev.*, in press.

[119] Tipton, C. L., Klun, J. A., Husted, R. R. and Pierson, M. D. (1967). Cyclic hydroxamic acids and related compounds from maize. *Biochemistry*, *6*, 2866-2870.

[120] Tohge, T., Nishiyama, Y., Hirai, M. Y., Yano, M., Nakajima, J., Awazuhara, M., Inoue, E., Takahashi, H., Goodenowe, D. B., Kitayama, M., Noji, M., Yamazaki, M. and Saito, K. (2005). Functional genomics by integrated analysis of metabolome and transcriptome of Arabidopsis plants over-expressing an MYB transcription factor. *Plant J.*, *42*, 218-235.

[121] Ton, J., D'Alessandro, M., Jourdie, V., Jakab, G., Karlen, D., Held, M., Mauch-Mani, B. and Turlings, T. C. J. (2007). Priming by airborne signals boosts direct and indirect resistance in maize. *Plant J.*, *49*, 16-26.

[122] Tunç, I. and Erler, F. (2003). Repellency and repellent stability of essential oil constituents against *Tribolium confusum. Z. Pflanzenkr. Pflanzenschutz*, *110*, 394-400.

[123] Turlings, T. C. J., Tumlinson, J. H., Heath, R. B., Proveaux, A. T. and Doolittle, R. E. (1991). Isolation and identification of allelochemicals that attract the larval parasitoid,

Cotesia marginiventris (Cresson), to the microhabitat of one of its hosts. *J. Chem. Ecol.*, *17*, 2235-2251.

[124] Turlings, T. C. J., Tumlinson, J. H. and Lewis, W. J. (1990). Exploitation of herbivore-induced plant odors by host-seeking parasitic wasps. *Science*, *250*, 1251-1253.

[125] Ververdis, F., Trantas, E., Douglas, C., Vollmer, G., Kretzschmar, G. and Panopoulos, N. (2007). Biotechnology of flavonoids and other phenylpropanoid-derived natural products. Part II: Reconstruction of multienzyme pathways in plants and microbes. *Biotech. J.*, *2*, 1235-1249.

[126] Wang, S. F., Ridsdill-Smith, T. J. and Ghisalberti, E. L. (1999). Levels of isoflavonoids as indicators of resistance of subterranean clover trifolates to redlegged earth mite *Halotydeus destructor*. *J. Chem. Ecol.*, *25*, 795-803.

[127] Wentzell, A. M., Rowe, H. C., Hansen, B. G., Ticconi, C., Halkier, B. A. and Kliebenstein, D. J. (2007). Linking metabolic QTLs with network and cis-eQTLs controlling biosynthetic pathways. *PLoS. Genet.*, *3*, 1687-1701.

[128] Widholm, J. M. (1972). Anthranilate synthetase from 5-methyltryptophan-susceptible and -resistant cultured *Daucus carota* cells. *Biochim. Biophys. Acta*, *279*, 48-57.

[129] Widholm, J. M. (1972). Cultured *Nicotiana tabacum* cells with an altered anthranilate synthetase which is less sensitive to feedback inhibition. *Biochim. Biophys. Acta*, *261*, 52-58.

[130] Woodhead, S. and Bernays, E. (1977). Changes in release rates of cyanide in relation to palatability of sorghum to insects. *Nature*, *270*, 235-236.

[131] Youn, H., Lee, S., Cho, J. H. and Oh, H. (1991). Insecticidal isoflavon glycoside from *Maackia amurensis*. *Arch. Pharm. Res.*, *14*, 105-108.

[132] Yu, O., Shi, J., Hession, A. O., Maxwell, C. A., McGonigle, B. and Odell, J. T. (2003). Metabolic engineering to increase isoflavone biosynthesis in soybean seed. *Phytochemistry*, *63*, 753-763.

[133] Zafra-Stone, S., Bagchi, M. and Bagchi, D. (2007). Health benefits of edible berry anthocyanins: novel antioxidant and anti-angiogenic properties. *ACS Symp. Series*, *956*, 337-351.

[134] Zalkin, H. (1985). Anthranilate synthase. *Methods Enzymol.*, *113*, 287-292.

[135] Zhang, Z., Ober, J. A. and Kliebenstein, D. J. (2006). The gene controlling the quantitative trait locus *EPITHIOSPECIFIER MODIFIER1* alters glucosinolate hydrolysis and insect resistance in *Arabidopsis*. *Plant Cell*, *18*, 1524-1536.

In: Genetically Modified Plants
Editors: Timm Wolf and Jonas Koch

ISBN 978-1-60456-696-3

Chapter 8

GENE UP-REGULATION BY DNA DEMETHYLATION IN 35S-*GSHI* TRANSGENIC POPLARS (*POPULUS* X *CANESCENS*)

***G. Gyulai*[1*], *Z. Tóth*[1], *A. Bittsánszky*[1,2], *G. Gullner*[2], *Z. Szabó*[1], *J. Kiss*[1], *T. Kőmíves*[2] *and L. Heszky*[1]**

[1]St. Stephanus University, Institute of Genetics and Biotechnology, Gödöllő, H-2103, Hungary

[2]Plant Protection Institute, Hungarian Academy of Sciences, Budapest, H-1525, Hungary

ABSTRACT

Gene expression levels of transgene 35S-*gsh*I (γ-glutamylcysteine synthetase) cloned from *E. coli*, and the endogenous gene *gsh*1 of poplar (*Populus* x *canescens*) were up-regulated by the DNA demethylating agent DHAC (5,6-dihydro-5'-azacytidine hydrochloride) (10^{-4} M for 7 days) in aseptic leaf discs cultures. Two 35S-*gsh*I-transgenic (6*lgl* and 11*ggs*) and wild type (WT) poplar clones were used. The efficiency of gene upregulation was also analyzed under herbicide paraquat stress (4 x 10^{-7} M). Levels of *gsh*I-mRNA and *gsh*1-mRNA were determined by RT-qPCR (reverse transcriptase quantitative PCR) after cDNA synthesis. For internal control, the constitutively expressed housekeeping poplar genes *α-tubulin* and *actin* were used, and the $2^{-\Delta\Delta Ct}$ method was applied for data analysis. In long term DHAC treatment (21 days), a morphogenetic response of *de novo* root development was observed on leaf discs in a wide concentration range of DHAC (10^{-8} to 10^{-6} M). Adventitious shoots (11*ggs* clone) also emerged from leaf discs after a combined treatment with DHAC (10^{-4} M) and paraquat (10^{-7} M). Shoots were dissected, rooted and transplanted in glass houses for further analyses for phytoremediation capacity. Since DNA methylation patterns are inherited (*epigenetic memory*), these poplar plants with increased gene expression levels of both transgene 35S-*gsh*I and endogenous gene *gsh*1 provide novel plant sources for *in situ* application.

* e-mail: gyulai.gabor@mkk.szie.hu.

ABBREVIATIONS:

5-mC:	*5-methylcytosine*;
AdoMet:	S-adenosyl methionine dependent methyltransferases;
C:	cytosine nucleotide of DNA;
BA:	benzyl adenine;
cDNA:	copia *DNA*;
DHAC:	5,6-dihydro-5'-azacytidine hydrochloride;
E. coli:	bacterium *Escherichia coli*;
GMO:	genetically modified organism;
γ-ECS:	γ-glutamylcysteine synthetase enzyme encoded by gene *gsh*;
GSH:	glutathione tripeptide (γ-L-glutamyl-L-cysteinyl-glycine);
MTases:	DNA methyltransferases;
NAA:	naphthalene acetic acid;
PCR:	polymerase chain reaction;
NTC:	non template control;
PQ:	paraquat (*N,N'-Dimethyl-4,4'-bipyridinium dichloride*; *syn.*: methylviologen);
RT-qPCR:	reverse transcription based quantitative PCR;
*rbc*S:	gene for RBCS (*syn.*: RuBPCase SSU, ribulose-1,5-bisphosphate carboxylase small subunit);
TGS:	transcriptional gene silencing;
PTGS:	post-transcriptional gene silencing;
6*lgl* and 11*ggs*:	*gsh*I-transformed poplar (*Populus* x *canescens*) clones;
WT:	wild type.

INTRODUCTION

Poplars (*Populus* ssp) are capable of removing and degrading toxic substances and heavy metals from polluted soils through phytoremediation due to the extensive root system, high water uptake capacity, rapid growth and large biomass production (Kőmíves *et al.* 1998; Burken and Schnoor 1998; Di Baccio *et al.* 2003; Gyulai *et al.* 2005). The oxidative stress tolerance capacity of *Populus* x *canescens* clones has been significantly increased recently by genetic transformation with the 35S-*gsh*I gene, which encodes for γ-glutamylcysteine synthetase (γ-ECS, EC 6.3.2.2) (Leple *et al.* 1992; Arisi *et al.* 1997; Noctor *et al.* 1998) cloned from *E. coli* (Watanabe *et al.* 1986; NCBI X03954). The transformed poplar clones (6*lgl* and 11*ggs*) showed higher levels of GSH and its precursor of γ-L-glutamyl-L-cysteine (γ-EC) than the WT, which led to an improved detoxification capacity against various environmental pollutants (Noctor *et al.* 1998; Peuke and Rennenberg 2005).

Transgenic clones (6*lgl* and 11*ggs*) studied have been maintained in aseptic shoot cultures for about a decade without 35S-*gsh*I transgene elimination (Gyulai *et al.* 2005; Bittsánszky *et al.* 2006). However, transgenes have been exposed to gene silencing processes either in the region of the constitutive CaMV-35S promoter, or in the coding region of the genes. The CaMV-35S promoter is considered to be a constitutive promoter; nevertheless its expression might be modulated by photoperiod, temperature, developmental stages and DNA methylation (Benfey and Chua 1990; Zardo *et al.* 1999; Schnurr and Guerra 2000; Obertello *et al.* 2005; Yang *et al.* 2005). By the application of a DNA-demethylation treatment, as in the study presented here, this natural gene silencing process can be reversed.

DNA Methylation:

DNA methylation of plant genes is a general endogenous process for gene silencing catalyzed by nuclear enzyme family MTases (DNMT - *DN*A *m*ethyl*t*ransferases in mammals; and DNMT-orthologue MET - *me*thyl*t*ransferases, CMT - *c*ytosine *m*ethyl*t*ransferases, and DRM in plants). These enzymes transfer a methyl group (CH_3) mainly from S-*ad*eno*s*yl *met*hionine (AdoMet-dependent methyltransferases) either to the position of cytosine-C_5 (EC 2.1.1.73), cytosine-N_4 (E.C. 2.1.113) or adenine-N_6 of DNA (E.C. 2.1.2.72) (Pósfai *et al.* 1989; Cheng and Roberts 2001; Vaucheret 2006). The cytosine methylation to 5-methylcytosine (5mC) is the most frequent process resulting in down regulation of both genes and transgenes. DNA methylation is frequently directed by RdDM (*R*NA-*d*irected *D*NA *m*ethylation) (Linn *et al.* 1990; Kumpatla *et al.* 1997; Castilho *et al.* 1999; Mathieu and Bender 2004).

Genetically, DNA methylation patterns are associated with two types of gene silencing in plants, the *t*ranscriptional *g*ene *s*ilencing (TGS) caused by methylation in the promoters of genes, and the *p*ost-*t*ranscriptional *g*ene *s*ilencing (PTGS; called RNA interference in animals, RNAi) caused by methylation in the coding region of genes (Elmayan and Vaucheret 1996; Kooter *et al.* 1999). DNA methylation is not universal, as in the insect fruit fly *Drosophyla* where is does not occur (Hirochika *et al.* 2000). The meiotically heritable TGS (Park *et al.* 1996) results in methylation '*inprints*', whereas PTGS, which is not heritable, affects the stability of the RNA transcripts, at a translational level under the control of RNAi (Napoli *et al.* 1990; Krol *et al.* 1990; Fire *et al.* 1998; Agrawal *et al.* 2003). The PTGS (RNAi) system allows cells to control endogenous nucleic acids (e.g. transposons) and exogenous (e.g. virus, transgenes) invaders (Vaucheret 2006).

DNA methyltransferases can be blocked exogenously by MTase-inhibitors, which act via covalent complex formation (Wolfe and Matzke 1999; Fan *et al.* 2005) when present either in the cytosol or when incorporated into DNA (as base analogues) resulting in gene reactivation (Bird 2002; Issa and Kantarjian 2005). Gene reactivation through the application of thymidine analogues can also occur in demethylation-independent gene up-regulations (Fan *et al.* 2005).

The study presented here aims to achieve the simultaneous analysis of gene up-regulation of both the prokarytic gene 35S-*gsh*I and the endogenous eukaryotic poplar gene *gsh*1 of two 35S-*gsh*I-transformant (*6lgl* and 11*ggs*) poplar clones (Leple *et al.* 1992; Arisi *et al.* 1997; Noctor *et al.* 1998) and the WT, in response to DHAC-treatment (10^{-4} M) combined with paraquat (4×10^{-7} M) stress.

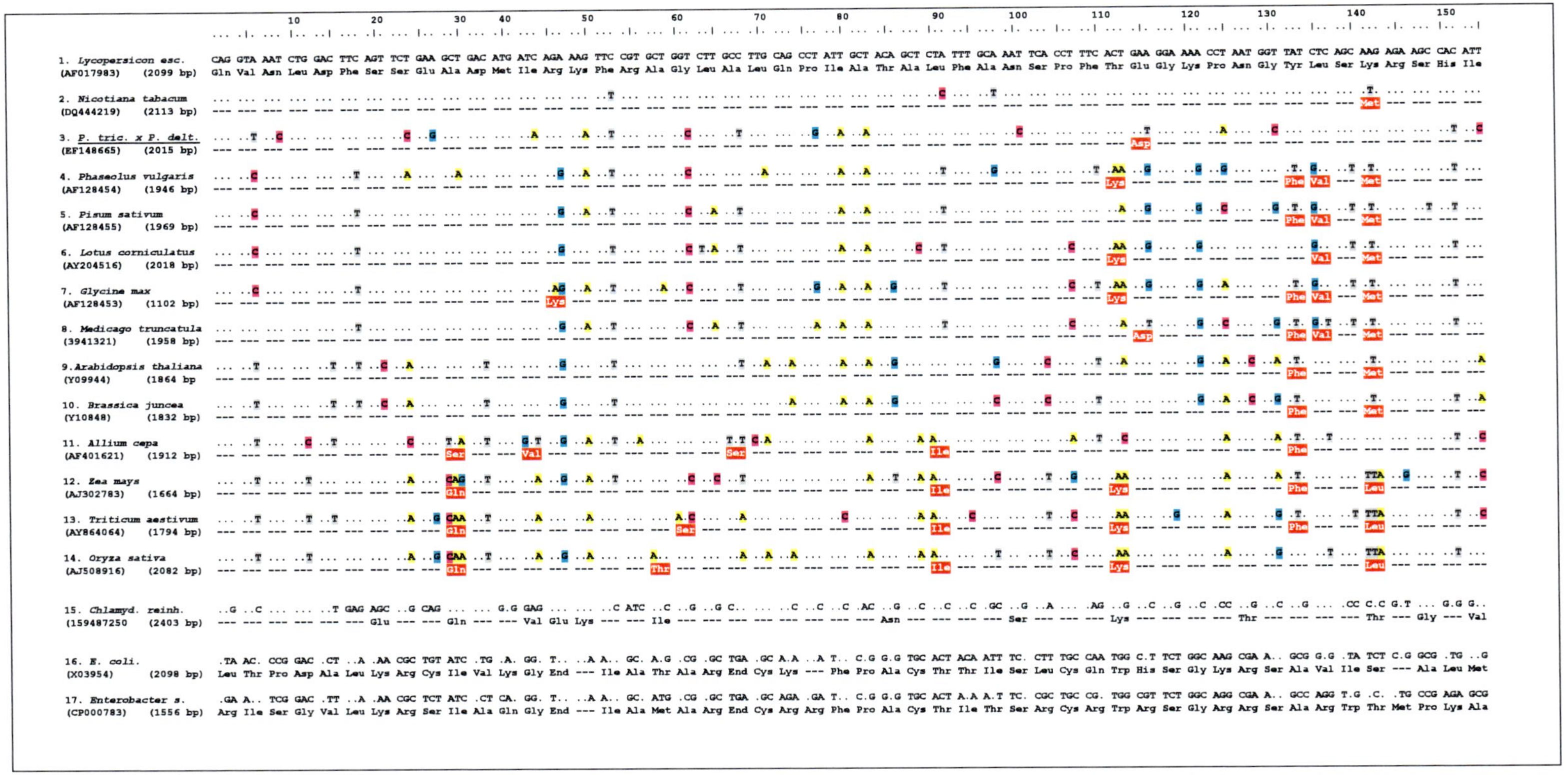

Figure 1. Multiple sequence alignments and box shade of parts (150 nt) of *gsh*1-cDNA triplets (top lines) of plants (1 – 14), alga *Chlamydomonas reinhardi* (15.); and the non-homologous *gsh*I-cDNAs of prokaryotes of *Escherichia coli* (16.) and *E. sakazaki* (17.) aligned with translated amino acid (50) sequences (below). Synonymous and non-synonymous (amino acid red box) nucleotide substitutions are indicated. Sequences found in NCBI databases (Altschul *et al.* 1997) were analyzed by BioEdit program (Hall 1999).

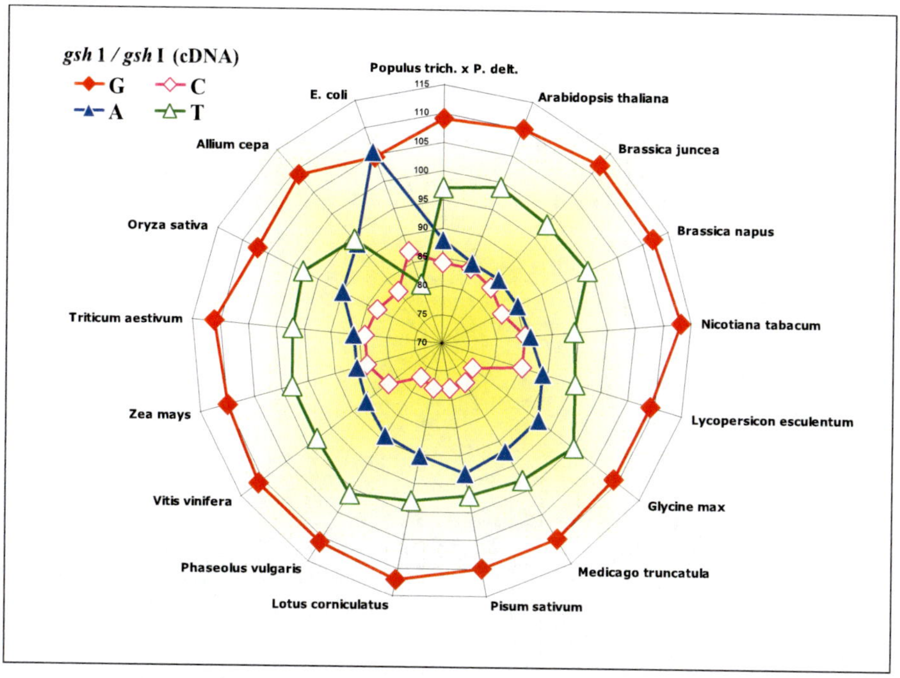

Figure 2. Nucleotide (G, C, A, T) compositions of *gsh*1-cDNAs of sixteen plant species and the non-homologous *gsh*I-cDNA of prokaryote *E. coli*. (accession # are indicated in Figure 1).

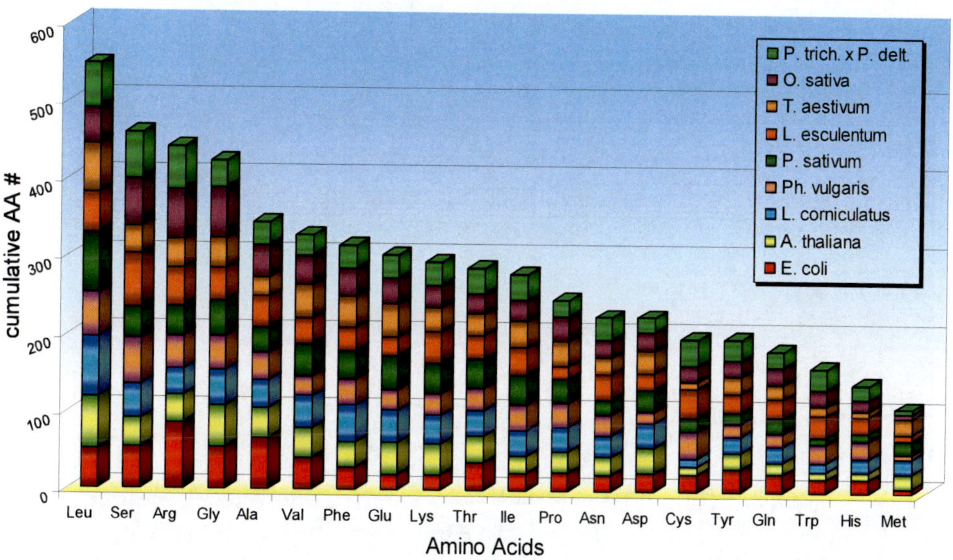

Figure 3. Comparative amino acid (AA) content of complete sequences of the leucine (Leu)-rich GHS1 enzymes of eight plant species compared to non-homologous GSHI of *E. coli* (accession # are indicated in Figure 1).

Figure 4. (Continued on next page.)

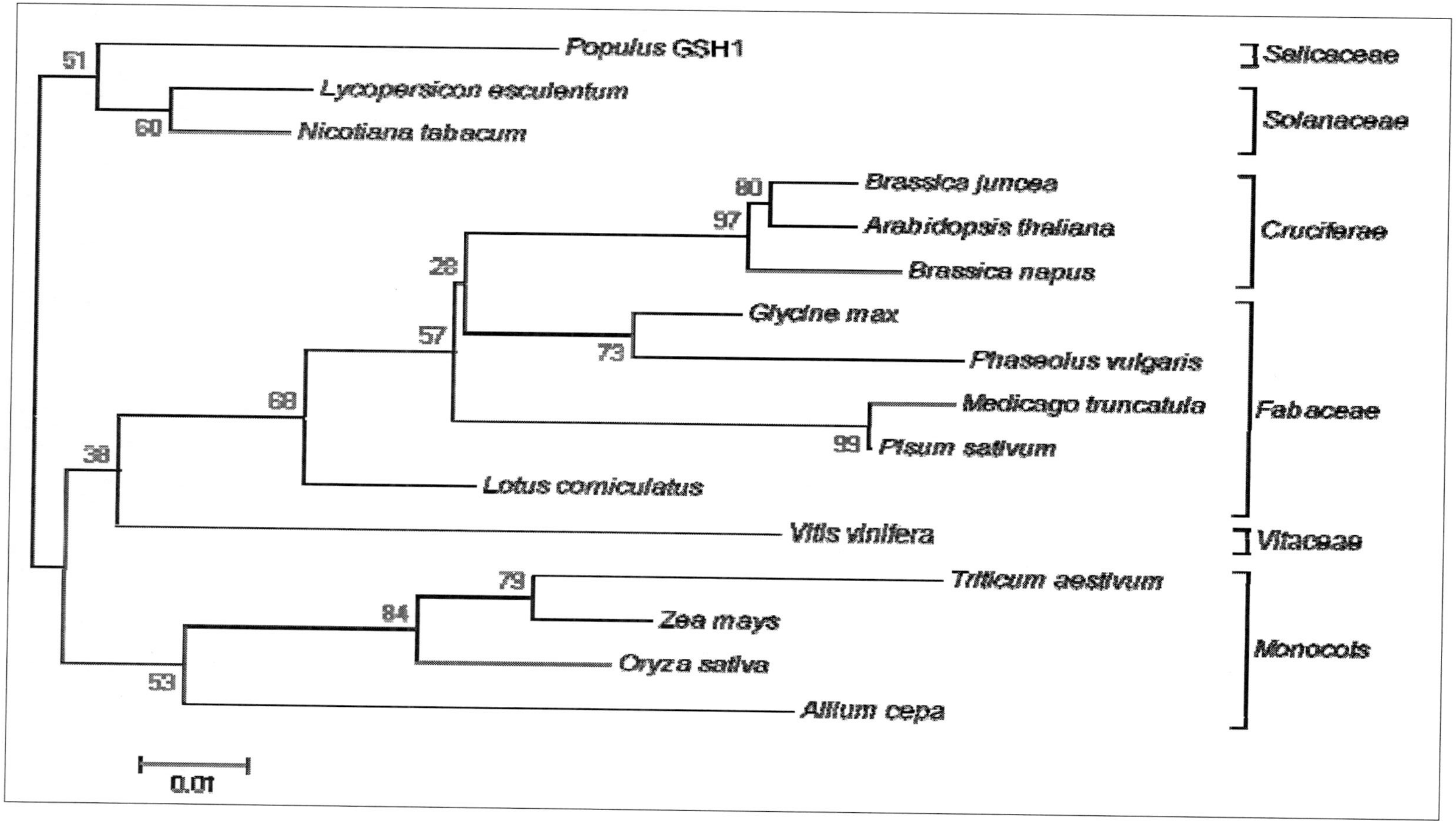

Figure 4. Distance trees (MEGA4; Tamura *et al.* 2007) of sixteen plant *gsh*1-cDNA (derived form *gsh*1-mRNAs) (left), and the translated GSH1 protein sequences (right) with boot strap analyses (1000). Size bars indicate numbers of substitutions per locus (accession numbers are indicated in Figure 1).

CONCLUSION

Sequence Analysis of *gsh* Genes:

Comparative analysis of sequence alignments (Figure 1), DNA nucleotide diversities (Figure 2), amino acid contents (Figure 3) and dendrogram analyses based on nucleotide and amino acid sequences of both *gsh*1 and *gsh*I genes of prokaryotic and eukaryotic plant species (Figure 4) reflects different rates of DNA nucleotide substitutions due to the consequences of evolution and adaptation (Baldwin 1896; Amirnovin 1997).

As a result of redundancy of genetic code, which provides protection against mutations, synonymous (*silent*) nucleotide substitutions (*point mutation*) in the protein coding triplets do not cause amino acid replacement in the protein sequences (Wong 1981 in 1988). Sample sections of the plant DNA and cDNA of *gsh*1 genes and the non-homologous prokaryotic *gsh*I gene of *E. coli* revealed several synonymous and non-synonymous nucleotide substitutions (Figure 1). In general, nucleotide substitution is a neutral processes as certain codons are translated more frequently (*codon use preferences*) which might be related to adaptation capacity (*fitness*) of the species (Palmer 1997); however, distance tree of *gsh*1 sequences of plant species showed correlation with taxonomical lineages, with dominance of *Brassicaceae* species (Figure 4). Both cDNA and protein sequences of poplar showed the closest linkage to *Solanaceae* species (tomato and tobacco), however nucleotide substitution scale (0.005) showed double level of that of amino acids (0.01) (Figure 4).

The Role of Reduced Glutathione (GSH) in Plant Cell:

Reduced glutathione (GSH), the major non-protein, tripeptide thiol compound (Meister and Anderson 1983), present in high concentrations (mM) of plant tissues, is regarded as one of the major determinants of cellular redox homeostasis (Rusznyák and Szent-Györgyi 1936) with a key role in the detoxification processes (Strohm *et al.* 1995; Foyer and Noctor 2005; Mullineaux and Rausch 2005; Wachter *et al.* 2005). Glutathione becomes reactive as the thiol group of cysteine residue releases electrons to unstable electron acceptor molecules, such as reactive oxygen species. The two reduced GSH molecules form glutathione disulfide (GSSG) which regenerates to GSH again by the catalysis of gluathione reductase. Cells under regular conditions ('healthy cells') have a GSH / GSSG ratio over about 90 %. Changes in the GSSG / GSH ratio indicates that plants are exposed to oxidative stress.

Up-Regulation of *gsh* Genes by DNA Demethylation:

Gene expression levels can be easily triggered *in vitro* by up-regulation (*syn.*: reactivation, hypomethylation and demethylation) or down-regulation (*syn.*: silencing, hypermethylation and suppression) of the gene of interest (Goffin and Eisenhauer 2002; Columbus 2006). For gene up-regulation, MTase-inhibitors such as the structurally modified cytosine analogues zebularine, 5-azacytidine (5-azaC), 5-aza-2'-deoxycytidine and DHAC have been shown to be highly effective (Chen and Pikaard 1997; Cao *et al.* 2000; Cheng and

Roberts 2001). Alternatively, the drug 3-aminobenzamide has been used for gene down-regulation in a series of genes by DNA hypermethylation (Zardo *et al.* 1999). In the study presented here, DHAC was applied due to its high chemical stability in aqueous solutions of long term treatments (Goffin and Eisenhauer 2002).

RT-qPCR:

Reverse transcription (RT) followed by qPCR analysis has proven to be an exceptionally sensitive method compared to RNA-DNA hybridization (*Northern blot*) (Alwine *et al.* 1977) for both absolute and relative quantification of gene expressions (Veres *et al.* 1987). In the study presented here, relative quantification was used as it is more relevant to compare expression levels of different treatments rather than absolute quantification (Pfaffl 2001; Livak and Schmittgen 2001; Tichopad *et al.* 2003).

The relative gene expression level of 35S-*gsh*I transgene in the 6*lgl* clone showed a 13.5-fold increase over the 11*ggs* clone (1.0) which was doubled (1.8-fold) in the DHAC-treated 6*lgl* samples (23.7) but not in the 11*ggs* clone (0.4-fold) (Figure 5). This expression pattern was contrary to the observation of relative copy numbers of the transgene 35S-*gsh*I as it was lower in the 6*lgl* clone (1.0) than in the 11*ggs* samples (1.69) (Table 1). These results might be due to the 35S-*gsh*I transgene construct of the 6*lgl* clone which included an additional targeting sequence (from 32nd nt to 202nd nt of the total 206 bp; NCBI M25614) of a transit peptide (TP) (57 amino acids) gene of *rbc*S (RuBPCase SSU: small subunit of RuBPCase, ribulose-1,5-bisphosphate carboxylase) (Leple *et al.* 1992; Arisi *et al.* 1997; Noctor *et al.* 1998; Bittsánszky *et al.* 2007). TPs as N-terminal extensions of transgene-products facilitate targeting and translocation of the cytosolically synthesized RBCS-GSH complex into plastids, which has also been found to be more effective in transgenic tobacco compared to regular transgene cassettes (Creissen *et al.* 1996; Bruce 2001; Wachter *et al.* 2005).

Gene expression levels of the endogenous poplar gene *gsh*1 also showed high responsiveness to DHAC-induced demethylation with an extremely high expression in the untransformed WT poplar clone (19.8-fold). A competition in the reactivation capacity between transgene 35S-*gsh*I and poplar *gsh*1 of 6*lgl* clone occurred as the gene expression of transgene 35S-*gsh*I increased from a high rel. expression level (13.5) up by about a two-fold (1.8 times) rate to 23.7 compared to the poplar *gsh*1 gene that increased by an 8.7-fold increment from a lower level (1.3) to 13.9 rel. unit (Figure 5). These results might indicate differences in DNA methylating capacity between transgenes and proper wild type genes as a type of cosuppression (Krol *et al.* 1990; Anand *et al.* 2003; Baumberger and Baulcombe 2005).

The phenomenon of co-suppression was originally discovered in plants (Napoli *et al.* 1990; Krol *et al.* 1990). Napoli *et al.* (1990) intended to up-regulate the activity of a gene for chalcone synthase (*chs*A), an enzyme involved in the production of anthocyanin pigments. Some of the transgenic petunia plants harboring the *chs*A coding region under the control of a CaMV-35S promoter lost both endogene and transgene chalcone synthase activity. As a result, many of the flowers were variegated in color or developed white sectors (Napoli *et al.* 1990). The molecular analysis discovered that transgene expression led to the formation of dsRNA, which, in turn, initiated PTGS (Metzlaff *et al.* 1997).

The interaction between transgenes and endogenous genes might also have resulted either from multicopy integration of transgene at the same locus (Assaad *et al.* 1993), the position effects due to random integration (Yang *et al.* 2005), the AT/CG composition of the transgene (Matzke and Matzke 1998), the presence of inverted repeats in the integration site (Stam *et al.* 1997), the overexpression of the transgene (Que *et al.* 1997), or the environmental conditions (Meyer and Heidman 1994).

Selective methylation capacity of transgene homologues has also been observed in supertransformants (a type of transgene pyramiding) where the resident transgene promoter *Ubi*1 from the first transformation remained unmethylated whereas the incoming 35S transgene promoter of the subsequent transformation was silenced in rice (Yang *et al.* 2005).

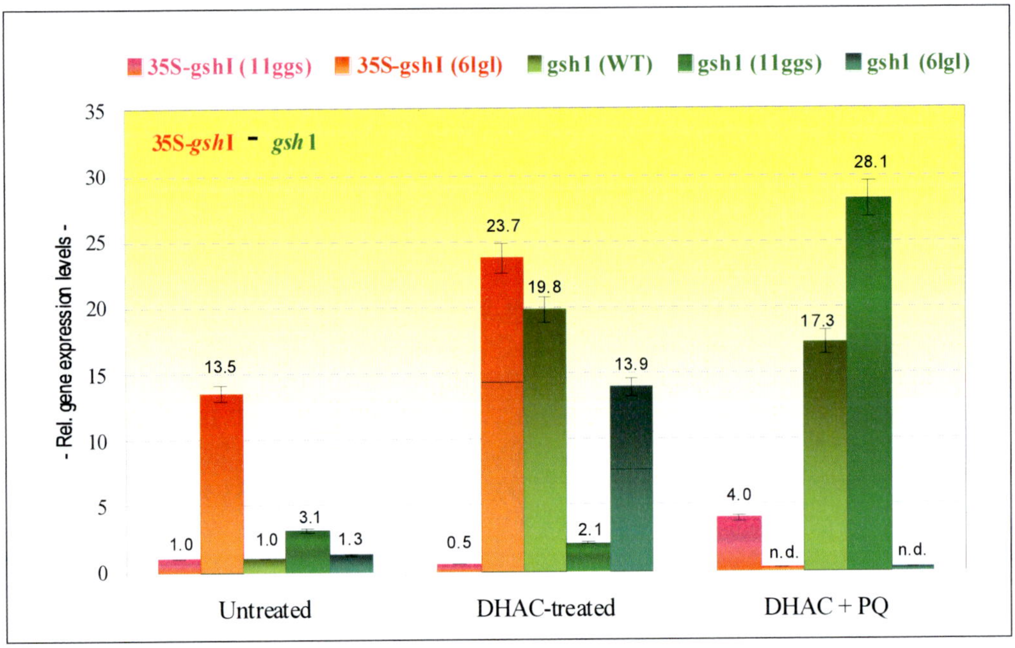

Figure 5. Relative gene expression levels (RT-qPCR) of transgene 35S-*gsh*I (*E. coli*) and endogenous poplar (*Populus x canescens*) gene *gsh*1 in the 35S-*gsh*I-transgenic clones of 11*ggs*, 6*lgl* and WT exposed to DHAC (10^{-4} M; 7 days) and DHAC combined with paraquat (PQ) (4×10^{-7} M; 7 days) (n = 6) (n.d. - not detected).

Table 1. Relative copy number of transgene 35S-*gsh*I (*E. coli*) in the transgenic poplar (*P. x canescens*) clones 6*lgl* and 11*ggs* determined by RT-qPCR analysis using the $2^{-\Delta\Delta Ct}$ method with the internal control of constitutively expressed poplar gene *actin* (three repetitions of two samples in each case, n = 6)

Poplar	35S-*gsh*I		*actin*		ΔCt		ΔΔCt		35S-*gsh*I copy # ($2^{-\Delta\Delta Ct}$)		
lines	Ct	SD	Ct	SD	Value	SD	Value	SD	-/+ range		Value
6*lgl*	23.75	0.14	23.41	0.24	0.34	0.21	0.0	0.21	0.86	1.16	1.00
11*ggs*	22.50	0.26	22.91	0.14	-0.41	0.13	-0.75	0.13	1.54	1.85	1.69

Paraquat Stress:

Paraquat (*N,N'-Dimethyl-4,4'-bipyridinium dichloride*; *syn.*: methylviologen) primarily acts as an electron acceptor in the electron transport chains located in chloroplasts (Gyulai 1984; Lehoczki *et al.* 1992; Szigeti *et al.* 2001). The paraquat, at bleaching concentrations totally impairs chloroplast function by reduction of the bleached tissues generating superoxide radicals that react with unsaturated lipids in membranes (Bittsánszky *et al.* 2005). This deleterious effect of paraquat was effectively eliminated by the DHAC treatment, as the endogenous poplar gene *gsh*1 of the 11*ggs* clone did not show DHAC responses but under paraquat stress it increased by 8-fold (from 3.1 to 28.1) along with co-expressing the transformed 35S-*gsh*I gene with 4-fold increment (from 1.0 to 4.0) (Figure 5). In contrast, endogenous *gsh*1 of the WT clone showed extreme upregulation in response to DHAC-induced DNA-demethylation with 20-fold increase (from 1.0 to 19.8) but no further responses to paraquat stress (17.3) (Figure 5).

Increased *gsh*-activity with elevated levels of *gsh*-mRNA (syn.: γ-ECS-mRNA) has also been reported in *Brassica napus* (Sun *et al.* 2005), *Brassica juncea* (Schäfer *et al.* 1998), and *Arabidopsis thaliana* under different stress conditions (Xiang and Oliver 1998; Harada *et al.* 2002). The moss *Physcomitrella patens* also showed a high level of γ-ECS overexpression (5.7 – 7.9-fold increase) in response to heavy metal (10 μM Cd^{2+}, for 3 days) stress (Rother *et al.* 2006). These results indicate a wide stress-response capacity of *gsh*1 genes not only to herbicides but also to heavy metals.

DHAC-Induced Morphogenesis:

A *de novo* root development was observed on DHAC-treated poplar leaf discs incubated for long term (21 days) treatment, which indicates a multi-target action site of DHAC at especially the auxin-related root initiating genes (Figure 6).

The morphogenetic capacity of DNA demethylating agents on plant development has been reported in different organs (Finnegan and Kovac 2000; Xiao *et al.* 2006). Flower induction of *Arabidopsis* (Finnegan *et al.* 1998) and shoot development of *Petunia* (Prakash and Kumar 1997; Prakash *et al.* 2003) were initiated by DNA-demethylating agents. Early flower bud development (vernalization, *remembering winter*) in *Arabidopsis* was found to be coupled with low levels of DNA methylation (Finnegan *et al.* 1998, 2005; Henderson and Dean 2004).

Low levels of methyl-cytosine were also associated with organogenetic capability in sugarbeet (*Beta vulgaris altissima*) (Causevic *et al.* 2005). Contrary to morphogenesis, incremental DNA methylation levels occurred during bud dormancy (Horvath *et al.* 2003; Law and Suttle 2003). Methylation also plays a key role in the chromosome modelling as it turnes out that the Pc-G (*P*oly*c*omb *G*roup) protein complexes encoded by *pcg-g* genes (*p*oly*c*omb *g*roup genes) control flowering in plants, which genes are similar to the PRC2 (Polycomb Repressive Complex 2) in animals, and functions as a histone methyltransferase (Chanvivattana *et al.* 2004).

Adventitious shoots were also developed from leaf discs of the 11*ggs* clone treated with DHAC (10^{-4} M) and paraquat (10^{-7} M) in long term cultur (Figure 7). Shoots (three in total of the four repetition experiments) were micropropagated and rooted *in vitro* according to

Gyulai *et al.* (2005), and transplanted in glass houses (86 lines) for further analyses for phytoremediation capacity. As DNA methylation patterns are inherited (*epigenetic memory*), these poplar plants with increased gene expression levels of both transgene 35S-*gsh*I and endogenous gene *gsh*1 provide novel plant sources with elevated stress capacity.

Figure 6. Root initiation capacity of the DNA demethylating agent DHAC (10^{-8} M to 10^{-6} M) on leaf discs of untransformed (WT) and 35S-*gsh*I-transformed poplar (*Populus* × *canescens*) clones (6*lgl* and 11*ggs*) incubated on long-term (21 days) aseptic agar media.

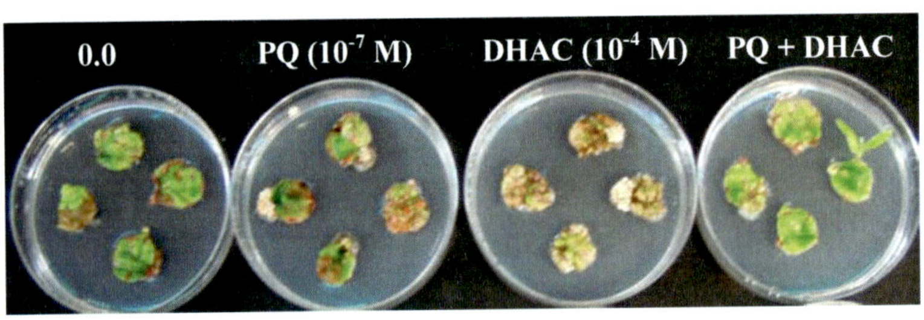

Figure 7. Samples of shoot regeneration from leaf discs of 35S-*gsh*I-transgenic poplar (*Populus* × *canescens*) clone (11*ggs*) treated with DNA demethylating agent DHAC (10^{-4} M) combined with paraquat (10^{-7} M) stress and incubated on aseptic agar media for 21 days.

MATERIALS AND METHODS

Plant material: Clones (INRA 717-1-B4) of the untransformed (WT) poplar (*Populus* x *canescens* = *P. tremula* x *P. alba*; 2n = 4x = 38; 4.5 to 5.5 x 10^8 bp) (Taylor 2002) and the genetically transformed lines overexpressing 35S-*gsh*I (γ-glutamylcysteine synthetase; EC 3.2.3.3; cloned from *Escherichia coli*; 1.557 bp) (Watanabe *et al.* 1986) (NCBI X03954) gene product, glutathione (GSH) either in the cytosol (line 11*ggs*) or in the chloroplasts (line 6*lgl*) were used. Gene constructs are driven by a copy of the CaMV-35S promoter (Leple *et al.* 1992; Arisi *et al.* 1997; Noctor *et al.* 1998).

Shoot culture in vitro: Nodal segments of poplar clones were micropropagated and maintained in aseptic shoot cultures *in vitro* (Gyulai *et al.* 1995; Kiss 1999; Gullner *et al.* 2005).

Detection of gene expression levels by RT-qPCR analysis, RNA isolation: Relative gene expression levels of 35S-*gsh*I-transgene (*E. coli*) and the endogenous poplar gene *gsh*1 were analyzed by RT-qPCR in the control of constitutively expressed housekeeping poplar gene *α-tubulin* and *actin*. Total RNA was extracted from 0.05 g leaf disc tissues using the Absolutely RNA Miniprep Kit (# 400800, Stratagene, USA - Biomedica, Hungary) following the manufacture's protocol. Three individual leaf discs were analyzed in duplicate measurements (n = 6) in each case. The quality and quantity of extracted RNA samples (2 μl) were measured by NanoDrop ND-1000 UV-Vis spectrophotometer (NanoDrop Technologies, Montchanin, DE, USA – BioScience, Budapest, Hungary).

cDNA synthesis: Reverse transcription of first strand cDNA was synthesized on the mRNA templates by RT (reverse transcriptase of Moloney Murine Leukemia Virus: M-MuLV) with oligo$(dT)_{18}$ (0.5 μg) as the primer following the manufacturer's protocol (# K1622; Fermentas – Biocenter, Szeged, Hungary).

Gene expression analysis: First strand cDNA samples (2.5 μl) were directly applied in RT-qPCR (25 μl) and probed by gene specific primers (400 nM). Primers were as follows: 35S-*gsh*I: 5'-aggtcaggacatcgaactgg-3' and 5'-gatgcaccaaacagataagg-3'; *gsh*1-poplar: 5'-agttccgaggctgacatgat-3' and 5'-cagcacggttgttgtcagta-3'; α-tubulin (poplar): 5'-taaccgccttgtttctcagg-3' and 5'-cctggggtatggaaccaagt-3'; and actin (poplar): 5'-aatggtaccggaatggtcaa-3' and 5'-cccaacatacgcatcctttt-3' according to Bittsánszky *et al.* (2007).

Kit of DyNAmo HS SybrGreenI qPCR kit (# F-410L, Finnzymes, Finland – Izinta, Hungary) was used. Reactions were performed in forty cycles (95 °C/20sec, 60 °C/20sec, 72 °C/20sec) prior to a hold at 95 °C for 10 min, and a final hold at 4°C. Reactions were run by Rotor Gene 6000 cycler (Corbett Research, Australia – Izinta, Hungary).

Detection of relative copy number of 35S-*gsh*I-transgene: Relative copy number of 35S-*gsh*I-transgene in clones of 11*ggs* and 6*lgl* were determined under the control of constitutively expressed housekeeping gene *actin* by using selective primer pairs used for RT-PCR. The levels of *gsh*I-mRNA were detected by the amplified fragment (273 nt) of the incorporated transgene (from 667 nt to 939 nt).

Data analysis of real-time RT-qPCR: For both calibration and quantification of reactions, ten-fold serial dilutions (1x, 10^{-1}x, 10^{-2}x, 10^{-3}x) of cDNAs were applied including controls of NTC (non DNA-template control) and ddH_2O. Data were analyzed by relative quantification of the $2^{-\Delta\Delta Ct}$ method (Livak and Schmittgen 2001). Ct values (threshold cycle): The threshold of fluorescence value (dR) of the amplified PCR products was determined manually above the background of fluorescence signals. Standard curve correlating Ct values to log amount of DNA were plotted at high R^2- ratio (0.976 to 0.987). ΔCt: ΔCt values were calculated as Ct_{gshI} minus $Ct_{\alpha\text{-}tubulin}$ and Ct_{gsh1} minus $Ct_{\alpha\text{-}tubulin}$ according to (Livak and Schmittgen 2001). ΔΔCt values: ΔΔCt values were determined as mean $Ct_{untreated}$ minus mean $Ct_{treated}$ (Livak and Schmittgen 2001).

Multiple sequence alignments were carried out by BioEdit Sequence Alignment Editor (NCSU, USA) (Hall 1999) and CLUSTALW EMBL-EBI (Thompson *et al.* 1994) software programs. BLAST (Basic Local Alignment Search Tool) analysis (Altschul *et al.* 1997) was carried out by a computer program of NCBI (National Center for Biotechnology Information). Distance trees based on *gsh*1 sequences were edited by either CLUSTALW EMBL-EBI (Thompson *et al.* 1994) or MEGA4 (Tamura *et al.* 2007). For MEGA4 the following steps were applied: Bootstrap Test of Phylogeny (1000); Neighbor-Joining; Gaps (Complete deletions); Substitution model (Nucleotide Maximum Composite Likelihood) according to Tamura *et al.* (2007). Diagrams were edited by Microsoft Office Excel program (9625 West 76th Street, Eden Prairie, MN 55344, USA).

Acknowledgments

Authors thank RP Malone (DIT, Dublin, Ireland) and L Waters Jr (AU, Alabama, USA), for their comments on the manuscript. Poplar clones were kindly provided by CH Foyer (Newcastle upon Tyne U, UK) and H Rennenberg (A-LU, Freiburg, G).

References

Agrawal, N; Dasaradhi, PVN; Mohmmed, A; Malhotra, P; Bhatnagar, RK; Mukherjee, SK. (2003) RNA Interference: Biology, Mechanism and Applications. *Microbiol. Mol. Biol. Reviews,* 67, 657-685.

Alwine, JC; Kemp, DJ; Stark, GR. (1977) Method for detection of specific RNAs in agarose gels by transfer to diazobenzyloxymethyl-paper and hybridization with DNA probes. *Proc. Natl. Acad. Sci. USA,* 74, 5350-5354.

Amirnovin, R. (1997) An analysis of the metabolic theory of the origin of the genetic code. *J. Mol. Evol.,* 44, 473-476.

Anand, A; Trick, HN; Gill, BS; Muthukrishnan, S. (2003) Stable transgene expression and random gene silencing in wheat. *Plant Biotechnology J.,* 1, 241-251.

Altschul, SF; Madden, TL; Schaffer, AA; Zhang, JH; Zhang, Z; Miller, W; Lipmand, DJ. (1997) Gapped BLAST and PSI-BLAST: a new generation of protein database search programs. *Nucleic Acids Res.,* 25, 3389-3402.

Arisi, A-CM; Noctor, G; Foyer, CF; Jouanin, L. (1997) Modification of thiol contents in poplars (*Populus tremula* x *P. alba*) overexpressing enzymes involved in glutathione synthesis. *Planta,* 203, 362-372.

Assaad, F; Tucker, KL; Signer, ER. (1993) Epigenetic repeat-induced gene silencing (RIGS) in *Arabidopsis*. *Plant Mol. Biol.,* 22, 1067–1085.

Baldwin, M.J. (1896) A new factor in evolution. *The American Naturalist*, 30, 441-451.

Baumberger, N. and Baulcombe, DC. (2005) *Arabidopsis* ARGONAUTE1 is an RNA slicer that selectively recruit microRNAs and short interfering RNAs. *Proc. Natl. Acad. Sci. USA,* 102, 11928-11933.

Benfey, P.N. and Chua, N.H. (1990) The cauliflower mosaic virus *35S* promoter: combinational regulation of transcription in plants. *Science,* 250, 959–966

Bird, A. (2002) DNA methylation patterns and epigenetic memory. *Genes and Dev.,* 16, 6-21.

Bittsánszky, A; Kőmíves, T; Gullner, G; Gyulai, G; Kiss, J; Heszky, L; Radimszky, L; Rennenberg, H. (2005) Ability of transgenic poplars with elevated glutathione content to tolerate Zinc (2+) stress. *Environment Int.*, 31, 251-254.

Bittsánszky, A; Gyulai, G; Humphreys, M; Gullner, G; Csintalan, Zs; Kiss, J; Szabó, Z; Lágler, R; Tóth, Z; Rennenberg, H; Heszky, L; Kőmíves T. (2006) RT-PCR analysis and stress response capacity of transgenic *gsh*I-poplar clones *(Populus* x *canescens*) in response to paraquat exposure. *Z Naturforschung,* 61c, 699-730.

Bittsánszky, A; Gyulai, G; Malone, RP; Gullner, G; Kiss, J; Czakó, M; Márton, L; Heszky, L; Kőmíves, T. (2007) Triggering of a plant molecular defense mechanism; gene expression levels of transgene *gsh*I and poplar gene *gsh*1 (*Populus* x *canescens*) in response to the DNA demethylating drug DHAC – an RT-qPCR analysis. *Acta Phytopathol. Entomol. Hung.,* 42, 235-243.

Bruce, B.D. (2001) The paradox of plastid transit peptides: conservation of function despite divergence in primary *structure. Biochimica Biophysica Acta,*1541, 2-21.

Burken, J.G. and Schnoor, J.L. (1998) Predictive relationships for uptake of organic contaminants by hybrid poplar trees. *Environ. Sci. Technol.,* 32, 3379-3385.

Castilho, A; Neves, N; Rufini-Castiglione, M; Viegas, W; Heslop-Harrison, JS. (1999) 5-Methylcytosine distribution and genome organization in *Triticale* before and after treatment with 5-azacytidine. *J. Cell Sci.,* 112, 4397-4404.

Causevic, A; Delaunay, A; Ounnar, S; Righezza, M; Delmotte, F; Brignolas, F; Hagège, D; Maury, S. (2005) DNA methylating and demethylating treatments modify phenotype and cell wall differentiation state in sugarbeet cell lines. *Plant Physiol. Biochem.,* 43, 681-691.

Cao, X; Springer, NM; Muszynskii, MG; Phillips, RL; Kaeppler, S; Jacobsen, SE. (2000) Conserved plant genes with similarity to mammalian de novo DNA methyltransferases. *Proc. Natl. Acad. Sci. USA,* 97, 4979–4984.

Chanvivattana, Y; Bishopp, A; Schubert, D; Stock, C; Moon, Y-H; Sung, ZR; Goodrich, J. (2004) Interaction of Polycomb-group proteins controlling flowering in Arabidopsis. *Development,* 131, 5263-5276.

Chen, Z.J. and Pikaard, C.S. (1997) Epigenetic silencing of RNA polymerase I transcription: a role for DNA methylation and histone modification in nucleolar dominance. *Genes and Development,* 11, 2124-2136.

Cheng, X. and Roberts, R.J. (2001) AdoMet-dependent methylation, DNA methyltransferases and base flipping. *Nucleic Acid Res.,* 29, 3784-3795.

Columbus, F.H. Focus on DNA research. Hauppauge, NY, USA: Nova Science Pub Inc. 2006. ISBN-10: 1594543437.

Creissen, G; Broadbent, P; Stevens, R; Wellburn, AR; Mullineaux, P, (1996) Manipulation of glutathione metabolism in transgenic plants. *Biochemical Society Transactions,* 24, 465-469.

Di Baccio, D; Tognetti, R; Sebastiani, L; Vitagliano, C. (2003) Responses of *Populus deltoides* x *Populus nigra* (*Populus* x *euramericana*) clone I-214 to high zinc concentrations. *New Phytol.,* 159, 443-452.

Elmayan, T. and Vaucheret, H. (1996) Expression of single copies of a strongly expressed 35S transgene can be silenced post-transcriptionally. *Plant J.,* 9, 787-797.

Fan, J; Kodama, E-I; Koh, Y; Nakao, M; Matsuoka, M. (2005) Halogenated Thymidine Analogues Restore the Expression of Silenced Genes without Demethylation. *Cancer Res.*, 65, 6927-6933.

Finnegan, E.J. and Kovac, K.A. (2000) Plant DNA methyltransferases. *Plant Mol. Biol.*, 43, 189–201.

Finnegan, EJ; Genger, RK; Kovac, KA; Peacock, WJ; Dennis, ES. (1998) DNA methylation and the promotion of flowering by vernalization. *Proc. Natl. Acad. Sci. USA*, 95, 5824 - 5829.

Finnegan, J; Kovac, KA; Jaligot, E; Sheldon, CC; Peacock, WJ; Dennis, ES. (2005) The downregulation of Flowering Locus C (FLC) expression in plants with low levels of DNA methylation and by vernalization occurs by distinct mechanisms. *Plant J.*, 44, 420–432.

Fire, A; Xu, S; Montgomery, MK; Kostas, SA; Driver, SE; Mello, CC. (1998) Potent and specific genetic interference by double-stranded RNA in *Caenorhabditis elegans*. *Nature,* 391, 806–811.

Foyer, C.H. and Noctor, G. (2005) Redox homeostasis and antioxidant signaling: a metabolic interface between stress perception and physiological responses. *Plant Cell,* 17, 1866–1875.

Goffin, J. and Eisenhauer, E. (2002) DNA methyltransferase inhibitors—state of the art. *Annals of Oncology,* 13, 1699–1716.

Gullner, G; Gyulai, G; Bittsánszky, A; Kiss, J; Heszky, L; Kömíves, T. (2005) Enhanced Inducibility of Glutathione S-Transferase Activity by Paraquat in Poplar Leaf Discs in the Presence of Sucrose. *Phyton,* 45, 39-44.

Gyulai, G. The effect of lipid metebolism regulators on the function of PS II of *Chlorella pyrenoidosa*. PhD Dissertation, Szeged University, Hungary, pp. 102, 1984.

Gyulai, G; Jekkel, Z; Kiss, E; Kiss, J; Heszky, LE. (1995) A selective auxin and cytokinin bioassay based on root and shoot formation in vitro. *J. Plant Physiol.*, 145, 379-382.

Gyulai, G; Humphreys, M; Bittsánszky, A; Skøt, K; Kiss, J; Skøt, L; Gullner, G; Heywood, S; Szabó, Z; Lovatt, A; Radimszky, L; Roderick, H; Abberton, M; Rennenberg, H; Kőmíves, T; Heszky, L. (2005) AFLP analysis and improved phytoextraction capacity of transgenic *gshI*-poplar clones (*Populus canescens* L.) *in vitro*. *Z Naturforschung,* 60c, 300-306.

Hall, T.A. (1999) BioEdit: a user-friendly biological sequence alignment editor and analysis program for Windows 95/98/NT. *Nucl. Acids Symp. Ser.*, 41,95-98.

Harada, E; Yamaguchi, Y; Koizumi, N; Hiroshi, S. (2002) Cadmium stress induces production of thiol compounds and transcripts for enzymes involved in sulfur assimilation pathways in *Arabidopsis*. *J. Plant Physiol.*, 159, 445–448.

Henderson, I.R. and Dean, C. (2004) Control of *Arabidopsis* flowering: the chill before the bloom. *Development,* 131, 3829-3838.

Hirochika, H; Okamoto, H; Kakutani, T. (2000) Silencing of retrotransposons in Arabidopsis and reactivation by the *ddm1* mutation. *Plant Cell,* 12, 357–368.

Horvath, DP; Anderson, JV; Chao, WS; Foley, ME. (2003) Knowing when to grow: signals regulating bud dormancy. *Trends Plant Sci.*, 8, 534-540.

Issa, J.P. and Kantarjian, H. (2005) Azacitidine. *Nature Reviews Drug Discovery*. May, Suppl., S6-7.

Kiss, J. Poplar germplasm improvement by means of biotechnology (PhD Dissertation), Gödöllő University, Hungary, pp. 98, 1999.

Kőmíves, T; Gullner, G; Király, Z. (1998) Role of glutathione and glutathione-related enzymes in response of plants to environmental stress. Stress of Life. *From Molecules to Man,* 851, 251-258.

Kooter, JM; Matzke MA; Meyer, P. (1999) Listening to silent gene: transgene silencing, gene regulation and pathogen control. *Trends Plant Sci.,* 4, 340–347.

Krol, AR; Mur, LA; Beld, M; Mol, JN; Stuitje, AR. (1990) Flavonoid genes in petunia: addition of a limited number of gene copies may lead to a suppression of gene expression. *Plant Cell,* 2, 291–299.

Kumpatla, SP; Teng, W; Buchholz, WG; Hall, TC. (1997) Epigenetic transcriptional silencing and 5-azacitidne-mediated reactivation of a complex transgene in rice. *Plant Physiol.,* 115, 361-73.

Law, R.D. and Suttle, J.C. (2003) Transient decreases in methylation at 50-CCGG-30 sequences in potato (Solanum tuberosum L.) meristem DNA during progression of tubers through dormancy precede the resumption of sprout growth. *Plant Mol. Biol.,* 51, 437–447.

Lehoczki, E; Laskay, G; Gaál, I; Szigeti, Z. (1992) Mode of action of paraquat in leaves of paraquat-resistant Conyza canadensis (L.) Cronq, *Plant Cell Environ.,* 15, 531-539.

Leple, JC; Brasileiro, ACM; Michel, BF; Delmotte, F; Jouanin, L. (1992) Transgenic poplars - Expression of chimeric genes using 4 different constructs. *Plant Cell Rep.,* 11, 137-141.

Linn, F; Heidmann, I; Saedler, H; Meyer, P. (1990) Epigenetic changes in the expression of the maize *A1* gene in *Petunia hybrida*: role of numbers of integrated gene copies and state of methylation. *Mol. Gen. Genet.,* 222, 329–336.

Livak, K.J. and Schmittgen, T.D. (2001) Analysis of Relative Gene Expression Data Using Real-Time Quantitative PCR and the $2^{-\Delta\Delta Ct}$ Method. *Methods,* 25, 402–408.

Mathieu, A. and Bender, J. (2004) RNA-directed DNA methylation. *J. Cell Sci.,* 117, 4881-4888.

Matzke, A.J.M. and Matzke, M.A. (1998) Position effects and epigenetic silencing of plant transgenes. *Current Opinion in Plant Biol.,* 1, 142-148.

Meister A. and Anderson, M.E. (1983) Glutathione. *Annu. Rev. Biochem.*, 52, 711–760.

Metzlaff, M; O'Dell, M; Cluster, PD; Flavell, RB. (1997) RNA-mediated RNA degradation and chalcone synthase A silencing in *Petunia. Cell,* 88, 845-854.

Meyer, P. and Heidmann, I. (1994) Epigenetic variants of a transgenic *petunia* line show hypermethylation in transgene DNA: an indication for specific recognition of foreign DNA in transgenic plants. *Mol. Gen. Genet.,* 243, 390-399.

Mullineaux, P. and Rausch, T. (2005) Glutathione, photosynthesis and the redox regulation of stress-responsive gene expression, *Photosynthesis Res.,* 86, 459-474.

Napoli, C; Lemieux, C; Jorgensen, RA. (1990) Introduction of a chimeric chalcone synthase gene into petunia results in reversible co-suppression of homologous genes *in trans. Plant Cell,* 2, 279–289.

Noctor, G; Arisi, A-CM; Jouanin, L; Kunert, KJ; Rennenberg, H; Foyer, CH. (1998) Glutathione: biosynthesis, metabolism and relationship to stress tolerance explored in transformed plants. *J. Exp. Bot.,* 49, 623–647.

Obertello, M; Santi, C; Sy, M-O; Laplaze, L; Auguy, F; Bogusz, D; Franche, C. (2005) Comparison of four constitutive promoters for the expression of transgenes in the tropical nitrogen-fixing tree *Allocasuarina verticillata. Plant Cell Rep.*, 24, 540-548.

Palmer, J.D. (1997) The mitochondrion that time forgot. *Nature,* 387, 454-455.

Park, Y-D; Papp, I; Moscone, EA; Iglesias, VA; Vaucheret, H; Matzke, AJM; Matzke, MA. (1996) Gene silencing mediated by promoter homology occurs at the level of transcription and results in meiotically heritable alterations in methylation and gene activity. *Plant J.,* 9, 183–194.

Peuke, A.D. and Rennenberg H. (2005) Phytoremediation with transgenic trees. *Z. Naturforschung,* 60c,:199-207.

Pfaffl, M.W. (2001) A new mathematical model for relative quantifcation in real-time RT-PCR. *Nucleic Acids Res.,* 29, E45:1-6.

Pósfai, J; Bhagwat, AS; Pósfai, Gy; Roberts, RJ. (1989) Predictive motifs derived from cytosine methyltransferases. *Nucleic Acids Res.,* 17, 2421-2435.

Prakash, A.P. and Kumar, P.P. (1997) Inhibition of shoot induction by 5-azacytidine and 5-aza-2′-deoxycytidine in Petunia involves DNA hypomethylation. *Plant Cell Rep.,* 16, 719-724.

Prakash, AP; Kush, A; Lakshmanan, P; Kumar PP (2003) Cytosine methylation occurs in a CDC48 homologue and a MADS-box gene during adventitious shoot induction in Petunia leaf explants. *J Exp. Bot.*, 54, 1361-1371.

Que, Q; Wang, H-Y; English, J; Jorgensen, RA. (1997) The frequency and degree of cosuppression by sense chalcone synthase transgenes are dependent on transgene promoter strength and are reduced by premature nonsense codons in the transgene coding sequence. *Plant Cell,* 9, 1357–1368.

Rother, M; Krauss, G-J; Grass, G; Wesenberg, D. (2006) Sulphate assimilation under Cd^{2+} stress in *Physcomitrella patens* – combined transcript, enzyme and metabolite profiling. *Plant Cell Environ.,* 29, 1801–1811.

Rusznyák S. and Szent-Györgyi A. (1936) Vitamin P flavonols as vitamins. *Nature*, 138, 27.

Schäfer, HJ; Haag-Kerwer, A; Rausch, T. (1998) cDNA cloning and expression analysis of genes encoding GSH synthesis in roots of the heavy-metal accumulator *Brassica juncea* L. evidence for Cd-induction of a putative mitochondrial gammaglutamylcysteine synthetase isoform. *Plant Mol. Biol.,* 37, 87–97.

Schnurr, J.A. and Guerra, D.J. (2000) The *CaMV-35S* promoter is sensitive to shortened photoperiod in transgenic tobacco. *Plant Cell Rep.,* 19, 279–282.

Stam, M; Mol, JNM; Kooter, JM. (1997) The Silence of Genes in Transgenic Plants. *Annals Bot.,* 79, 3-12.

Strohm, M; Jouanin, L; Kunert, KJ; Provost, C; Polle, A; Foyer, CH; Rennenberg, H. (1995) Regulation of glutathione synthesis in leaves of transgenic poplar (*Populus tremula×Populus alba*) overexpressing glutathione synthetase. *Plant J.,* 7,141-145.

Szigeti, Z; Rácz, I; Lásztity, D. (2001) Paraquat resistance of weeds – the case of *Conyza canadensis* (L.) Cronq, *Z. Naturforschung,* 65c, 319-328.

Sun, X; Sun, XM; Yang, ZM; Li, SQ; Wang, J; Wang, SH. (2005) Expression of *Brassica napus* L. γ-glutamylcysteine synthetase and low-and high-affinity sulfate transporters in response to excess cadmium. *J. Integrative Plant Biol.,* 47, 243–250.

Tamura, K; Dudley, J; Nei, M; Kumar, S. (2007) MEGA4: Molecular Evolutionary Genetics Analysis (MEGA) software version 4.0. *Mol. Biol. Evol.,* 24, 1596-1599.

Taylor, G. (2002) *Populus*: Arabidopsis for forestry. Do we need a model tree? *Annals of Bot.,* 90, 681-689.

Thompson, JD; Higgins, DG; Gibson, TJ. (1994) CLUSTAL W: improving the sensitivity of progressive multiple sequence alignment through sequence weighting, positions-specific gap penalties and weight matrix choice. *Nucleic Acids Res.,* 22, 4673-4680.

Tichopad, A; Dilger, M; Schwarz, G; Pfaffl, MW. (2003) Standardized determination of real-time PCR effciency from a single reaction set-up. *Nucleic Acids Res.,* 31, N°. 20 e122, 1-6.

Vaucheret, H. (2006) Post-transcriptional small RNA pathways in plants: mechanisms and regulations. *Genes and Dev.,* 20, 759–771.

Veres, G; Gibbs, RA; Scherer, SE; Caskey, CT. (1987) The molecular basis of the sparse fur mouse mutation. *Science,* 237, 415-417.

Wachter, A; Wolf S; Steininger, H; Bogs, J; Rausch, T. (2005) Differential targeting of GSH1 and GSH2 is achieved by multiple transcription initiation: implications for the compartmentation of glutathione biosynthesis in the *Brassicaceae*. *Plant J.*, 41, 15-30.

Watanabe, K; Yamano, Y; Murata, K; Kimura, A. (1986) The nucleotide sequence of the gene for gamma-glutamylcysteine synthetase of *Escherichia coli*. *Nucleic Acids Res.,* 14: 4393-4400.

Wolfe, A.P. and Matzke, M.A. (1999) Epigenetics: Regulation through repression. *Science,* 286, 481- 486.

Wong, J.T.F. (1988) Evolution of the genetic code. *Microbiol.,* 5, 174-181.

Yang, G; Lee, YH; Jiang, Y; Kumpatla, SP; Hall, TC. (2005) Organization, not duplication, triggers silencing in a complex transgene locus in rice. *Plant Mol. Biol.*, 58, 351-66.

Xiang, C. and Oliver, D.J. (1998) Glutathione metabolic genes coordinately respond to heavy metals and jasmonic acid in *Arabidopsis*. *Plant Cell,* 10, 1539–1550.

Xiao, W; Brown, RC; Lemmon, BE; Harada, JJ; Goldberg, RB; Fischer, RL. (2006) Regulation of Seed Size by Hypomethylation of Maternal and Paternal Genomes. *Plant Physiol.,* 142, 1160–1168.

Zardo, G; Marenzi, S; Perilli, M; Caiafa, P. (1999) Inhibition of poly(ADP-ribosyl)ation introduces an anomalous methylation pattern in transfected foreign DNA. *The FASEB J.,* 13, 1518-1522.

In: Genetically Modified Plants
Editors: Timm Wolf and Jonas Koch

ISBN 978-1-60456-696-3
2008 Nova Science Publishers, Inc.

Chapter 9

THE EFFECTS OF ORGANIC AND GENETICALLY MODIFIED SOY ON CHOLESTEROL AND TRIGLYCERIDES SERUM LEVELS. STUDY IN RATS

***Lavínia Leal Soares*[*1], *Ana Maria Martins Lucas*[1] *and Gilson Teles Boaventura*[2]**

[1] Medicine College, Fluminense Federal University, Niterói, Rio de Janeiro,Brazil

[2] Laboratory of Experimental Nutrition, Nutrition College, Fluminense Federal University, Niterói, Rio de Janeiro, Brazil

ABSTRACT

Soy is a food rich in proteins, fibers, vitamins, minerals and isoflavones, compounds with a very similar chemical structure to the estrogen, being able to perform similar activities to this hormone. Studies report that its use decreases the risk of cardiovascular diseases by the reduction of cholesterol and triglycerides serum levels. 24female Wistar rats were divided into three groups (N = 8): organic soy group (OSG) receiving an organic soy-based diet, genetically modified soy group (GMSG) receiving a transgenic soy-based diet, and a control group (CG) receiving a casein-based. All animals with water, isocaloric and isoproteic diets *Ad libtum* during all their lives until old age. It was observed that the consumption of organic and transgenic soy seems to promote a reduction in the levels of cholesterol and triglycerides compared with control group. It was also observed that the genetically modified soy behaved in a similar way to the organic soy in relation to lipids levels, no inserted DNA or any other component of the transgenic soy had some adverse effect that could imply the transgenic food is not adequate for consumption.

Keywords: *Soybean, transgenic soybean, isoflavones, cholesterol, triglycerides.*

---

* Lavínia Leal Soares Adress: Rua das Palmas, 1/22, Iguabinha, Araruama, Rio de Janeiro, 28970-000, Brazil. e-mail: lavinialeal@yahoo.com.br.

INTRODUCTION

Soy (*Glycine Max* (L) *Merril*) is an important plant for human and animal nutrition, as well as for industrial purposes. Approximately 60% of all the processed food products contain ingredients derived from soy. It contains large amounts of protein despite its deficiency of sulfur-containing amino acids such as methionine and cystine. Its grain contains approximately 38% protein, 18% oil, 30% carbohydrates. It is also an important source of various minerals (Fe, Zn, Mg, K, Ca, Mn and Se), vitamins (A, B1, B2, B6 and folic acid), phytoestrogens and fiber [1].

Organic soy is grown in an ecological manner without chemical products, and does not pollute the soil, does not contaminate the producer nor does it modify the product. This process, however, implies a significant loss of productivity and profit [2].

Today there is a tendency to use soy in substitution of animal protein, as traditionally done in countries like Japan and China. Because of the increased soy consumption, transgenic soy has been created by genetic engineering. This is a genetically modified organism (GMO) to which three foreign genes were added, one of them from a virus and the others from a bacterium found in soil. The advantage of this modification is that the plant becomes resistant to glyphosate herbicides used to destroy weeds, which end up by being harmful to the crop itself. With the genetic modification, this problem does not occur, with consequent increased production and reduced costs [3]. However, some studies report that this modification could cause toxicity or allergic reactions, resistance to antibiotics and risks to the environment [4]. There are still many doubts about the possible risks of transgenic soy regarding consumption and the environment and more studies are necessary to elucidate such questions.

Besides its high nutricional value, soy is the vegetable species which presented highest concentration of isoflavones, compounds with chemical structure similar to the estrogen, a hormone which has its production decreased in menopause. Due to this similarity, these compounds may bind to estrogen receptors, being able to perform similar activites to the estrogen, this way being called phytoestrogens. According to researches, the isoflavones are capable of acting in the prevention of cardiovascular diseases, in the reduction of cholesterol and triglycerides, in reduction of the osteoporosis, in the relief of menopause symptoms and, in the prevention of the types of breast and uterus cancer which increase under the influence of hormones. Nonetheless, they are considered weak estrogens, because their potency is from 1.000 to 10.000 times inferior to the estrogen, which would not cause so many side effects [5].

According to studies, the consumption of soy besides helps in the reduction of cholesterol levels, due it contents in isoflavones, increasing HDL and decrease LDL serum levels, reducing the risk of cardiovascular diseases [6]. Such fact may be explained by the increase of activity in the LDL receptors. A research made with rats which were deficient in LDL receptors helped to support this hypothesis. When normocholesterolemic C57BL/6J mice were fed with high cholesterol, isoflavone-poor diet, they developed hyperlipidemia and lesions in aorta. However, those animals fed and isoflavone-rich diet exhibited a decrease in total cholesterol and VLDL cholesterol. This benefit of an isoflavone-rich diet, however, was not observed in rats deficient in LDL receptors [7].

In another study, in which women were supplemented with 20g of soy protein per day, it was observed a significant decrease in the diastolic pressure (5mm Hg), in the total cholesterol (6%) and in the levels of LDL (7%) [8].

The isoflavones present antioxidant action, preventing the formation of free radicals due to the increase of the superoxide dismutase synthesis, responsible for elimination of the superoxide free radical, which promotes the formation of lesions and trombus [9]. They also promote a decrease in the development of the arteriosclerotic lesion since it decreases the cellular adhesion, reducing the platelet aggregation. The isoflavones also act promoting vasodilatation which will interfere in the decrease of arterial pressure, probably due to the increase in the production of nitric oxide [10].

Postmenopausal female monkeys feeded with diets based on phytoestrogens during six months presented normal vasodilatation with the administration of acetylcholine. However, vasoconstriction was verified in males and females which ingested low quantity of phytoestrogens [11].

Nonetheless, there are studies which do not approve the benefits in the use of isoflavones in relation to their cardiovascular effects. Twenty, healthy, postmenopausal women, aging from 50 to 70 years old, with evidences of endothelial disfunction, were treated with 80mg/day of isoflavones. Placebo was used for the control group. After three weeks ingesting a standard fat-reduced diet, subjects received isoflavones or placebo during eight weeks. Comparing with the control group, no significant effects were found in relation to arterial pressure, lipids levels, lipoprotein concentration or vasodilatation [12].

Nestel *et all* [13] demostrated that pills containing from 40 to 150 mg of isolated isoflavones per day, compared to the placebo, did not presented the reducing effects in cholesterol already verified in a diet rich in soy and in soy protein supplementation.

There are still many questions to be answered in relation to the benefits caused by isoflavones, which demands more studies in order to prove their real effects.

The objective of this study is to evaluate organic and transgenic soy and their effects in lipids serum levels when used by female rats during all their lives.

Methods and Materials

Animals and Groups

The biological assay was conducted on 24 female Wistar rats aged 21 days (after lactation) at the beginning of the experiment from the Laboratory of Experimental Nutrition (LABNE) of the Department of Nutrition and Dietetics, Nutrition College, Fluminense Federal University. The rats were divided into three groups of eight animals each, which received the experimental diets, as follows: control group (CG) fed a casein-based diet, organic soy group (OSG) fed an organic-soy based diet supplemented with 0.3g cystine, and a genetically modified soy group (GMSG) receiving a transgenic soy-based diet.

All animals were fed the above diets exclusively, until they were 1 year and 3 months old (which is considered as old age for this kind of animal) and were the offspring of parents (preceding generation) who also received the same diet during all their lives. At the end of this period, the animals were sacrificed. Blood collection was made through cardiac puncture, in order to determine the cholesterol, triglycerides serum levels. During all the time the rats were kept in polypropylene cages, in an environment with controlled temperature at 22°C and a 12-h light/dark period. Water and diets were offered A*d libitum*. Food consumption and animal weight were recorded daily.

The research was approved by Committee of Ethics from Fluminense Federal University.

Diets

Transgenic soy was supplied by Jasmine Integral Foods (Curitiba, PR, Brazil) and organic soy was supplied by Bunge Foods (Porto Alegre, RS, Brazil). The suppliers of the other components of the diets were: Maizena starch by Refinements of Maize Ltda (Granhuns, Recife, PE, Brazil), refined sugar by União (Rio de Janeiro, RJ, Brazil), Liza soy oil by Cargill Agricultural Ltda (Mairinque, SP, Brazil), Microcell cellulose by Blanver Ltda (Cotia, SP, Brazil) and cysteine, choline bitartrate, casein and mixtures of vitamins and minerals by Rhoster Commerce and Industry (Vargem Grande Paulista, SP, Brazil).

The soybeans used for the preparation of flour were manually selected, weighed on a Toledo digital scale (São Paulo, SP, Brazil) with 0.1 g accuracy and dried in a ventilated Fabbe-Prima (São Paulo, SP, Brazil) stove at 60°C for 30 min. Boiling water was then added to the grains for 3 min and discarded, to permit husk removal under running water. The grains were then left in water for 4 h and the water was discarded. The soybeans were cooked in water for 30 min, the water was discarded again, and the beans were placed in an oven for 12 h at 60°C. After this procedure, the beans were ground and the flour obtained was used as the protein source for diet preparation.

All diets were prepared in the LABNE and contained 10% protein (1.75% nitrogen) and 363.95 kcal per 100g, added to the mixtures of vitamins and minerals according to the rules of the Committee on Laboratory Animal Diets, 1979, modified according to the recommendations of the American Institute of Nutrition-93 [14]. The ingredients of the diets (Table 1) were homogenized in an industrial mixer with boiling water. The mass obtained was transformed into tablets, which were dried in a ventilated oven at 60°C for 24 h, properly identified and stored refrigeration until the time for use

Table 1. Composition of the diets used for the assay

Foods 100g	Casein(g)	Organic Soy(g)	Transgenic Soy(g)
Proteic Source	11.54	20.96	20.28
Starch	61.41	57.23	58.04
Refined Sugar	10.00	10.00	10.00
Minerals Mix	3.50	3.50	3.50
Vitamins Mix	1.00	1.00	1.00
Soy Oil	7.00	2.81	2.94
Celulose	5.00	3.95	3.99
C. Bitartrate	0.25	0.25	0.25
Cistein	0.30	0.30	0.00
Total	100.00	100.00	100.00

Collected Material

The animals were anaesthetized by inhalation of eter and the blood was collected through cardiac puncture using needles and vacuum tubes (Vaccueete) in a volume of 10ml. The blood collected in test tubes was centrifuged in a centrifuge (Sigma) during 30 minutes at 3.500 RPM, in order to obtain the serum for cholesterol and triglycerides analysis.

The cholesterol and triglycerides analysis were determined through kits BIOCLIN (Indústria Quibasa- Química Básica Ltda, Belo Horizonte, MG, Brazil) in the LABNE. The reading was made in a spectrophotometer (Spec 20MV) in 500 and 490 nanometers (nm) respectively.

Data are reported as means $\pm$ SD and the results were analyzed statistically by multiple comparison One-way analysis of variance, with the level of significance set at $P \leq 0.05$. When a statistically significant difference was detected between variables, the Sheffé test was applied using the Bonferroni coefficient for multiple comparisons. All statistical analyses were performed using the Stratigraphics Software plus 6.0.

Results and Discussion

Observing the means related to cholesterol levels (table 2), the experimental groups were lower than the CG, even though no statistical difference was detected. OSG (104.2 ± 11.1 mg/dl) and GMSG (86.8 ± 6.1 mg/dl) had a reduction of 8.7 and 23.9%, respectively, in relation to the CG (114.2 ± 7.3 mg/dl). All the groups, including CG, presented cholesterol serum levels within the normality, 40 to 130 mg/dl, according Harkness e Wagner [16].

In relation to triglycerides serum levels, the groups which received soy based diet were inferior to the CG. The GMSG had the lowest levels (60.3 ± 4.6 mg/dl), with a significant difference ($p<0.04$) from the CG. The OSG (72.3 ± 10.1 mg/dl) was similar to the GMSG and although it was numerically inferior to CG, these difference was not significant (table 2). A hypothesis to explain this fact is that the supplementation of the OSG with L-cystine probably made that the protein of the organic soy behaved in a similar way to casein. The CG had the higher levels of triglycerides (97.8 ± 13.1). According to Harkness e Wagner [16] all the groups obtained triglycerides levels within normality, which is up to 145 mg/dl.

Kern *et all* [17] obtained results similar to the ones found in our research. Rats which consumed isolated soy protein (with 20% of protein, supplemented with methionine) during 28 days presented a reduction in the levels of cholesterol and triglycerides, compared to the control group (with a casein-based diet). The group which consumed isolated soy protein presented levels of cholesterol of 59.3 ±11.9 mg/dl against 73.8 ± 17.8 mg/dl from the casein group ($p<0.05$). In relation to the triglycerides levels, the following results were found: 93.4 ±37.4 mg/dl against 96.3 ± 41.8 mg/dl from the control group. The group which received the isolated soy protein presented a reduction in the levels of triglycerides even though it did not present a significant difference. It was also observed that the addition of methionine did not decrease the hypocholesterolemic effect of soy. As well as in our study, the supplementation with cystine did not alter the hypocholesterolemic effect of organic soy, even though the group which consumed this kind of diet did not present significant difference in relation to the CG.

Table 2. Serum Levels of Cholesterol and Triglycerides at the end of the experiment

GROUPS	*Cholesterol (mg/dl)*	*Triglycerides (mg/dl)*
GC	114.2 ± 7.3^{a}	97.8 ± 13.1^{a}
GSO	104.2 ± 11.1^{a}	72.3 ± 10.1^{ab}
GSGMO	86.8 ± 6.1^{a}	60.3 ± 4.6^{b}

Numbers followed by different superscript letter are statistically significant ($p \leq 0{,}05$, ANOVA followed by Scheffé and Bonferroni test). CG= control group fed a casein-based diet; OSG= organic soy group fed an organic diet supplemented with 0.3g cysteine; GMSG= genetically modified soy group receiving a transgenic soy based diet.

In a study, after inducing diabetes in rats, the animals were supplied with diets with different concentrations of isoflavones, in quantities 1, 2 and 8 times higher than the consumption suggested by the manufacturer, during 24 days. A reduction in the cholesterol levels was observed only in the group with higher concentration of isoflavones: 190.1±154.1mg/dl against 224.4 ±77.8 mg/dl from the casein group. In relation to the triglycerides levels there was a reduction also in the groups with higher levels of isoflavones (190.1 ± 154.1mg/dl). The control group presented higher levels (224.4 ± 77.8). According to these data, only a quantity of isoflavones much higher that the one indicated for consumption was efficient in the reduction of total cholesterol and triglycerides levels [18].

Hodgson *et all* [19] verified that men and postmenopausal women who ingested 55 mg of isoflavones/day during 8 weeks did not present reduction in the levels of cholesterol and triglycerides. No effects were found regarding the lipid levels in healthy persons.

In a research were applied either isolated isoflavones (114 mg/dl) or placebo to postmenopausal women during 3 months. It was observed that the isoflavones did not present any effect in the levels of cholesterol and triglycerides [20].

According to Potter [21], in a study with postmenopausal women who received different concentrations of isoflavones, it was observed that small quantities of isoflavones presented effects in the lipid profile earlier than in the groups which received high concentrations. This fact suggests that the cholesterol lowering component of soy is not or is only partially related to the isoflavones.

In the present study, isoflavones were not supplied to the rats in the isolated form but, instead, in the form of soy protein with isoflavones. In the researches quoted above we verified a reduction in these lipid levels only in the cases when soy protein was used, or isoflavones in such high concentrations that would be inadequate for human consumption. In the groups which received isoflavones in the isolated form it was not verified reduction in the lipid levels. Such fact questions if the isoflavones are, really, the soy component capable of acting in the reduction of the cholesterol and triglycerides levels. The reduction verified among the experimental groups in our study, may have occurred not solely due to the isoflavones, but due to the soy protein itself, or because it is a kind of food rich in fibers, what helps to decrease the cholesterol levels.

A research helped to confirm this hypothesis. 60 *Syrian* female hamsters, from 6 to 8 months old were used. During 10 days, they were offered either soy protein, with different concentrations of isoflavones, or soy protein, from which the content of isoflavones was extracted. There was a reduction in the levels of cholesterol and triglycerides in all the groups, including in the one from which the content of isoflavones was extracted and no significant difference was found among them. Based on these results, the authors report that the hypocholesterolemic effect of soy cannot be entitled solely to the isoflavones but to the aminoacids composition of soy protein itself, or to other factors such as the presence of fibers, saponins and phenolic acids [22].

Tovar-Palacio *et all* [23] found similar results. They offered gerbils either isolated soy protein or isolated soy protein supplemented with different concentrations of isoflavones (2.1; 3.6 e 6.2 mg / g of protein). All the groups presented reduction in the levels of cholesterol in relation to the control group, based on casein. The supplementation of isoflavones did not promote higher reduction of the above mentioned levels.

It was also observed that the genetically modified soy behaved in a similar way to the organic soy in relation to lipids levels, no inserted DNA or any other component of the transgenic soy had some adverse effect that could imply the transgenic food is not adequate for consumption. There are, however, many other questions in relation to the possible risks caused by genetic modification, leading to the conclusion that more studies are necessary regarding the transgenic soy.

Based on the results found, we may conclude that the consumption of organic or transgenic soy in female rats, during all their lives until old age, seems to promote the reduction in the plasma levels of cholesterol and triglycerides, a fact which - so far - cannot be imputed solely to the isoflavones. And the genetically modified soy behaved in a similar way to the organic soy in relation to lipids levels.

REFERERENCES

[1] Anderson J.W, Johnstone B.M, and Cook-Newell ME. Meta-analysis of the effects of soy protein intake on serum lipids. *N. Engl. J. Med..* 1995;333 (5): 276-282.

[2] Magkos F, Arvaniti F, and Zampelas A. Putting the safety of organic Food into perspective. *Nutrition Research Reviews.* 2003;16:241-252.

[3] Rott M.E, Lawrencw T.S, Wall E.M, and Green M.J. Detection and quantification of roundup ready soy in foods by conventional and real-time polymerase chain reaction. *J. Agric. Food. Chem.* 2004;52: 5223-5232.

[4] Nodari R.O, and Guerra M.P. Plantas transgênicas e seus produtos: impactos, riscos e segurança alimentar (biossegurança de plantas transgênicas).*Rev. Nutr.* 2003;6 (1):105-116.

[5] Lissin W.L, and Cooke J.P. Phytoestrogens and cardiovascular health. *J. Am. Coll. Cardiol,* 2000; 35 (6):1403-1410.

[6] Nilausen K, and Meirertz H. Lipoprotein (a) and dietary proteins: casein lowers lipoprotein (a) concentrations and compared with soy protein. *Am. J. Clin. Nutr.* 1999; 69 (3):419-425.

[7] Kirk E.A, Sutherland P, and Wang S.A. Dietary isoflavones reduce plasma cholesterol and atherosclerosis in C57BL/6 mice but not LDL receptor-deficient mice. *J. Nutr.* 1998; 128: 954-959.

[8] Washburn S, Bruke G.L, Morgan T, and Anthony M. Effect f protein supplementation on serum lipoproteins, blood pressure, and menopausal symptoms in perimeopausal women. *Menopause.* 1999; 6 (1): 7-13.

[9] Bazzoli D.L, Hill S, and DiSilvestro R.A. Soy protein antioxidant actions in active, young adult women. *Nutr. Res*. 1996; 233 (2): 3186-3190.

[10] Okigami H. Fitoestrógenos. *Science News*. 2005; 1 (2): 8-11.

[11] Honore E.K, Williams J.K, Anthony M.S, Clarkson T.B. Soy isoflavones enhance coronary vascular reactivity in atherosclerotic female macaques. *Fertil. Steril.* 1997;67:148-154.

[12] Simons L.A, Konigsmark M, Simons J, and Celermajer D.S. Phytoestrogens do not influence lipoprotein levels or endothelial functions in healthy, postmenopausal women. *Am. J. Cardiol.* 2000; 85:1297-1301.

[13] Nestel P.J, Yamashita T, Pomeroy S, Dart A, komesaroff P, and Owen A. Soy isoflavones improve systemic arterial compliance bu not plasma lipids in menopausal and perimenopausal women. *Arterioscler. Thromb. Vasc. Biol.* 1997; 17: 3392-3398.

[14] Reeves P.G, Nielsen F.H, Fahey Jr. G.C.F. AIN-93 purified diet laboratory rodents: final report of the American Institute of Nutrition ad hoc Writing Committee on the Reformulation of the AIN -76A rodents diet. *J. Nutr.* 1993;123: 1939-1951.

[15] Ramos C.F, Lima A.P.S, Teixeira C.V, Brito P.D, Moura E.G. Thyroid function in post-weaning rats, wich dams were fed a low protein diet during suckling. *Braz. J. Med. Biol. Res.* 1997; 30:133-137.

[16] Harkness J. E, Wagner J.E. Biologia e Clínica de Coelhos e Roedores. São Paulo, Editora Roca, 1993.

[17] Kern M, Ellison D, Marroquin Y, Ambrose M, and Mosier K. Effects of soy protein supplemented with metionine on blood lipids and adiposity of rats. *Nutrition.* 2002; 18: 654-656.

[18] Hsu C, Chiu W, and Yeh S. Effects of soy supplementation on plasma glucose, lipids, and antioxidant enzyme activities in streptozotocin-induced diabetic rats. *Nutrition Research*. 2003; 23:67-75.

[19] Hodgson J.M, Puddley I.B, Beilin L.J, Mori T.A, Croft K.D. Supplementation with isoflavonoid phytoestrogens does not alter serum lipid concentration: a randomized controlled trial in humans. *J. Nut. r*1998; 128(4): 728-732.

[20] Nikander E, Tiitinen A, Laitinen K, Tikkanen M, and Ylikorta O. Effects of isolated isoflavonids on lipids, lipoproteins, insulin sensivity, and ghrelin in postmenopausal women. *J. Clin. Endocrinol. Metab.* 2004; 87 (7): 3567-3572.

[21] Potter S.M, Baum J.A, Teng H, Stilman R.J, Shay N.F, and Erman J.W. Jr. Am. *J. Clin. Nutr.* 1998; 68, 1375.

[22] Song T, Lee S, Murphy P. A, and Hendrich S. Soy protein with or without isoflavones, soy germ and soy germ extract, and daidzein lessen plasma cholesterol levels in golden Syrian hamsters. *Exp. Biol. Med*. 2003; 228(9):1063-1068.

[23] Tovar-Palacio C, Potter S.M, Hafermann J,C, and Shay. Intake of soy protein extracts influences lipid metabolism and hepatic gene expression in gerbils. N.F. *J. Nutr*. 1998; 28:839-842.]

Funding Sources: PIBIC/ CNPq/UFF, FAPERJ (n^0 E-26/171/173/2002).

In: Genetically Modified Plants
Editors: Timm Wolf and Jonas Koch

ISBN 978-1-60456-696-3

Chapter 10

QUALITATIVE AND QUANTITATIVE PCR METHODS FOR BIOTECH CANOLA GT73 AND MS8/RF3

Seong-Hun Lee[1]
Gene Analysis Laboratory, Experiment Research Institute of
National Agricultural Products Quality Management Service, Seoul, 150-043, Korea

ABSTRACT

Canola is one of the biotechnologically developed crops in the world. Biotech canola is a GMO (genetically modified organism) in food and an LMO (living modified organism) in environment. Several kinds of biotech canola have been commercialized in United States and Canada. Among them, two kinds of biotech canola, GT73 and Ms8/Rf3, were mainly cultivated in the world.

The detection method is an essential element for the GMO labeling system or LMO management about the biotech crops. In this study, the specific primer pairs and probes were developed to the qualitative and quantitative PCR methods for GT73 and Ms8/Rf3. The designed primers were confirmed in specificity and sensitivity. The single PCR product was obtained from GT73 or Ms8/Rf3 by the specifically designed primer pair, respectively. Limits of detection (LODs) of the qualitative method were all 0.03% for these biotech canola events.

On the other hand, the quantitative method was developed by using a TaqMan real-time PCR. As a reference molecule, a synthetic plasmid was constructed from a taxon-specific DNA sequence of canola and construct-specific DNA sequences of GT73 and Ms8/Rf3. The validation of this quantitative method was performed using six levels of mixing samples, 0.1 to 10.0%, with two biotech canola. As a result, the biases from the true value and the relative standard deviations were almost within the range of ±30%. Limits of quantitation (LOQs) of the quantitative method were all 0.1%. Consequently, we report that the proposed detection methods were applicable for qualitative and quantitative analysis for the biotech canola GT73 and Ms8/Rf3.

---

[1] correspondence should be addressed. Tel: 82-2-2165-6080, FAX: 82-2-2165-6005, E-mail: starlee@naqs.go.kr / starslee7@hanmail.net.

Keywords: biotech canola; genetically modified organism (GMO); living modified organism (LMO); biotech crops; limits of detection (LODs), real-time PCR; reference molecule; limits of quantitation (LOQs),

INTRODUCTION

A lot of biotech or genetically modified (GM) crops have been developed and commercialized since the first born of GM tomato, Flavr Savr™, in 1994. A living GMO is also called as a living modified organism, LMO. The global area for biotech crops has continuously increased from 1.7 million hectares in 1996 to 114.3 million hectares in 2007. There are four major biotech crops which are now being cultivated worldwide; soybean 64%, maize 24%, cotton 43% and canola 20% [1].

Biotech crops may affect human health and ecological environment. Most of consumers continuously concern about potential risk of GMO. Thus a number of countries have made their own labeling system for the biotech products. The purpose of the labeling system is to inform consumers of the presence of the biotech products in the crops or derived-products, and therefore to help the consumers to choose their preferred products. Many countries have established labeling systems based on their own criteria, with thresholds for adventitious presence (AP) of biotech crops defined as 0.9% in the EU [2], 3% in Korea [3] and 5% in Japan [4]. In addition, biotech crops are also treated as LMOs in the area of environment according to the Cartagena Protocol on Biosafety (BSP). There are the specific provisions for handling, transport, packaging and identification of LMOs between countries in BSP. In Korea, the mandatory GMO labeling system has been performed since 2001 and the consolidated notice on transboundary movement of LMO started on January 1st in 2008. As a management tool for these systems, the detection methods for GMOs or LMOs have been required from the regulatory authorities in many countries.

The detection method is an essential element for the GMO labeling system or LMO management about the biotech crops. The polymerase chain reaction (PCR) is the most commonly used DNA-based detection method for the identification of biotech crops. Qualitative methods using conventional PCR have been developed to detect the presence of the biotech crops [5-9]. Quantitative methods using real-time PCR have been also developed to analyze contents of biotech crops [10-16]. There are several kinds of biotech canola in the world; Ms1/Rf1, Ms1/Rf2, Ms8/Rf3, GT73, T45, and Topas 19/2. However, only GT73 and Ms8/Rf3 canola events are currently cultivated. Others are not produced from companies now. The detection methods for some biotech canola have been developed [17-19].

Canola (*Brassica napus* L.) is one of four major genetically modified crops; soybean, maize, cotton and canola. Canola is mainly grown in U.S.A, Canada and India, and used as a vegetable oil in food for human and feed for livestock. Several biotech canola events have been commercialized. GT73 canola expresses glyphosate tolerant CP4 EPSPS (5-enolpyruvylshikimate-3-phosphate syntease) and gox v247 (glyphosate oxidoreductase variant) for glyphosate degradation proteins, which allow GT73 canola to be protected against herbicide glyphosate damage [20].

On the other hand, Ms8/Rf3 canola has bar gene encodes phosphinothricin aceyl transferase (PAT) enzyme to allow the use of herbicide glufosinate ammonium. Ms8/Rf3 canola is gene stacked hybrid between Ms8 and Rf3. For the pollination control system, the parental line Ms8 has barnase gene encodes ribonuclease for male sterility and another parental line Rf3 has barstar gene encodes ribonuclease inhibitor for fertility restorer [21].

In this study, we report the qualitative and quantitative detection methods for biotech canola GT73 and Ms8/Rf3 by using the specific primer pairs, probes, and one standard plasmid, and confirmed the applicability for practical use by in-house validation experiment.

Materials and Methods

Plant Materials. Ground seeds of non-biotech canola and biotech canola, GT73 and Ms8/Rf3, were provided from Monsanto (St. Louis, USA) and BayerCropScience (Gent, Belgium) companies, respectively. Biotech canola, T45 and Topas 19/2, were also provided from BayerCropScience. Other biotech canola Ms8 and Rf3, biotech soybean Roundup Ready^{TM}, biotech maize Bt176 and biotech cotton RR1445 were purchased from Institute for Reference Materials and Measurements (IRMM, Geel, Belgium).

Genomic DNA Extraction and PCR Inhibitor Check. Canola and other plant materials were ground by an electric mill (Fritsch pulverisette 14, Germany). According to the modified manufacturer's protocol [22], the genomic DNA was extracted from about 1 g of the ground sample by using the DNeasy Plant Maxi kit (Qiagen, Germany). The concentration of the extracted DNA was measured by a UV spectrophotometer ND-1000 (Nano Drop Technologies., USA) and confirmed on agarose gel by electrophoresis. As a good quality of DNA for PCR, the absorption ratios at 260/280 and 260/230 nm wavelengths should be all more than 1.7.

The extracted DNA was checked to identify if the PCR inhibitor is present or not by quantitative PCR. Two levels of concentration (50ng/$\mu\ell$ and 10ng/$\mu\ell$) with GT73 or Ms8/Rf3 genomic DNAs were prepared and quantitated by Hmg-F/R and GT-F/R or MS-F/R primers/probe sets, respectively. The absence of PCR inhibitor was confirmed by comparison with copy numbers between two levels of concentrations.

Primers and Probes. As a canola-specific reference system, the Hmg-F/R primer pair and Hmg-Taq probe were designed to qualitatively or quantitatively detection for the high mobility group protein (Hmg) gene (GenBank Accession No. AF127919), which is a taxon-specific reference gene for canola.

On the other hand, as the construct-specific target systems for two kinds of biotech canola, GT-F/R primer pair and a GT-Taq probe were designed to qualitatively or quantitatively detection for GT73 based on the junction region between the gox v247 gene (GenBank Accession No. AR016595) and E9 gene (GenBank Accession No. X00806).

MS-F/R primer pair and an MS-Taq probe were designed to qualitative or quantitative detection for Ms8 and Ms8/Rf3 based on the junction region between the barnase gene (GenBank Accession No. X12871) and nopaline synthetase (NOS) gene (GenBank Accession No. V00087). RF-F/R primer pair and RF-Taq probe were also designed to qualitative or quantitative detection for Rf3 and Ms8/Rf3 based on the junction region between the barstar gene (GenBank Accession No. X15545) and NOS gene. Both of MS-F/R and RF-F/R primers are specific to Ms8/Rf3 because Ms8/Rf3 is gene stacked hybrid between Ms8 and Rf3 events. We selected the MS-F/R primers and MS-Taq probe for quantitation of Ms8/Rf3.

All primers and probes were designed by using the Primer Express software 3.0 (Applied Biosystems, USA). They were synthesized and purified by Genotech Company (Daejeon, Korea). The probes were labeled with 6-carboxy-fluorecein and 6-carboxyteramethyl-rhodamine at the 5′ and 3′ ends, respectively. The locations of primers and probes are shown in Figure 1, and their nucleotide sequences are listed in Table 1.

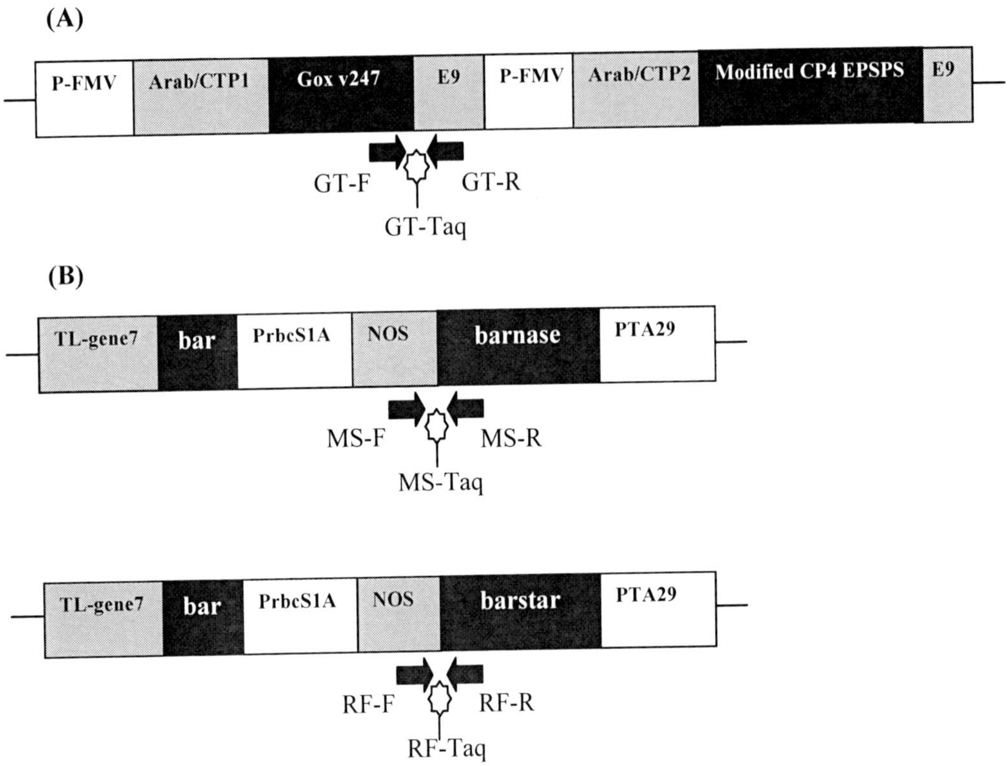

Figure 1. Schematic diagram of PCR strategy to detect biotech canola GT73 and Ms8/Rf3. The arrows indicate the specific primers and stars indicate the specific probes for qualitative or quantitative detection for biotech canola GT73 (A) and Ms8/Rf3 (B).

Conventional PCR for Qualitative Analysis. Conventional PCR was performed to confirm the specificity of the designed primers on the genomic DNAs extracted from non-biotech canola, six biotech canola including GT73 and Ms8/Rf3, biotech soybean, biotech maize, and biotech cotton. In order to determine the sensitivity of the qualitative PCR method, DNA mixtures were prepared from two biotech canola GT73 and Ms8/Rf3 by the following levels: 0, 0.03, 0.1, 0.3, 1.0, 3.0, 5.0, and 15.0% (w/w).

The conventional PCR was run with a 25$\mu\ell$ volume of reaction solution using a GeneAmp PCR System 9700 (Applied Biosystems, USA). A reaction solution contained 50ng of genomic DNA, 2.5$\mu\ell$ 10×PCR buffer, 0.2mM dNTPs, 1.5mM $MgCl_2$, 0.5μM of each primer, and 1.25 unit of AmpliTaqGold DNA polymerase (Applied Biosystems). The PCR was performed according to the following program: 1 cycle of 10 min at 94°C, 40 cycles of 30 sec at 94°C, 30 sec at 60°C, 30 sec at 72°C, and a final extension cycle of 7 min at 72°C. The PCR product was resolved on 2% (w/v) agarose gel by electrophoresis.

Table 1. Primers and probes used for qualitative and quantitative PCRs.

Target		Name	Sequence (5′ → 3′)	Specificity	Length (bp)
Endogenous					
High mobility group protein (Hmg) gene		Hmg-F	CCG TTT CAT ATG GTA GAT CTG ATT TTT T	Hmg	111
		Hmg-R	GAT CGC CGT CTT GTT GCA A	Hmg	
		Hmg-Taq	FAM-ATC GAA GCA TCG AAC GAC GCG AAC-TAMRA	Hmg	
Construct Specific					
GT73		GT-F	CAT TCG CAC CAA ACC GTT T	gox v247	115
		GT-R	GAT GAT ACG AAC GAA AGC TCT AGC T	E9	
		GT-Taq	FAM-TAT TGG TAA GTC CAA GCA AAC TGG TCC TGC-TAMRA	gox v247	
Ms8/Rf3	Ms8	MS-F	GAG GCC GTT TTT TTC AGC TTT AC	barnase	118
		MS-R	TGC CAA ATG TTT GAA CGA TCT G	NOS	
		MS-Taq	FAM-CGG TCA ATT TCA CTT TCC GGC TCT AGA GG -TAMRA	barnase/NOS	
	Rf3	RF-F	GCG ACA TCA CCA TCA TAC TTT CTT AAT A	barstar	113
		RF-R	CCA AAT GTT TGA ACG ATC TGC TT	NOS	
		RF-Taq	FAM-AGA TGA ACA ATA TGG AAA CAC AAA CCC GCA-TAMRA	barstar/NOS	

Reference Molecule for Quantitative PCR. As a reference molecule, a standard plasmid was constructed based on a pCR2.1 vector (Invitrogen, USA) integrated with four PCR products, which were amplified from the specific primer pairs for canola Hmg gene, GT73, Ms8 and Rf3.

The standard plasmid was constructed according to Kuribara *et al. (10)*. Two kinds of PCRs were performed for connection and cloning of PCR products. Firstly, to make a blunt-end PCR product for connection, a 50$\mu\ell$ volume of reaction solution was composed of 5$\mu\ell$ of 10×PCR buffer, 0.2 mM dNTP, 1 mM $MgSO_4$, 0.3 μM primer pair, 1 unit KOD-Plus-DNA polymerase (Toyobo Co., Japan), and 2ng of plasmid DNA as a reaction template. Secondly, to connect the first PCR products for cloning, a 25$\mu\ell$ reaction solution contained 2.5$\mu\ell$ of 10×PCR buffer, 0.2 mM dNTP, 1.5 mM $MgCl_2$, 0.5 μM primer pair, 1.25 unit of AmpliTaqGold DNA polymerase, and 2$\mu\ell$ of the two first PCR products as reaction templates. All PCRs were performed by GeneAmp PCR system 9700 (Applied Biosystems, USA) according to the following PCR program: 1 cycle of 10 min at 94°C, 40 cycles of 30 sec at 94°C, 30 sec at 60°C, 30 sec at 72°C, and a final extension cycle of 7 min at 72°C.

The final-connected PCR product was ligated into the pCR2.1 plasmid vector by TOPO TA cloning kit (Invitrogen, USA), and this recombinant plasmid transformed *Escherichia coli* strain TOP10 cell (Invitrogen, USA). The cloned plasmid was selected by *Bst*X I digestion

(Roche Applied Science, Germany) and confirmed through DNA sequencing analysis by Solgent Company (Daejeon, Korea).

The cloned plasmid was extracted by the Qiagen Plasmid Midi kit (Qiagen) and the extracted plasmid DNA was cut by *Not* I restriction endonuclease (Roche). The linearized plasmid DNA was purified from 2% agarose gel by the QIA Quick Gel Extraction kit (Qiagen). The concentration of the plasmid DNA was measured by the UV spectrophotometer ND-1000 (Nano Drop Technologies., USA). As a calibrant for quantitation, the standard plasmid was serially diluted to 2×10, 1.25×10^2, 1.5×10^3, 2×10^4 and 2.5×10^5 copies per 2.5$\mu\ell$ with ColE1 DNA (5ng/$\mu\ell$, Wako, Japan).

Real-Time PCR for Quantitative Analysis. For quantitation of biotech canola DNA, a real-time PCR was performed in duplicates using two wells. A 25$\mu\ell$ volume of one well contained 50ng of sample DNA, 0.5 μM primer pair, 0.1 μM probe, and 12.5$\mu\ell$ Universal Master Mix (Applied Biosystems, USA). The real-time PCR was carried out by ABI PRISM 7900(Applied Biosystems, USA). The PCR amplification was run with the following program: uracil-N-glycosylate treatment at 50°C for 2 min, denaturation at 95°C for 10 min, 45 cycles of denaturation at 95°C for 30 sec, and annealing or extension at 59°C for 1 min.

Five levels of standard plasmid were used for the standard curve as calibrants, which are reference molecules. ColE1 DNA (5ng/$\mu\ell$) was used as no template control (NTC). To calculate the biotech cotton content, the conversion factor (C_f) was required. It defines the ratio of the DNA copy numbers between an introduced gene and an endogenous gene, and it is used to change the experimental value into real value.

In order to validate the quantitative method, the test samples were prepared by mixing DNAs from conventional cotton, two biotech cotton GT73 and Ms8/Rf3 according to the six levels of 0.1, 0.5, 1.0, 3.0, 5.0, and 10.0%.

RESULTS AND DISCUSSION

Specific Detection of Qualitative PCR Method. The Hmg-F/R primer pair was designed to differentiate canola from other crops such as soybean, maize and cotton. There are several reference genes in canola. The Hmg gene also presents as a single copy gene in the canola genome [23-25]. Hmg-F/R primer pair is specific to a high mobility group protein (Hmg) gene of the canola. The newly designed primer pairs, GT-F/R and MS-F/R or RF-F/R were specific to GT73 and Ms8/Rf3, respectively.

We used the qualified genomic DNAs, which had no PCR inhibitor, extracted from GT73 and Ms8/Rf3(data not shown). When the GT73 and Ms8/Rf3 DNA was amplified as templates, amplified fragment of 115 and 118 or 113 bp was observed, respectively. On the other hand, no amplification was observed from the non-biotech canola, biotech canola T45 and Topas 19/2, biotech soybean, biotech maize and biotech cotton (Figure 2).

This specificity is basically attributable to the specific primer pair designed to amplify the junction region containing two genes of each biotech event. These results showed that the developed qualitative PCR detection methods could be applicable for the identification of GT73 and Ms8/Rf3.

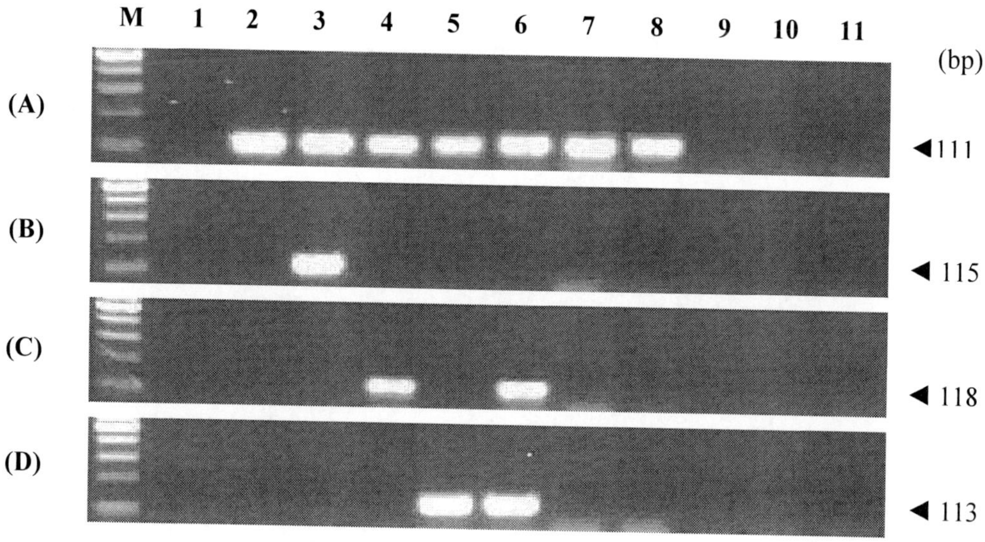

Figure 2. Specificity of the primer pairs designed for biotech canola GT73 and Ms8/Rf3. PCR products were electrophoresed on 2% agarose gel. Arrowheads indicate the expected PCR amplification products. The respective primer pairs for detecting Hmg (A), GT73 (B), Ms8 and Ms8/Rf3 (C), Rf3 and Ms8/Rf3 (D) were used. Template DNAs for each lane were as follows: Lane1-11; No template control, Non-biotech canola, biotech canola GT73, Ms8, Rf3, Ms8/Rf3, T45, Topas 19/2, biotech soybean RRS, biotech maize Bt176 and biotech cotton RR1445. M: 100 bp size ladder.

Limit of Detection of Qualitative PCR Method. The sensitivity test was performed to determine the limit of detection (LOD) of the qualitative PCR method by three times repeat. The test DNA samples, mixed with GT73 and Ms8/Rf3, were prepared to eight levels; 0, 0.03, 0.1, 0.3, 1.0, 3.0, 5.0 and 15.0% (w/w). As a result, the LODs for GT73 and Ms8/Rf3 were all 0.03% (Figure 3). The lowest threshold value is 0.9% of EU in the world. Thus, this result indicated that the qualitative PCR method developed from this study could be used in monitoring work for the GMO labeling systems in every country.

Standard Plasmid as a Reference Molecule used in Quantitative Method. A plasmid pCano2 was constructed by the integration of four PCR amplicons, which were from one endogenous gene, Hmg and three introduced genes of GT73, Ms8 and Rf3, into a pCR2.1 vector (Invitrogen). The nucleotide sequences of the integrated PCR amplicons of pCano2 are shown in Figure 4.

As a reference molecule, five levels of standard plasmid were set to 2×10, 1.25×10^2, 1.5×10^3, 2×10^4, and 2.5×10^5 copies per reaction for the quantitative real-time PCR. It was sufficient to quantitate 0.1~ 100% of biotech canola using 50 ng of genomic DNA for one reaction ranging from 2×10 to 2.5×10^5 copies. The linearity of the standard curves for GT73 and Ms8/Rf3 was confirmed in the quantitative PCR using the designed primer pair, probe and the standard plasmid, and calculated to 0.999 of R^2 value (Figure 5).

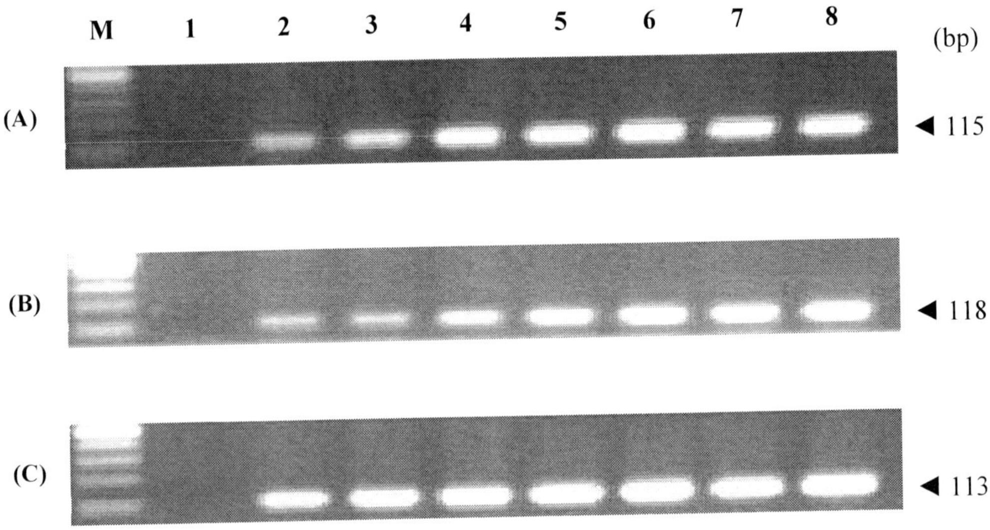

Figure. 3. Sensitivity of the primer pairs designed for biotech canola GT73 and Ms8/Rf3. PCR products were electrophoresed on 2% agarose gel. Arrowheads indicate the expected PCR amplification product. The primer pair for detection of biotech canola GT73(A, GT-F/R) and Ms8/Rf3(B, MS-F/R; C, RF-F/R) was used. Template mixing DNAs for each lane were as follows; Lane1-8; 0, 0.03, 0.1, 0.3, 1.0, 3.0, 5.0 and 15.0%, respectively. M; 100bp size ladder.

As for the five levels of the standard plasmid (pCano2), the repeatability of the copy number at each level of standard plasmid was confirmed from the data of duplicate reactions. The values of relative standard deviation (RSD) of the duplicate reactions ranged from 0.00 to 23.39% (Table 2). Most of RSD values were below 25%. The variation within this range was not significant, so the standard plasmid was confirmed to be stable and reliable reference material.

In Europe, the genomic DNAs from the certified reference materials (CRMs) have been used as reference molecules. Compared with the genomic DNA, the standard plasmid can be supplied as unlimited quantities with a consistent quality and a single plasmid can be used as a single reference molecule for several biotech events. As a result of this study, we propose that this standard plasmid can be used as a credible reference molecule for quantitating GT73 and Ms8/Rf3 in biotech canola.

Conversion Factors for Quantitation. Biotech canola has different copy numbers of recombinant DNA, so it is necessary to know the original copy numbers of a recombinant DNA of each biotech canola event for standard-plasmid quantitative method. The conversion factor (C_f) is a ratio of the copy numbers between a recombinant DNA and a taxon-specific DNA in a biotech canola event.

The biotech canola content of a canola sample can be calculated from the following formula; (copy numbers of a construct-specific recombinant DNA sequence / copy numbers of a taxon-specific DNA sequence) $\times$ 1/ C_f $\times$ 100(%). All experiments were repeated three times and the mean value was determined to be C_f for real content.

(A)

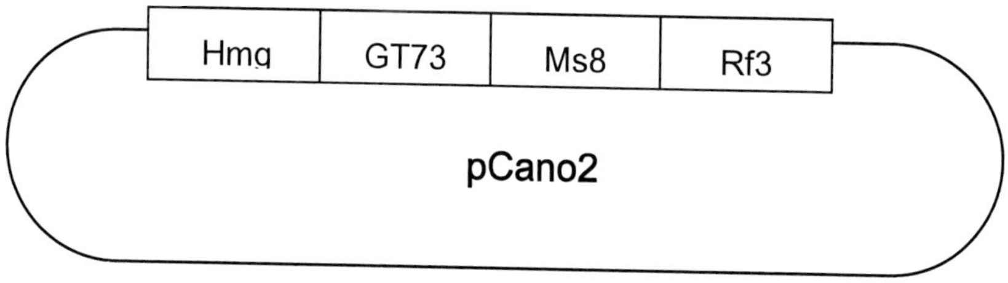

(B)

ccgtttcata tggtagatct gatttttCG GTTTTGTGG AAATCAGATG 50
Hmg-F

ATCATGGAAG CG***ATCGAAGC ATCGAACGAC GCGAAC***GG**tt gcaacaagac** 100
Hmg-Taq Hmg-R

ggcgatccat tcgcaccaaa ccgtttCGG***T ATTGG TAAGT CCAAGCAAAC*** 150
GT-F *GT-Taq*

TGGTCCTGCA TCCTA AGTGG GAATT CGAGC TCGGT ACCGG ATCCT CT**agc** 200

tagagctttc gttcgtatca tcGGTTTC**ga ggccgttttt ttcagcttta** 250
GT-R MS-F

c ATAAAGTGT GTAATAAATT TTTCTTCAAA CTCTGAT***CGG TCAATTTCAC*** 300
MS-Taq

TTTCCGGCTC TAGAGGATC CGAAG**cagatc gttcaaacat ttggca**GAAG 350
MS-R

GCT**gcgacat caccatcata ctttcttaat a**CGATCAATG GG***AGATGAAC*** 400
RF-F

AATATGGAAA CACAAACCCG CAAGCTTGGC TCTAGAGGAT CCG**aagcaga** 450
RF-Taq RF-R

tcgttcaaac atttgg 466

Figure 4. Standard plasmid pCano2 as a reference molecule. (A) Schematic diagram of pCano2. (B) Sequence of the integrated PCR amplicon of Hmg gene, GT73, Ms8 and Rf3 in the pCano2. The lowercase letters in bold indicate the sequence of primers with arrows indicating the orientation. The capital letters and italics in bold indicate the sequence of TaqMan probes.

(A)

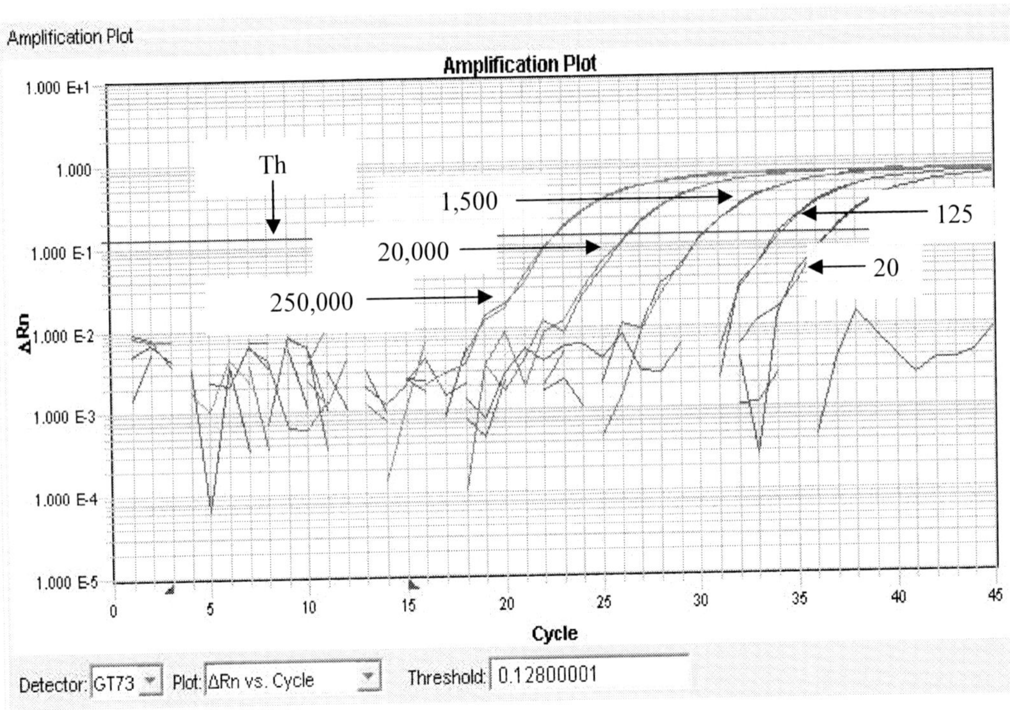

(B)

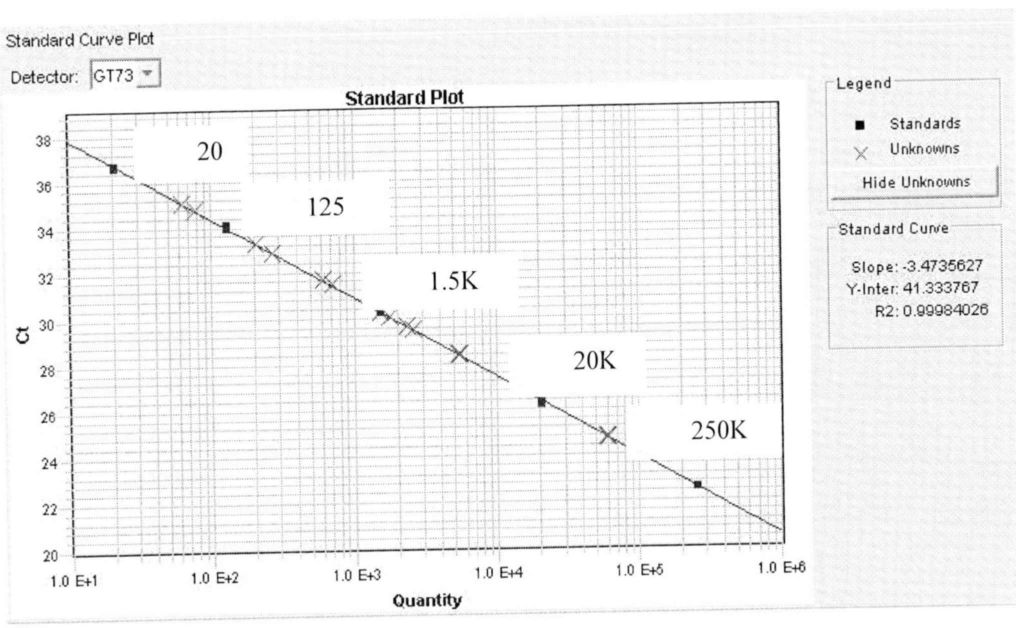

Figure 5. Amplification plots and standard curve of real-time PCR. (A) Amplification plots were generated from PCR of GT-F/R primers and GT-Taq probe for the detection of biotech cotton GT73 by the five levels of pCano2. (B) The standard curve from the data of the above amplification plots.

Table 2. Repeatability of the copy numbers of pCano2.

Target	Copy No. True value	Mean value	RSD[a]
Hmg	20	22.20	23.39
	125	127.01	10.37
	1500	1459.46	1.89
	20000	20397.37	10.84
	250000	248852.61	2.36
GT73	20	20.75	3.28
	125	119.06	5.62
	1500	1476.58	0.00
	20000	20470.54	7.02
	250000	248770.99	0.47
Ms8/Rf3	20	22.14	13.91
	125	113.21	2.89
	1500	1463.00	12.95
	20000	19273.49	2.89
	250000	263016.86	5.77

[a]RSD = an average of the relative standard deviations of the duplicate reactions in a single experiment for Hmg, GT73 and Ms8/Rf3

Canola (*Brassica napus* L.) is primarily self-pollinating plant. GT73 is a homozygous trait, and so the F1 seed of GT73 contains four GT73 copies per four haploid genomes. In case of gene stacked heterozygous hybrid Ms8/Rf3, their parental line Ms8 is a heterozygous trait, while Rf3 is a homozygous trait. But only Ms8/Rf3 (Ms8 x Rf3) F1 hybrid seeds are currently used for cultivation in the market. The F1 seeds of Ms8/Rf3 contain one copy of Ms8 and two copies of Rf3 per four haploid genomes. Thus, theoretical C_f of canola F1 seed with one or four copies transgenes per four haploid genome will be 0.25 or 1.0 because the Hmg gene, which is an endogenous gene, is a single copy per haploid genome of canola. The experimental mean C_f values are 0.85 for GT73 and 0.23 for Ms8/Rf3 (Table 3).

Table 3. Conversion factors (C_fs) of quantitative PCR for GT73 and Ms8/Rf3

Target	Mean	SD[a]	RSD[b]
GT73	0.85	0.04	4.71
Ms8/Rf3	0.23	0.03	13.04

[a]SD = Standard deviation.

[b]RSD (Relative standard deviation) was calculated by dividing the standard deviation by the mean value, and given in %. Experiments were performed three times each.

Table 4. Accuracy and precision of the quantitative PCR method

		Accuracy		Precision		
		Mean	Bias			Below 20 copies[c]
Biotech event	True value (%)	Experimental Value (%)	True value (%)	SD[a]	RSD[b]	
GT73	0.1	0.10	0.0	0.02	20.0	0/3
	0.5	0.43	-14.0	0.03	7.0	0/3
	1.0	1.06	6.0	0.03	2.8	0/3
	3.0	2.79	-7.0	0.26	9.3	0/3
	5.0	4.27	-14.6	0.08	1.9	0/3
	10.0	9.89	-1.1	1.49	15.1	0/3
Ms8/Rf3	0.1	0.09	-10.0	0.03	33.3	0/3
	0.5	0.46	-8.0	0.13	28.3	0/3
	1.0	0.99	-1.0	0.18	18.2	0/3
	3.0	2.79	-7.0	0.57	20.4	0/3
	5.0	4.79	-4.2	0.31	6.5	0/3
	10.0	9.41	-5.9	0.98	10.4	0/3

[a]SD = Standard deviation.
[b]RSD = Relative standard deviation. Experiments were repeated three times.
[c]Below 20 copies = the number of experiments below 20 copies/total number of experiments.

Validation of the Quantitative PCR Method. The accuracy of the method was evaluated as the bias (%) of the experimental mean value from the theoretical value. The precision was evaluated by the relative standard deviation (RSD).

To evaluate the accuracy and precision, the test DNA samples were prepared as six mixing levels; 0.1, 0.5, 1.0, 3.0, 5.0 and 10.0%. According to the GMO labeling systems of some countries, thresholds for the adventitious presence (AP) level of biotech products are 0.9% in the EU, 3.0% in Korea and 5.0% in Japan.

As a result at 1.0, 3.0, and 5.0% mixing levels, in case of GT73, the biases (mean vs. true value) were 6.0, -7.0 and –14.6%, and their RSDs were 2.8, 9.3 and 1.9%, respectively. In case of Ms8/Rf3, the biases (mean vs. true value) were –1.0, -7.0 and –4.2%, and their RSDs were 18.2, 20.4 and 6.5%, respectively. Overall result, the values of biases and relative standard deviations (RSDs) of the three times repeated tests ranged from 1.9 to 33.3% at six levels (Table 4). Except the lowest level, most of variations were within 30%, therefore the accuracy and precision of this quantitative method was credible for the method's practical application [26].

On the other hand, the limits of quantitation (LOQs) of the detection methods for biotech canola GT73 and Ms8/Rf3 were all 0.1%. Therefore, our method can be applicable for GMO labeling systems in the world because the lowest AP threshold is 0.9% of the EU. In conclusion, we report that the qualitative and quantitative PCR methods for biotech canola GT73 and Ms8/Rf3 were developed and validated in house. Consequently, we confirmed these methods could be used as a practical monitoring method for the biotech canola.

ACKNOWLEDGMENTS

I am very grateful to the Monsanto and BayerCropScience companies for providing the ground seeds of biotech canola GT73 and Ms8/Rf3, respectively.

LITERATURE CITED

[1] Clive, J. Preview: Global status of commercialized Biotech/GM crops. *ISAAA Brief.* 2007, No.37.

[2] The European Parliament and The Council of the European Union. Regulation (EC) No.1829/2003. Office J. Euro. Uni. 2003, Sec2, Article 24.

[3] Ministry of Agriculture and Forestry (July 3, 2007) Notification No. 2007-43.

[4] Food and Marketing Bureau, Ministry of Agriculture, Forestry and Fisheries of Japan (June 10, 2000) Notification No. 1775.

[5] Vollenhofer, S.; Burg, K.; Schmidt, J.; Kroath, H. Genetically modified organisms in food-screening and specific detection by polymerase chain reaction. *J. Agric. Food Chem.* 1999, 47, 5038-5043.

[6] Matsuoka, T.; Kawashima, Y.; Akiyama, H.; Miura, H.; Goda, Y.; Kusakabe, Y.; Isshiki, K.; Toyoda, M.; Hino, A. A method of detecting recombinant DNAs from four lines of genetically modified maize. *J. Food Hyg. Soc. Japan.* 2000, 41, 137-143.

[7] Matsuoka, T.; Hideo, K.; Akiyama, H.; Miura, H.; Goda, Y.; Kusakabe, Y.; Isshiki, K.; Toyoda, M.; Hino, A. A multiplex PCR method of detecting recombinant DNAs from five lines of genetically modified maize. *J. Food Hyg. Soc. Japan.* 2001, 42, No.1, 24-32.

[8] Hernandez, M.; Rodriguez-Lazaro, D.; Zhang, D.; Esteve, T.; Pla, M.; Prat, S. Interlaboratory transfer of a PCR multiplex method for simultaneous detection of four genetically modified maize lines: Bt11, MON810, T25 and GA21. *J. Agric. Food Chem.* 2005, 53, 3333-3337.

[9] Lee, S. H.; Min, D. M.; Kim, J. K. Qualitative and quantitative polymerase chain reaction analysis for genetically modified maize MON863. *J. Agric. Food Chem.* 2006, 54, 1124-1129.

[10] Kuribara, H.; Shindo, Y.; Matsuoka, T.; Takubo, K.; Futo, S.; Aoki, N.; Hirao, T.; Akiyama, H.; Goda, Y.; Toyoda, M.; Hino, A. Novel reference molecules for quantitation of genetically modified maize and soybean. *J. AOAC Int.* 2002, 85, 1077-1089.

[11] Yoshimura, T.; Kuribara, H.; Matsuoka T.; Kodama, T.; Iida, M.; Watanabe, T.; Akiyama, H.; Maitani, T.; Furui, S.; Hino, A. Applicability of the Quantification of Genetically Modified Organisms to Foods Processed from Maize and Soy. *J. Agric. Food Chem.* 2005, 53, 2052-2059.

[12] Akiyama, H.; Watanabe, T.; Wakabayashi, K.; Nakade, S.; Yasui, S.; Sakata, K.; Chiba, R.; Spiegelhalter, F.; Hino, A.; Maitani, T. Quantitative detection system for maize sample containing combined-trait genetically modified maize. *Anal. Chem.* 2005, 77, 7421-7428

[13] Yang, L.; Pan, A.; Zhang, K.; Guo, J., Yin, C.; Chen, J.; Huang, C.; Zhang, D. Identification and quantification of three genetically modified insect resistant cotton lines using conventional and TaqMan real-time polymerase chain reaction methods. *J. Agric. Food Chem*. 2005, 53, 6222-6229.

[14] Lee, S. H.; Kang, S. H.; Park, Y. H.; Min, D. M.; Kim, Y. M. Quantitative analysis of two genetically modified maize lines by real-time PCR. *J. Microbiol. Biotechnol.* 2006, 16(2), 205-211.

[15] Baeumler, S.; Wulff, D.; Tagliani, L.; Song, P. A real-time quantitative PCR detection method specific to Widestrike transgenic cotton (event 281-24-236/3006-210-23). *J. Agric. Food Chem*. 2006, 54, 6527-6534.

[16] Lee, S. H.; Kim, J. K ; Yi, B. Y. Detection methods for biotech cotton MON 15985 and MON 88913. *J. Agric. Food Chem*. 2007, 55, 3351-3357

[17] James D.; Schmidt A.; Wall E.; Green M.; Masri S. Reliable detection and identification of genetically modified maize, soybean and canola by multiplex PCR analysis. *J. Agric. Food Chem*. 2003, 51, 5829-5834.

[18] Yang L.; Pan A.; Zhang H.; Guo J.; Yin C.; Zhang D. Event-specific qualitative and quantitative Polymerase Chain Reaction analysis for genetically modified canola T45. *J. Agric. Food Chem*. 2006, 54, 9735-9740

[19] Wu Y.; Wu G.; Xiao L.; Lu C. Event specific qualitative and quantitative PCR detection methods for transgenic rapeseed hybrids MS1 x RF1 and MS1 x RF2. *J. Agric. Food Chem*. 2007, 55, 8380-8389.

[20] U.S. Department Agriculture, Animal and Plant Health Inspection Service, Roundup Ready Canola RT73. *http://www.agbios.com.*

[21] U.S. Department Agriculture, Animal and Plant Health Inspection Service, Hybrid Canola Transformation Events Ms8/Rf3. *http://www.agbios.com.*

[22] National Agricultural Products Quality Management Service. Testing manual for genetically modified organism. 2007. *http://enr.naqs.go.kr*

[23] Hernandez M.; Rio A.; Esteve T.; Prat S.; Pla M. A rapeseed-specific gene, Acetyl-CoA carboxylase, can be used as a reference for qualitative and real-time PCR detection of transgenes from mixed food samples. *J. Agric. Food Chem*. 2001, 49, 3622-3627

[24] Tang J.; Scarth R.; Fristensky B. Effects of genomic position and copy number of Acyl-ACP thioesterase transgenes on the level of the target fatty acids in *Brassica napus* L. *Molecular Breeding*. 2003, 12, 71-81

[25] Weng H.; Yang L.; Liu Z.; Ding J.; Pan A.; Zhang D. Novel reference gene, high-mobility-group protein I/Y, used in qualitative and real-time quantitative polymerase chain reaction detection of transgenic rapeseed cultivars. *J. AOAC International*. 2005, Vol. 88, No. 2.

[26] Shindo, Y.; Kuribara, H.; Matsuoka, T.; Futo, S.; Sawada, C.; Shono, J.; Akiyama, H.; Goda, Y.; Toyoda, M.; Hino, A. Validation of real-time PCR analyses for line-specific quantitation of genetically modified maize and soybean using new reference molecule. *J. AOAC Int*. 2002, 85, 1119-1126.

INDEX

A

B

C

D

E

F

G

H

I

J

K

L

M

N

O

P

Q

R

S

T

U

V

W